International Conference on e-Health 2019

Held at the Multi-Conference on Computer Science and Information Systems (MCCSIS 2019)

Porto, Portugal
16-19 July 2019

Editors:

Mário Macedo
Piet Kommers
Pedro Isaías
Luís Rodrigues

ISBN: 978-1-5108-9239-2

Printed from e-media with permission by:

Curran Associates, Inc.
57 Morehouse Lane
Red Hook, NY 12571

CURRAN ASSOCIATES INC.
proceedings.com

Some format issues inherent in the e-media version may also appear in this print version.

Printed by Curran Associates, Inc. (2019)

For permission requests, please contact International Association for Development of the Information Society (IADIS) at the address below.

International Association for Development of the Information Society (IADIS)
Rua Sao Sebastiao Da Pedreira
No. 100, 30, 1050-209
Lisbon, Portugal

Phone: 351 21 3151373
Fax: 351 21 3151244

secretariat@iadis.org

Additional copies of this publication are available from:

Curran Associates, Inc.
57 Morehouse Lane
Red Hook, NY 12571 USA
Phone: 845-758-0400
Fax: 845-758-2633
Email: curran@proceedings.com
Web: www.proceedings.com

MULTI CONFERENCE ON COMPUTER SCIENCE AND INFORMATION SYSTEMS

MCCSIS

16-19 JULY **2019** PORTO, PORTUGAL

Proceedings of the International Conference

e-Health 2019

Edited by
Mário Macedo

iadis
international association for development of the information society

Volume Editor:
Mário Macedo

Computer Science and Information Systems Series Editors:
Piet Kommers and Pedro Isaías

Associate Editor: Luís Rodrigues

PROCEEDINGS OF THE INTERNATIONAL CONFERENCE

E-HEALTH 2019

Porto, Portugal
JULY 17 - 19, 2019

Organised by

iadis
international association for development of the information society

Co-Organised by

isep Instituto Superior de Engenharia do Porto

INTERNATIONAL CONFERENCE

E-HEALTH 2019

part of the

MULTI CONFERENCE ON COMPUTER SCIENCE

AND INFORMATION SYSTEMS 2019

TABLE OF CONTENTS

FULL PAPERS

SHORT PAPERS

REFLECTION PAPERS

POSTERS

FOREWORD

These proceedings contain the papers and posters of the 11th International Conference on e-Health 2019, which was organised by the International Association for Development of the Information Society and co-organised by ISEP – Instituto Superior de Engenharia do Porto, from 17 - 19 July, 2019. This conference is part of the Multi Conference on Computer Science and Information Systems 2019, 16 - 19 July, which had a total of 926 submissions.

The use of ICTs (Information and Communication Technologies) in Healthcare Services is the main mechanism to improve efficiency and effectiveness. Nowadays ICTs are being developed to achieve the following objectives:
– To integrate heterogeneous systems;
– To develop frameworks to make all data meaningful, accessible and available everywhere and permanently;
– To develop AIDC (Automatic Identification and Data Collectors) systems;
– To develop intelligent systems to support clinical and management decisions;
The use of these technologies also improves the quality of patient care and reduces clinical risk. At the same time, the patient will be part of the healthcare process, having more information about diseases and access to his/her electronic health record.

The e-Health (EH) 2019 conference aims to draw together information systems, practitioners and management experts from all quadrants involved in developing computer technology to improve healthcare quality.

Submissions were accepted under the following 3 main areas in the field of e-health within specific topics:

-Research Issues
-Management Issues
-Applications

The EH 2019 received 127 submissions from more than 26 countries. Each submission has been anonymously reviewed by an average of three independent reviewers, to ensure that accepted submissions were of a high standard. Consequently, only 23 full papers were approved which means an acceptance rate of 18%. A few more papers were accepted as short papers, reflection papers and posters. An extended version of the best papers may be published in the IADIS International Journal on WWW/Internet (ISSN: 1645-7641) and also in other selected journals, including journals from Inderscience.

Besides the presentation of full, short and reflection papers and posters, the conference also included one keynote presentation from an internationally distinguished researcher. We would therefore like to express our gratitude to Prof. Dr. Miguel Coimbra, Faculdade de Ciências da Universidade do Porto (*Porto University*), Portugal, for accepting our invitation to be our keynote speaker.

This volume has taken shape as a result of the contributions from a number of individuals. We are grateful to all authors who have submitted their papers to enrich the conference proceedings. We wish to thank all members of the organizing committee, delegates, invitees and guests whose contribution and involvement are crucial for the success of the conference.

Last but not the least, we hope that everybody had a good time in Porto, and we invite all participants for the next years' edition of this conference.

Mário Macedo, Universidade Atlântica, Portugal
e-Health 2019 Program Chair

Piet Kommers, University of Twente, The Netherlands
Pedro Isaías, The University of Queensland, Australia
MCCSIS 2019 General Conference Co-Chairs

Bertil Marques, ISEP, Portugal
Local Organising Chair

Porto, Portugal
July 2019

PROGRAM COMMITTEE

E-HEALTH PROGRAM CHAIR

Mário Macedo, Universidade Atlântica, Portugal

MCCSIS GENERAL CONFERENCE CO-CHAIRS

Piet Kommers, University of Twente, The Netherlands
Pedro Isaias, The University of Queensland, Australia

MCCSIS LOCAL ORGANISING CHAIR

Bertil Marques, ISEP, Portugal

E-HEALTH

COMMITTEE MEMBERS

Abdel-Badeeh M.Salem, Ain Shams University, Egypt
Andreas Schrader, University of Lübeck, Germany
Antonio Lanatà, University of Pisa, Italy
Asa Smedberg, Stockholm University, Sweden
Bert-Jan van Beijnum, University of Twente, Netherlands
Bobak Jack Mortazavi, Texas A&M University, USA
Bridget Kane, Karlstad University, Sweden
David Isern, University Rovira I Virgili, Spain
Edward de Quincey, Keele University, United Kingdom
Eric Campo, Laas-cnrs, France
Frederic Bousefsaf, Université de Lorraine, France
Gayo Diallo, University of Bordeaux, France
Georg Duftschmid, Medical University of Vienna, Austria
George Okeyo, Jomo Kenyatta University of Agriculture and Techno, Kenya
Guillaume Lopez, Aoyama Gakuin University , Japan
Jan Vejvalka, Charles University, Czech Republic
Jaspaljeet Singh, Universiti Tenaga Nasional, Malaysia
Jesuk Ko, Gwangju University, South Korea
Jolanta Masiak, Medical University of Lublin, Poland
Jose Luis Sanchez , Laboratorio de Computadoras y Control, Spain
Laurent Billonnet, Ensil - Université de Limoges, France
Lenka Lhotska, Czech Technical University in Prague, Czech Republic
Malina Jordanova, Bulgarian Academy of Sciences, Bulgaria
Mario Cannataro, University , Italy
Maurice Mars, University of KwaZulu-Natal, South Africa
Miguel Angel Rodriguez-Florido, Chair of Medical Technology - ULPGC, Spain
Mohy Uddin, King Abdullah International Medical Research Cente, Saudi Arabia
Nathalie Souf, University Paul Sabatier, France
Ovidiu Stan, Technical University Of Cluj Napoca, Romania
Paivi Hamalainen, The National Institute for Health and Welfare, Finland

KEYNOTE LECTURE

AI KNOCKING ON HEALTHCARE'S DOOR - A FEW GREAT EXAMPLES

Prof. Dr. Miguel Coimbra

ABSTRACT

Artificial Intelligence has promised a revolution in nearly all areas of our society. It is already knocking on healthcare's door and in some fields we already let it in. Research in Portugal has been particularly active in this field, with a rich set of successful research, development and innovation stories. In this talk, this will be illustrated by a set of projects coordinated by Instituto de Telecomunicações and by INESC Tec, inspiring young researchers to embark into the fascinating challenge of transforming healthcare using artificial intelligence.

Full Papers

ELECTRONIC HEALTH RECORD (EHR) ADOPTION IN SOUTH AFRICAN HEALTHCARE CENTRES: A CASE OF NW PROVINCE

Thatoyaone Modise[1], Nehemiah Mavetera[1] and Mmaki Jantjies[2]
[1]North West University P Bag X 2046 Mmabatho 2735 +27-76151868
[2]University of Western Cape Department of Information Systems South Africa

ABSTRACT

Electronic Health Records (EHRs) enable the healthcare sector to enjoy increased efficiency and throughput, whilst cutting overhead costs in healthcare centres. The aim of this study was to investigate the adoption of EHRs in South Africa by identifying the factors that influence their rate of adoption within healthcare institutions. The Theory of Planned Behaviour (TPB), Diffusion of Innovation (DoI) Theory and the Technology Acceptance Model 2, (TAM2) are used as the theoretical lenses through which this problem was viewed.

Results showed support for some of the factors of TPB, TAM2 and DoI, namely Relative Advantage, Output Quality, Result Demonstrability, Computer Self-Efficacy, System Complexity and Enjoyment/Job Satisfaction. A new variable – Patient Safety Endangerment – was also found to have a significant influence on the healthcare worker's decision to use a particular EHR. It is with this information that a better understanding of how EHRs are used in the North West Province can be established and this information can be used by decision makers when implementing similar systems to maximise their adoption.

KEYWORDS

Electronic Health Records, Technology Acceptance Model, Theory of Planned Behaviour, Diffusion of Innovations Theory, National Health Insurance, South Africa

1. INTRODUCTION

Patient records in South African healthcare facilities are semi-electronic (with the exception of public clinics) (South African Department of Health, 2012). The electronic systems used in public hospitals tend to mostly cater for Hospital Administrative duties as patient medical records are still kept in filing cabinets. The majority of public clinics have no electronic system at all.

South Africa is making efforts to increase access to quality healthcare for all its citizens, and doing away decentralized health records by introducing a National Health Insurance (NHI) fund [South African Department of Health, 2012; Mayosi and Benatar, 2014; Mayosi et. al, 2009]. At the heart of this fund is a national e-Health system that will handle all patient information and reporting (South African Department of Health, 2012). Electronic systems of any kind have been known to increase efficiencies and throughput, as well as reduce overhead costs of doing business, however these systems add no value if they are not utilized. This negates all the efforts that were taken to develop such systems. EHRs in particular have the potential to improve availability, continuity of care, empowerment, patient safety, and quality of care [Chib et al, 2008; Gartner, 2009; Ruxwana, Herselman, and Conradie, 2010].

This study provides an investigation into the adoption of technologies used in healthcare facilities. This is done to help identify possible barriers impeding the use of a healthcare patient administration system once developed, to ensure that new e-health systems do not suffer from low adoption. Factors identified as impeding adoption can be addressed in future e-Health systems to ensure maximal adoption by system users and ultimately provide the patient with the intended benefits resulting from the use of the technology.

2. REVIEW OF RELATED LITERATURE

TPB is a revision of the Theory of Reasoned Action (TRA) constructed by Ajzen & Fishbein, (1980). According to TRA, an individual's intention to perform a behaviour is a combination of their attitude towards performing the behaviour and subjective norms (Ajzen & Fishbein, 1980). The construction of TPB includes an extra construct namely Perceived Behavioural Control (PBC) which is based on two beliefs: control beliefs and their perceived facilitation. The Technology Acceptance Model (TAM) is also a derivative of TRA. According to Holden and Ben-Tzion (2009: p2) "TAM was developed in the 1980's, in light of concern that workers were not using IT made available to them".

TAM dictates that a person's acceptance of an Information System (IS) lies in their intention to use it. The intention, can be determined by taking into account a person's attitude toward the IS in question, and their perceptions concerning its usefulness. Attitudes towards the IS are a result of the beliefs that a person holds about the IS. Mathhews (2012: p1) further elaborates that "External variables, such as the task, user characteristics, political influences, organizational factors, and the development process, are expected to influence technology acceptance behaviour indirectly by affecting beliefs, attitudes, or intentions."

TRA/TPB have been used to predict adoption of a broad range of activities including but not limited to smoking, exercise, condom use, and moral behaviour [Paula, Modia, and Patel, 2016; Kangas, et al., 2015; Yzer et al., 2015]. TAM on the other hand was created solely to predict use in the IT context. In an article by Chuttur, (2009), the author explains how TPB provides more details that can be used to explain the intention of participants to use a computer application when compared to TAM. The added predictive power of TPB stems from the addition of the construct PBC. This construct is what allows for the identification of barriers that impede system use such as limitations on user's skills.

TAM2 (Venkatesh and Davis, 2000), a derivative of TAM identifies several variables including those already identified in TAM. TAM2 takes into consideration the individual as well as organizational context, thus making it better suited (when compared to TAM), to explain software use in the workplace. When compared to TPB, TAM2 takes into account factors that an employee or employer would take into consideration before adopting a technology in the workplace. However, researchers have found that facilitating conditions have a more direct influence on the behaviour itself. This makes sense as no matter how high the intention to use a technology, that action cannot be carried out if there is a barrier standing in the way of the potential user.

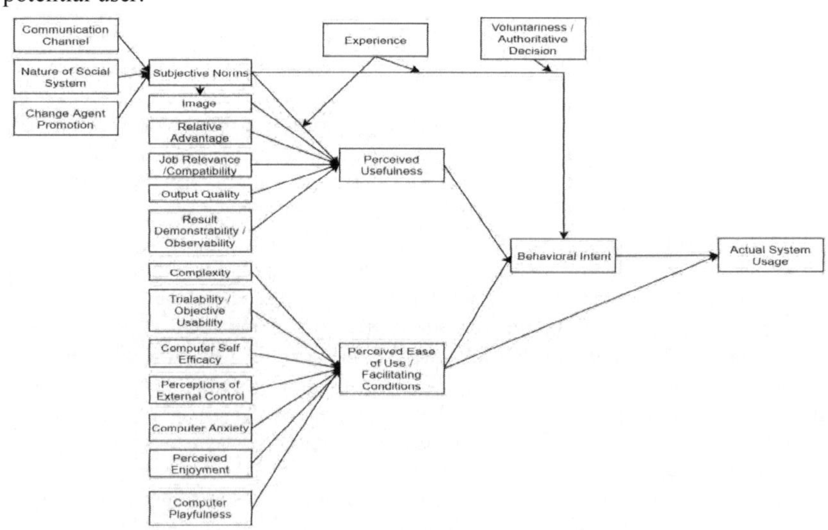

Figure 1. Conceptual model of this study

Another aspect of adoption that is explored in this study is the innovativeness of the innovation. By this we refer to attributes (features) of the innovation itself that persuade the potential user to adopt the innovation. To explain this, we turn to the Diffusion of Innovations (DoI) theory. The theory identifies five (5) innovation design variables namely: relative advantage, compatibility, complexity, trialability, and observability that, according to the theory have an influence on whether an innovation is adopted and how quickly its adoption will spread. These factors from DoI, TAM2 as well as TPB have been combined to form a conceptual model that was utilised by this study. The diagram shown depicts this conceptual model.

Structurally, it can be seen that the conceptual model bears many similarities to TAM2 with the addition of DoI innovation attributes. The review of the determinants of TRA, TPB, TAM, TAM2 & DoI as well as their provided definitions revealed several overlapping attributes. Such attributes included Job Relevance & Compatibility, Result Demonstrability & Observability, Voluntariness & Authoritative Decisions, and Trialability & Objective Usability. To avoid measuring attributes that seemed to be duplicates, these attributes were then grouped together to produce only one attribute. Unique attributes from DoI and TPB were added to TAM2 under the existing constructs of Perceived Usefulness and Perceived Ease of Use so as to supplement the predictive power of these constructs.

3. RESEARCH QUESTIONS

The following are research questions that this study aims to answer.

i. What are the factors that influence the adoption of EHRs in the North West Province?
ii. How can the adoption of EHRs in South Africa be improved?

4. RESEARCH METHODOLOGY

It has become best practice that when conducting any form of TPB research, the researcher must identify key factors having an influence on the subject under discussion (Ajzen, 2002). For quality and reliability, the factors identified must be independent of researcher input. To ensure this, this research study utilised a mixed methods approach. This means that both the qualitative and quantitative methodology were utilised to collect the data required for this study in a specific order.

The strategy employed in our case is referred to as a sequential exploratory research strategy and ensures that the salient beliefs responsible for influencing the individual's decision to use EHRs, were determined in the initial phase of the study using interviews and confirmed by respondents through the use of a questionnaire at the last stage.

Data from the interviews with the participants was recorded and later transcribed. Using thematic analysis, the transcribed interviews were analysed and coded according to the coding framework made up of the innovation attributes from the conceptual model. New themes were identified from the data and added to the framework.

5. RESULTS

5.1 Qualitative Investigation

The main intention of the qualitative phase of the study was to identify the salient beliefs for the main innovation attributes of our conceptual model. The table below shows the beliefs that were identified for the main constructs of the study.

Table 1. Salient beliefs identified from Interviews with participants

Construct	**Behavioural Intent**	**Subjective Norms**	**Perceptions of External Control (Facilitating Conditions)**
Salient Beliefs Identified	• Patient Safety Endangerment	• Co-workers Influence • Family Influence • Supervisor Influence	• Incomplete Functionality • Electricity Outages • Network Outages • Change Management

The interviews with the participants helped identify new factors that were not initially in the conceptual model. Some factors were identified as not being applicable to the study. The Construct Image, defined as "the degree to which an innovation is perceived to enhance one's status in one's social system" (Venkatesh, and Bala, 2008), was not supported and was subsequently removed from the conceptual model. Respondents highlighted the fact that there was no prestige associated with using the patient administration system at their workplace as all administrative staff had the same access to it. It is for this reason that the Image Attribute was discarded from the model.

5.2 Hypotheses and Questionnaire Construction

The conceptual model of this study is made up of a number of factors identified from TPB, TAM2 and DoI. Following the qualitative investigation that resulted in a revised conceptual model, hypotheses, and subsequently sets of questions, were documented for each innovation attribute of the model to construct the hypotheses of this study, as well as the questionnaire to be used by this study.

5.3 Instrument Reliability

The table below contains the alpha values of those measures. Of the 21 measures, reliability was computed for eleven (11) of the measures. The other ten (10) measures had a single question determining that measure and as such, reliability could not be computed. Of the 11 measures whose reliability was computed, two (2) displayed unacceptable reliability: Experience (N = 2, alpha = .109) and Communication Channels (N = 3, alpha = -.204). The other nine (9) measures showed an alpha value between .648 and .991, demonstrating fair to high reliability.

The Experience measure presented a low alpha of (N = 2, alpha = .109), this was attributed to users who indicated that they had no prior encounters with systems similar to the one at their workplace. The same users also indicated that a lack of experience would not let it deter them from using the system in their workplace. The two contrasting responses were not anticipated at the time of instrument design and thus resulted in the low alpha value.

Communication Channels (N = 3, alpha = -.204) also presented a low alpha as well. An inspection of the question items revealed that the questions evaluated "which of the two communication mediums (e-mail or contact sessions) did users prefer to be informed about new systems in their workplace". These questions evaluated two different forms of communication and as such the responses to these question items were not consistent with one another.

The low alphas observed in Experience and Communication Channels were as a result of a lack of unidimensionality in the questions measuring those scales. Unidimensionality in Cronbach's Alpha assumes that the questions are only measuring one latent variable or dimension (StatisticsHowTo, 2014) These two measures were, however, not discarded from the instrument based on their alpha values as they were still valid measures of adoption, their questions were simply rectified to ensure unidimensionality.

5.4 Hypotheses Testing

Correlational analysis is a technique that deals with relationships among variables. According to Cohen, West and Aiken, (2014) multivariate correlation analysis or sometimes multiple correlation analysis is a technique that may be used "whenever a quantitative variable, the dependent variable (Y), is to be studied as a function of, or relationship to, any factors of interest, the independent variables (IVs)". Multivariate correlational analysis was used in this case to evaluate the nature of the relationship between the identified innovation attributes and the usage of EHRs at healthcare facilities of the North West Province.

The table below contains the hypotheses of this study and the results from the testing of those hypotheses. Although the direction of the relationship between the influencing components on the influenced components could be estimated, a two-tailed test of significance was taken to try to remove the probability of error in this assumption. A value of .05 was used as the significance value as is widely used in academic research (Gelman, 2012). Correlation analysis of the hypothesis of this study were as follows.

Table 2. Correlation Analysis of Study

(H)#	Correlation		Supported
	R	Sig. (2-tail)	
1	.091	.704	No
2	.076	.749	No
3	.269	.252	No
4	.163	.481	No
5	.132	.570	No
6	.433	.050	Yes
7	.496	.022	Yes
8	.558	.009	Yes
9	.490	.024	Yes
10	-.268	.241	No
11	.279	.221	No
12	.045	.848	No
13	.343	.127	No
14	-.051	.825	No
15	.304	.181	No
16	.132	.570	No
17	-.250	.274	No
18	.400	.072	No
19	.638	.002	Yes
20	-.512	.018	Yes
21	.596	.004	Yes
22	.668	.001	Yes
23	.725	< .000	Yes

The results of the correlations revealed that Relative Advantage, Output Quality and Result Demonstrability were statistically significant in determining Perceived Usefulness. Experience (E) and Voluntariness (V) were hypothesized to have a mediating effect on Subjective Norm and Perceived Usefulness. Direct correlations between Subjective Norm (SN) and Perceived Usefulness (PU) yielded a strong correlation (SN-PU: r=. 533; p=.013).

The Perceived Ease of Use component was hypothesized to be influenced by Change Management, Complexity, Trialability, Computer Self-Efficacy, Perceptions of External Control, Perceived Enjoyment and Computer Anxiety. For all of these components of Perceived Ease of Use, none of them were found to be statistically significant.

A new component, Patient Safety Endangerment, was found to have a statistically significant and negative effect on Behavioural Intention (r= -.512, p= .018). Essentially, the higher the chances of the system endangering the safety and privacy of the patient, the less likely the intention of potential users to use such a system.

Further exploration of the correlation data found relationships between influencing variables and the main variables (Perceived Usefulness, Perceived Ease of Usefulness, Behavioural Intention and Actual System Usage). Complexity (r= -.663, p= .001), Computer Self-Efficacy (r= .593, p= .007) and Computer

Playfulness (r= .634, p= .002) were found to have a relationship that was statistically significant with Actual System Use. This relationship was strange as it had been expected that these factors would contribute towards the ease of use of the system and not the actual usage. That being said, the effects of these factors cannot be overlooked as they hold some merit. The implication is that these factors have the ability to impede actual usage of a system regardless of the attitudes the user may have towards the usage of the system.

Actual System Use was defined in this study as the frequency in which healthcare workers utilised the patient administration at their workplace daily. It then follows that the usability of a system will most likely suffer a loss in terms of the usage frequency if that system is substantially or highly complicated.

Table 3. Correlation Values between Actual System Use and other factors

	Complexity	Computer Self-Efficacy	Computer Playfulness
Actual System Use (r)	-.663	.593	.634
S ig. (2-tailed)	.001	.007	.002

The table above contains the correlation between factors that were hypothesized to have a relationship with Perceived Ease of Use but were instead found to have a statistically significant relationship with Actual System Use.

6. DISCUSSION

6.1 What are the Factors that Influence the Adoption of EHRs in Healthcare Facilities?

Amongst all of the innovation attributes investigated, there were some whose definitions seemed to be referring to the same thing, or overlapping each other. These attributes were combined into a single attribute instead. Those attributes included: Job Relevance and Compatibility, Result Demonstrability & Observability, Voluntariness & Authoritative Decisions, and Trialability & Objective Usability.

From the hypothesized relationships between the adoption attributes of this study, some attributes were found to be supported by this study, while others were not. Those attributes whose hypothesized relationship was not found to be significant were subsequently dropped from the model. Having analysed all the above, the initial conceptual model is then refined to produce the following model in Figure 3 below:

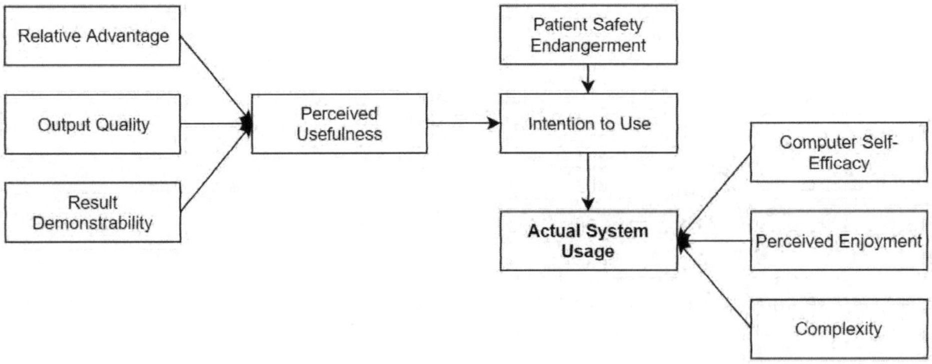

Figure 2. Revised Conceptual Model

The revised conceptual model has less factors than those of the initial model which is acceptable, as not all of the hypothesized factors were found to be supported by the study. Complexity, Computer Self-Efficacy and Perceived Enjoyment are seen having a more direct influence on Actual System Usage. A new factor, Patient Safety Endangerment, is also identified as having a direct influence on the intention of a user to use the patient administration system at their workplace. Relative Advantage, Output Quality and Result

Demonstrability are the only determinants that were identified as having a significant influence on the perceived usefulness of the patient administration system. Subjective Norms and its determinants, Voluntariness, Experience, and Job Relevance were found to be non-significant. The hypothesized determinants of Perceived Ease of use: Trialability, Perceptions of External Control, Computer Anxiety, and Perceived Enjoyment were also found to be non-significant.

6.2 How can the Adoption of EHRs be Improved in South African Healthcare Facilities?

This section proposes recommendations on ways in which the use of EHRs in healthcare facilities in South Africa can be maximized. The recommendations stem from literature and empirical data that was collected as part of this study.

i. Involvement of Staff

Management needs to interact with their staff before proposing upgrades or new implementations, particularly staff that utilizes the EHR system on a daily basis, as these users have in-depth knowledge of any pitfalls, the system they work with may have, when it comes to the work they do.

This form of involvement and transparency in the workplace decreases resistance and has been found to increase productivity levels as well as a range of other workplace and worker characteristics (Walden et al., 2015).

Staff are also positioned to provide insights that could assist in the design, testing and customization of the proposed system and its workflow (Jeffrey, 2016). Staff can also be utilized to provide feedback of their daily usage to aid in the optimization and modification of the proposed system and workflow (Blavin al., 2013).

ii. System Functionality and User Interfaces

Ensure that the systems being procured provide functionality that is relevant to the job functions of the staff that will use it, produces results that are in-line with expectations of management as well as ensures that benefits that arise from the use of the system are clear to the user – as per the Perceived Usefulness attributes put forward by this model. Functionality that is currently not available in the systems in healthcare facilities needs to be included.

iii. Change Management

Change Management is defined as the process of continually renewing the direction of an organisation, its structure as well as capabilities in order to serve the ever-changing needs of its customers (Nassif, Capretz, and Ho, 2010).

As part of the change management plan, training is one of the most popular interventions to increasing the ease of use of a system. "Training offers a good opportunity to help users adjust to the change that has been introduced by [any] system, and helps build positive attitudes toward the [new system]" (Munassar, and Govardhan, 2010: p271). Training introduces the new users of a technology to how the new system looks and functions. This helps reduce the view that the target group may have regarding the complexity of the system.

iv. Creating and Hiring Digital Natives

Seeing how the use of technology is increasing in the healthcare sector, it follows that healthcare facilities must ensure that their staff are familiar with the role that computers play in their lives. Training courses on end-user computing are widely available in South Africa, which staff members with no computing experience can be made to attend. The integration of introductory computing courses in the curriculum of students pursuing careers in the healthcare sector could also add to maximising their efficacy and playfulness around computers.

7. CONCLUSION

Several factors from the conceptual model were not supported by the data collected and were thus not considered for the revised model. Of the factors that emerged from the qualitative phase of the study, Patient Safety Endangerment – defined in this study as the degree to which a system could potentially endanger and/or compromise the safety and privacy of the patient - was found to be statistically significant to the study and thus, was included in the revised model. Interventions to reduce or alleviate the problematic factors and maximise those factors that increase adoption have been put forward with the hopes that management will take these into consideration.

REFERENCES

Ajzen, I., & Fishbein, M. (1980). Understanding attitudes and predicting social behaviour. EnglewoodCliffs, NJ: Prentice-Hall

Ajzen, I., 2002. Perceived behavioral control, self-efficacy, locus of control, and the theory of planned behavior 1. Journal of applied social psychology, 32(4), pp.665-683

Blavin, F., Ramos, C., Shah, A., & Devers, S. 2013. Lessons from the Literature on Electronic Health Record Implementation. Urban Institute. Retrieved 04 April 2016 from https://www.healthit.gov/sites/default/files/hit_lessons_learned_lit_review_final_08-01-2013.pdf

Chib, A., Lwin, M.O., Ang, J., Lin, H. 2008. Midwives and mobiles: using ICTs to improve healthcare in Aceh Besar, Indonesia. Asian Journal of Communication. Volume 18, Issue 4, 2008. Availability http://www.tandfonline.com/doi/citedby/10.1080/01292980802344182

Chou, H., Huang, Y., Wu, M., & Weng, Y. 2011. TAM2-based Study of Website User Behavior—Using Web 2.0 Websites as an Example.

Chuttur M.Y. (2009). "Overview of the Technology Acceptance Model: Origins, Developments and Future Directions," Indiana University, USA. Sprouts: Working Papers on Information Systems, 9(37). http://sprouts.aisnet.org/9-37

Cohen, P., West, S.G. and Aiken, L.S., 2014. Applied multiple regression/correlation analysis for the behavioral sciences. Psychology Press.

Department of Health. 2012. E-Health strategy South Africa: 2012-2016. Pretoria: Government Printer.

Gartner, 2009, eHealth for a Healthier Europe, Swedish Ministry of Health and Social Affairs

Gelman, A., 2012. What do statistical p-values mean when the sample= the population? Statistical Modeling, Causal Inference, and Social Science, 26.

Holden, Richard J., and Ben-Tzion Karsh. "A theoretical model of health information technology usage behaviour with implications for patient safety." Behaviour & Information Technology 28.1 (2009): 21-38.

Jeffrey, D.A., 2016. Testing the technology acceptance model 3 (TAM 3) with the inclusion of change fatigue and overload, in the context of faculty from Seventh-day Adventist universities: A revised model. Andrews University.

Kangas, J. L., Baldwin, A. S., Rosenfield, D., Smits, J. A. J., Rethorst, C. D. 2015. Examining the moderating effect of depressive symptoms on the relation between exercise and self-efficacy during the initiation of regular exercise. Health Psychology, Vol 34(5), May 2015, 556-565. http://dx.doi.org/10.1037/hea0000142

Mathhews, S. 2012. Technology Acceptance Model (TAM). Viewed on 17 March 2016 from http://smatts.blogspot.co.za/2012/08/technology-acceptance-model-tam.html

Mayosi, B.M. and Benatar, S.R., 2014. Health and health care in South Africa—20 years after Mandela. New England Journal of Medicine, 371(14), pp.1344-1353.

Mayosi BM, Lawn JE, van Niekerk A, Bradshaw D, Abdool Karim SS, Coovadia HM. Health in South Africa: changes and challenges since 2009. Lancet 2012; 380:2029-43.

Munassar, N.M.A. and Govardhan, A., 2010. A comparison between five models of software engineering. IJCSI, 5, pp.95-101.

Nassif, A.B., Capretz, L.F. and Ho, D., 2010, January. Software estimation in the early stages of the software life cycle. In International conference on emerging trends in computer science, communication and information technology pp. 5-13.

Paula, J., Modia, A.; Patel, J. 2016. Predicting green product consumption using theory of planned behaviour and reasoned action. Journal of Retailing and Consumer Services. Vol 29, March 2016, pp 123–134

Ruxwana, N.L., Herselman, M.E., and Conradie, D.P. 2010. ICT Applications as e-Health Solutions in Rural Healthcare in the Eastern Cape Province of South Africa [online]. Health Information Management Journal, Vol. 39, No. 1, 2010: 17-29. Availability: http://search.informit.com.au/documentSummary;dn=961281654549031;res=IELAPA

StatisticsHowTo. 2014. Cronbach's Alpha: Simple Definition, Use and Interpretation. Retrieved 15 January 2018 from http://www.statisticshowto.com/cronbachs-alpha-spss/

Venkatesh, V. and Bala, H. 2008. Technology acceptance model 3 and a research agenda on interventions. Decision sciences, 39(2), pp. 273-315.

Venkatesh, V. and Davis, F.D., 2000. A theoretical extension of the technology acceptance model: Four longitudinal field studies. Management science, 46(2), pp.186-204.

Walden, D.D., Roedler, G.J., Forsberg, K., Hamelin, R.D. and Shortell, T.M., 2015. Systems engineering handbook: A guide for system life cycle processes and activities. John Wiley & Sons.

Yzer, M., Weisman, S., Mejia, N., Hennrikus, D., Choi, K., & DeSimone, S. 2015. Informing Tobacco Cessation Benefit Use Interventions for Unionized Blue-Collar Workers: A Mixed-Methods Reasoned Action Approach. Prevention Science. August 2015, Volume 16, Issue 6, pp 811-821

SECONDARY USE OF CLINICAL DATA FOR SCIENTIFIC RESEARCH PURPOSES IN THE DITAS PROJECT

Paola Aurucci, Ilio Catallo, Mariet NouriJanian, Andrea Micheletti and Alberto Sanna
Center for Advanced Technology in Health and Wellbeing, IRCCS San Raffaele Hospital, Milan, Italy

ABSTRACT

Thanks to the advent of the digital revolution, coupled with advances in computing power, medical research is becoming increasingly data intensive. Extracting meaningful information from data sets and continuous flood of data is a challenge but holds unparalleled potential for observational studies on, e.g., the use and safety of drugs and rare diseases, epidemiological studies and digital disease detection. At the same time Cloud computing is used for two main reasons in secondary health research: first, to allow large-scale digital health data processing using readily scalable, external infrastructure; and second, to allow the sharing of digital health data with collaborators via a jointly usable IT environment. In this context of the DITAS project (https://www.ditas-project.eu/), to reconcile the often-competing values of data protection and innovation, aims at providing developers with processes and tools for the creation of GDPR compliant data-intensive applications in cloud computing. The aim of this paper is to show how in DITAS context researchers can lawfully further process for research purposes sensitive data originally collected for clinical activity when anonymization (Section 2) and pseudonymization (Section 3) techniques are applied.

KEYWORDS

Data Protection, Cloud, DITAS, Medical Research, Anonymization, Pseudonymization

1. INTRODUCTION

Over the past twenty-five years, the digital revolution has made it possible for healthcare facilities and health agencies to use massive digital databases for the storage of administrative and health service data gathered from routine clinical practice [Stenardo et al. 2013].

These databases, despite being designed and alimented for service-related organizational purposes, have proved to be useful in observational biomedical research [Romio et al. 2013].

Indeed, the number of studies that are based on these "secondary" data sources increased [Thiese 2014] [Sorense 1996]. In the last ten years, the availability of such a large group of heterogeneous data-sets, coupled with advances in computing power, is driving researchers to create sophisticated algorithms for the analysis of pre-existing and real-time data valuable to medical research (not previously combinable through matching techniques), to look for patterns, correlations, and links of potential significance [Mostert et al. 2016] and even to predict the likelihood of events based on continuous data stream through the application of machine learning techniques (Big Data Analytics) [Mantelero 2017]. Extracting meaningful information from data sets and continuous flood of data is a challenge but holds unparalleled potential for observational studies on, e.g., the use and safety of drugs and rare diseases, epidemiological studies and digital disease detection. Cloud computing is used for two main reasons in secondary health research: first, to allow large-scale digital health data processing using readily scalable, external infrastructure; and second, to allow the sharing of digital health data with collaborators via a jointly usable IT environment [Molnár-Gábor et al. 2017]. Yet, whilst the possibilities in terms of innovative research springing from "general analysis"[1] continue to expand, developments in IT and the utilization and reutilization of sensitive data have led to increasing concern about the effectiveness of data protection regulations still based on "notice and consent or anonymization paradigm", which is challenged in the context of data intensive medical research [Mostert et al. 2016]. On the other hand, concerns regard also the utilization of clouds, since their limitations in data security and in

[1] Recital 29, GDPR, available at: <http://data.europa.eu/eli/reg/2016/679/oj>.

respecting the rights of patient donors and research participants [Molnár-Gábor et al. 2017]. In the wake of a growing need for consistent protection of personal data that should not hinder responsible technological research of Big Data through manifold techniques, such as machine learning [Pagallo 2017], the European Parliament and the Council in 2016 adopted the much-awaited "Regulation on the protection of natural persons with regard to the processing of personal Data and on the free Movement of such data" (from now on "GDPR").[2] This act repealed Directive 95/46/EC and entered in force on May 2018. This two-year grace period was supposed to allow Member States to revise or adapt their legislation in order to come into compliance with the GDPR [Piciocchi et al. 2017]. The GDPR carves out numerous derogations for "historical or scientific purposes" [Maldoff 2016], also called "research exemption", to avoid restrictions on secondary uses of sensitive data (Article 5(1)(b), Recital 50). However, these derogations depend on the appropriate safeguards set up by the data controller in keeping with recognized ethical standards [Maldoff 2016], which lower the risks of discrimination or stigmatisation. Article 89(1) specifies that one way for the controller to comply with the duty to apply appropriate safeguards - where anonymization contrasts with the purposes of the processing [Bolognini and Bistolfi 2017] - concerns the use of "pseudonymization". The use of this strong de identification technique however is not enough and must be combined with additional technical and organizational measures such as security measures for the conservation of the secret key used for the encryption or hash function, the principle of privacy-by-design, privacy-by-default, transparency and the presence of a data protection officer [Bolognini and Bistolfi 2017]. For the sake of illustration, in the following sections we consider how such data protection aspects have been addressed in the context of the DITAS project (https://www.ditas-project.eu/), a H2020 European project that aims at providing developers with processes and tools for the development of GDPR compliant data-intensive applications in cloud computing. In particular, we examine the case of a medical research institute that would like to provide its researchers with a digitalized data sharing platform for the access to clinical data. Ideally, the platform should assist researchers in expressing their need for research data and provide an immediate way of downloading data of interest. At the same time, researchers should be prevented from accessing patients' personal data while carrying out their research activities. DITAS attempts to reconcile these two contrasting objectives. It does so by mediating the access to the clinical data, so as to put in place pseudonymization (or – when necessary, anonymization) techniques. Hence, researchers never get direct access to clinical data. Instead, DITAS provides a uniform entry-point to the data, and decides, depending on the person requesting the access, what kind of transformation should be applied to the data prior to their delivery. The aim of this paper is to show how in DITAS context researchers can lawfully further process for research purposes sensitive data originally collected for clinical activity when anonymization (Section 2) and pseudonymization (Section 3) techniques are applied. Since GDPR crows back the power to EU Member States, allowing them to introduce further safeguards with regard to the processing of genetic data, biometric data or data concerning health, in the Conclusion the paper aims to restrict the focus of analysis to the Italian legal framework that on this specific subject is a real "Dedalus"[3]of different inputs and rules [Piciocchi et al. 2017] . The Italian scenario shows that delegation of powers back to the national legal systems of the Member States entails a number of critical drawbacks, like hampering the progress of cross-border medical research, and preventing the achievement of the second – but of equal importance – goal of the GDPR: to guarantee the "free movement of personal data".

2. ANONYMIZATION CHOICE: USE CASE 1

GDPR applies to "personal data," which is defined, in part, as "any information relating to an identified or identifiable natural person".[4]

---

[2] Regulation (EU) 2016/679 of the European Parliament and of the Coucil of 27 April 2016 on the protection of natural persons with regard to the processing of personal data and on the free movement of such data, and repealing Directive 95/46/EC (General Data Protection Regulation), L 119/1, available at: <http://data.europa.eu/eli/reg/2016/679/oj>.

[3] In Greek Mythology Deadalus was a skillful craftsman which was imprisoned within Daedalus the labyrinth he created.

[4] Article 4(1), GDPR, available at: <http://data.europa.eu/eli/reg/2016/679/oj>.

The final text of the GDPR should be a cause for celebration by the medical research community for two reasons: first, scientific research is defined broadly in the final text, including fundamental research, applied research and privately funded research and – second, it grants some exemptions from its requirements when personal data are processed for scientific research [Dove 2018].

However, GDPR continues to clearly favor scientific research that uses anonymized data[5] [Penasa 2018]. Indeed, Article 89(1) states that, if the purposes of the scientific research can be achieved using these data, *then it can be fulfilled by further processing which does not permit or no longer permits the identification of data subjects".*[6] Datum which have been anonymized is no longer "personal data" and is therefore not subject to the requirements of data protection law. In the context of the use case n. 1 of the DITAS project, researchers can access anonymized nutritional data, which have been collected as part of the day-to-day clinical activity, and that can also serve as a valuable source of information for conducting, e.g., statistical studies on the population. Researchers can then use this data collected during the clinical practice for new research purposes, unknown at the time of the collection, without the need to respect GDPR provisions, usually trigger by the use of personal data or personal identifiable information [Vayena and Gasser, 2016]. Furthermore, in the scientific research context the use of anonymization technique is then functional not only for the storage of data, but also for enabling the use of historically collected data for secondary research, their eventual sharing with third parties, and potential linkage to other datasets. For this reason, use case n. 3 of DITAS project provides that researchers external to the research institute can only access anonymized data, dispensing them to respect GDPR requirements governing data sharing [Vayena and Gasser, 2016]. The Article 29 Working Party (WP29) 2014 in his Opinion on Anonymization Techniques[7] set a high bar for what qualifies as fully anonymized data. WP29 states that, taking into account all means "likely reasonably" to be used to re-identify the data, anonymization must be "irreversible" and "as permanent as erasure."[8] At the same time, according to Article 29 WP29 there are many anonymization techniques[9] but it also recognized that the risk of re-identification of each of these technique is never zero, since they present risk in terms of inference and *linkability*.[10] It is fundamental then to establish what constitutes "an acceptable low level of risk". Recital 26 GDPR states that "account should be taken of all means reasonably likely to be used (…) to identify the natural person either directly or indirectly". This language suggests a "reasonable acceptable standard of anonymization" should be preferred to the "impossible standard". In any case, to anonymize data, it is necessary to proceed with an ex ante risk of identification assessmen [Bolognini and Bistolfi 2017] that must be context and technology dependent. According to Recital 26 the risk re-identification should be analyzed considering different factors such as time, effort or researches needed to re-identify data, the purpose of the processing, the risk of external attack and the available technology at the time of the processing. Furthermore, even if the risk of re-identification of an anonymized dataset is calculated to be acceptably small at the time of data disclosure, potential linkage of that dataset to others at an unknown future point in time may increase the possibility of identifying the person or of inferring information about him/her connecting different data and trace them back to a single subject. This is why, as required in the resents Guidelines on the protection of individuals with regard to the processing of personal data in a world of Big Data[11], data controllers "shall regularly review the assessment of the risk of re-identification". Much of the literature have, in fact, shows that in the era of big data with potentially multiple data linkage opportunities and new analytical methods, e.g. data mining, increase the risk of re-identification [de Montjoye et al. 2015].

[5] Recital 26, GDPR, available at: <http://data.europa.eu/eli/reg/2016/679/oj>.
[6] Article 89(1), GDPR, available at: <http://data.europa.eu/eli/reg/2016/679/oj>.
[7] Article 29 Working Party, Opinion 05/2014 on anonymization techniques, 10 April 2014, available at: https://ec.europa.eu/justice/article-29/documentation/opinion-recommendation/index_en.htm.
[8] Id., at 6.
[9] Id., at 12.
[10] Id., at 12-17
[11] Council of Europe, Guidelines on the protection of individuals with regard to the processing of personal data in a world of Big Data, available at: <https://rm.coe.int/16806ebe7a>.

3. THE PSEUDONIMIZIATION CHOICE: USE CASE 3

As seen in the previous paragraph GDPR continues to clearly favor scientific research that uses anonymized data.

However, when this could also not be possible, since currently popular method of redaction of any information which may increase the risk of re-identification of the individual can greatly compromise the scientific utility of the data.

Furthermore, potential linkage of that dataset to others and the use of analytics [Mantalero 2017] will increase the number of attributes will need to be considered in the assessment of risk, rendering any calculation of this risk of reidentification difficult or potentially unknowable. In these cases scientific research may be carried out using personal data. To lawfully process personal data for scientific research purposes at least the following legal elements must be considered in advance: a) the respect of principles set by Articles 5 and b) The presence of legal grounds for data processing c) the obligation to inform data subjects according to Articles 13-14 GDPR. In order to reconcile the often-competing values of data protection and innovation, in all this three set of rules, the GDPR carves out numerous derogations (also called "research exemptions") for "historical or scientific purposes" [Chassang 2017], allowing researchers to avoid restrictions on secondary uses of health data. These "research exemptions" will be analyzed in the following paragraphs. In the particular case of the DITAS project, researchers internal to the research institute are allowed to access pseudonymized data, whereas external researchers can only access anonymized data. DITAS transparently applies the proper transformation depending on the kind of person that initiates the data request, thus greatly simplifying the provisioning of a data sharing platform for a research institute

3.1 GDPR Research Exemptions to the General Principle

Starting from the general principles, the "purpose limitation" principle of Article 5 (1)(b) - which is of fundamental importance in the general application of data protection rules and compliance with these rules - requires that specified, explicit and legitimate purposes should be known and defined at the moment of data collection and that data must not be further processed in a way that is incompatible with those purposes. In the contest of modern observational medical research – where sensitive data are constantly shared, aggregated and reused - it is for the controller increasingly hard to predict at the time of data collection all the specific future purposes of investigations [Mittelstadt et al. 2016]. To re-contact each individual data subject to obtain explicit consent or to inform her/him on the new research purpose is often impractical, prohibitively expensive and could undermine the validity of outcomes [Vayena and Gasser, 2016]. The possibility to predict at the time of the collection future possible purposes of research it becomes impossible when researches imply the use of AI techniques - like machine learning - that are hypothesis-generating rather than hypothesis-testing research. This is because the aim of these techniques is to discover unforeseen connections between data points, that cannot be accurately predicted prior to using the data. In order to facilitate of secondary observational research - where health data are further processed for purposes that are different from those for which they were originally collected – Article 5(1)(b) also provides that further processing for archiving purposes in the public interest, scientific or historical research purposes shall not be considered incompatible with the initial purposes. This language is reflected in Recital (50) that goes even further, stating that further processing for scientific research should in any case be considered "compatible lawful processing". This recital also provides the possibility of using the original data subject's consent, or another legal basis that has allowed the original collection when the purpose of the new processing (further processing or secondary use) is "compatible" with the purposes initially declared. This presumption of compatibility with the purposes announced at the time of the collection is related to the exemption to the principle of "storage limitation" which allows personal data to be stored for longer periods insofar as the personal data will be processed solely for archiving purposes in the public interest, scientific or historical research purposes or statistical purposes. Both articles, however, specify that this presumption of compatibility with the initial purposes is not fully automatic, but requires a data controller and a data processor to implement appropriate safeguards designed to protect the data subject and to reduce risks in accordance with Article 89(1). Article 89(1) specifies that the purpose of this safeguards is to ensure that specific technical and organizational measures are in place in particular in order to ensure respect for the principle of data minimization. The use of "pseudonymization" techniques allows the controller to ensure respect for the principle of data minimization (where anonymization contrasts with the purposes of the processing as seen at the beginning of the paragraph). Recital 29 specify that when pseudonymization is used to make "general analysis" legitimate it must be combined with additional security measures to reduce the risk of reidentification [Bolognini and Bistolfi, 2017]. Just to name a few: technical measures to store the cryptographic system or hash function separately from the data in order to prevent the pseudonym from being attributed to the original data, measure to protect the places where they are stored from undue access, authentication mechanisms (ID and password biometrics or both) to to allow (researchers) or prevent (unauthorized persons) access to the keys or to correlation tables. This presumption of compatibility has been subject to restrictive interpretation the European Commission which in the draft Q&A analyzing the interplay between the EU Clinical Trials Regulation ("CTR") and the EU General Data Protection Regulation ("GDPR") excluded the presumption of compatibility provided under Article 5(1)(b) in all circumstances. However, following the interpretation of the European Data Protection Board (from now on EDPB) on this draft, the EC has highlighted in the new Q&A that secondary use of clinical trial data for scientific research purposes is by default compatible with its original use, in accordance with Article 5(1)(b) of the GDPR. As a result, it should not be necessary to obtain a new consent in order to engage in additional secondary research.

3.2 GDPR Research Exemptions to the Legal Basis for Processing

Respecting fundamental principles of data protection listed under Article 5 is not enough for lawfully processing personal data. The controller – before starting the processing - must find one or more legal grounds (legitimate basis) for data processing, listed under Article 6. However, primary and secondary medical research implies the use of genetic data, biometric data, data concerning health, which are traditionally held to be particularly sensitive type of personal data, necessitating stricter condition for its processing. For this reason, GDPR requires that for the processing of these special categories of data, the legal basis identified under Article 6 shall be applied only if Article 9 provides for a specific derogation from the general prohibition to process this data (i.e. exceptions to the general prohibition). While patient consent provides both a lawful basis and a permitted condition for processing personal data collected in the context of a health, it is one only of the available options, and others may be more appropriate. The EU's data protection advisory body, the Article 29 Working Party (WP29, now the European Data Protection Board), issued its finalised Guideline on the concept of consent under the GDPR in mid-April 2018 (the EU Guidelines)[12]. The EU Guidelines explain that under the GDPR, consent is not always required to process data for the purpose of scientific research—for example, processing may be lawful on the basis that it is necessary for the performance of a task carried out in the public interest, for compliance with a legal obligation, or for the legitimate interests of the data controller or a third party, as long as appropriate safeguards are in place and the processing is fair, lawful, transparent and accords with data minimisation standards and individual rights, and that similar considerations also apply to the processing of special category data under the scientific research condition (Article 9(1)(j)). However, s seen in the previous paragraph further processing of clinical data for scientific research purposes is by default compatible with its original use, in accordance with Article 5(1)(b) of the GDPR. As a result, it should not be necessary to obtain a new consent in order to engage in additional secondary research

3.3 GDPR Research Exemptions to the Obligation to Inform Data Subjects

To respect general fundamental principles and to ensure a valid legal grounds/lawfulness conditions of data processing could not be enough. A general rule, coming from Articles 13-14 of the GDPR, imposes to adequately inform data subjects about their personal data processing, even in case of further processing, no matter whether a new legal basis is required or not. This is the "privacy notice" or "privacy policy" that we are used to receive as data subjects and users, almost in all cases personal data are going to be collected and processed (some exceptions occur, for instance, just in justice/law enforcement scenarios). Collecting data from the (dark, deep, surface) web, such as collecting data from publicly available records and documents, or

[12] Article 29 Working Pary, Guidelines on consent under Regulation 2016/679, available at: https://ec.europa.eu/newsroom/article29/item-detail.cfm?item_id=623051

from another hospital, implies that data are not obtained directly from the data subject. In this case (collection not from the data subject, but from third parties), Article 14.5.b) of the GDPR applies and justify at certain conditions even the exclusion of the obligation for the data controller to inform data subjects, where and insofar as "the provision of such information proves impossible or would involve a disproportionate effort, in particular for processing for archiving purposes in the public interest, scientific or historical research purposes or statistical purposes, subject to the conditions and safeguards referred to in Article 89(1) or in so far as the obligation referred to in paragraph 1 of this Article is likely to render impossible or seriously impair the achievement of the objectives of that processing. In such cases the controller shall take appropriate measures to protect the data subject's rights and freedoms and legitimate interests, including making the information publicly available. However, the only exception to a controller's Article 13 information obligations, where it has collected personal data directly from a data subject, occurs "where and insofar as, the data subject already has the information " (Article 13(4)). The WP29 in its Guidelines on transparency under Regulation 2016/679[13] notes that this exception should be construed narrowly, and the principle of accountability requires controllers to demonstrate and document precisely what information the data subject has, how and when they received it and that no changes have occurred to that information that would render it out of date. This is a problem Scientific Institute for Research, Hospitalization and Health Care, like those involved in DITAS use case 3. In this case due to the instrumental nature of the activity of health care provided by the aforementioned institutions with respect to research, the further processing for research purposes is carried out not by a third party, but by the same controller that collected the data in the first place for providing medical assistance. This imply that controller should recontact and inform the patient each time he/her datum are going to be used for a scientific research purpose that was not specified at the moment of the collection. It seems here to be room for the use Recital 33, which recognize the difficulty to predict all possible specific research purposes at the time of data collection and seems to allow for a broader consent to certain areas of scientific research and for participants to select those areas. However, the Article 29 Working Party in its Guideline on consent down play Recital 33 to the point of becoming nearly non-existent. More clarity must be provided at European level on this topic.

4. CONCLUSION

The previous Sections focus on which GDPR requirements should be respected in order to lawfully further processing sensitive data in medical observational studies.

This is fundamental in order to reach the aim of the DITAS project: providing developers with processes and tools for the development of GDPR compliant data-intensive applications in cloud computing.

However, although the GDPR is intended favor scientific research and to reduce inconsistencies in the implementation of data protection across the EU, it crows back powers to EU Member States. In line with Article 9(4) of the GDPR - that allows Member States to introduce further safeguards with regard to the processing of genetic data, biometric data or data concerning health [Pagallo 2017] - the recently reformed Italian Data Protection Code ("IDPC"), as amended by Legislative Decree no. 101/2018, adds and requires further conditions in order to lawfully process secondary health related data for scientific purposes without consent or information to data subjects. At present terms, the relevant legal framework applying to processing data for medical research in Italy is a complex combination of IDPC's provisions, old and new deontological rules, and old and "updated" general Authorizations of the Italian Data Protection Authority (from now on "IDPA" or "Garante"). Furthermore, Article 22(11) of the Legislative Decree no. 101/2018 says that rules concerning the processing of genetic data, biometric data and data concerning health, of the previous Code of Privacy (Legislative Decree no. 196/2003) continue to apply, if compatible with the GDPR, until the biannual guidance on the processing of these data, foreseen in in Article 2 *septies* of the IDPC, will be provided by the Garante. This document (biannual guidance) will identify the safeguard measures (including security measures, such as encryption and pseudonymization) to apply for the processing of biometric, genetic and health-related data, having regard to the specific purposes of processing. This provision of the IDPC appears in contradiction with the principle of self-accountability of data controllers

[13]Article 29 Working Pary, Guidelines on transparency under Regulation 2016/679, available at: https://ec.europa.eu/newsroom/article29/item-detail.cfm%3Fitem_id%3D622227+&cd=4&hl=it&ct=clnk&gl=it

(i.e., no *ex ante* identification of the proper safeguards to apply for a specific research purposes but *ex post* onus on data controllers to demonstrate compliance with the GDPR in case of breach). However, until this document won't be issued by the Garante, it is not clear which rules actually applies to the secondary processing of sensitive data for scientific research purposes. The only thing that is sure is that fragmentation of data protection legislation inside EU, and particularly strict requirements for lawfully processing of data for health research (which creates further burdens for data controllers and processors) may hamper the responsible data sharing that DITAS project aims to provide, and consequently, the free movement of personal data within the internal market.[14] The hope is that emerging code of conducts at European level, encouraged by the GDPR in Articles 40 and 41, may enable harmonization in the application of data protection law in this specific field [Marelli and Testa, 2018]. Initiatives already exist to promote code of conducts on the large-scale re-use and linkage of health-related and genomic data.[15] On the other hand, code of conducts at National level may also constitute a valid point of reference for the scientific community to understand how to be compliant and a possibility for research organizations, patient advocacy groups, and industrial representative [Marelli and Testa, 2018] to agree on practical data protection rules that will meet the needs of the sector as well as the requirements of the GDPR.

ACKNOWLEDGEMENT

This work is supported by European Commission H2020 Programme through the DITAS Project no. 731945.

REFERENCES

Barfield et. al., 2018. *Research handbook on the law of artificial intelligence. Title (in italics).* Edward Elgar Publishing, Cheltenham, UK. Northampton, MA, USA.

Bodorik P. et al, 1991. Deciding to Correct Distributed Query Processing. *In IEEE Transactions on Data and Knowledge Engineering*, Vol. 4, No. 3, pp. 253-265.

Bolognini L. et al., 2017. Pseudonymization and impacts of Big (personal / anonymous) Data processing in the transition from the Directive 95 / 46 / EC to the new EU General Data Protection Regulation. *Comput Law Secur Rev Int J Technol Law Pract,* Vol. 33, No. 2, pp. 171–81.

Chassang G., 2017, The impact of the EU general data protection regulation on scientific research. Ecancermedicalscienc, Vol. 11, No. 709, pp. 1-12.

De Montjoye Y. A. et al., 2015. Unique in the shopping mall: on the reidentificability of credt card data. *Science*, Vol. 347, No. 6221, pp. 536-539.

Dove E. S., 2019, The EU General Data Protecion Regulation: *Implication for International Scientific Research in the Digital Era.* Journal of Law, Medicin & Ethics, Vol. 46, No. 4, pp. 1-32.

Mantelero A., 2017. Regulating Big Data. The guidelines of the Council of Europe in the Context of the European Data Protection Framework. *Computer Law and Security Review*, Vol. 33, No 5, pp. 584-602.

Marelli L. and Testa G., 2018, Scrutinizing the EU General Data Protection Regulation, Science Vol. 360, No. 6388, pp. 496-498.

Molnár-Gábor F. et al., 2017. Computing patient data in the cloud: practical and legal considerations for genetics and genomics research in Europe and internationally. *Genome Medicine*, Vol. 9, No. 58, pp. 1-12.

Mondschein C.F. et al. 2019, The EU's General Data Protection Regulation (GDPR) in a Research Context. In: Kubben P. et al. (eds) *Fundamentals of Clinical Data Science*. Springer, Cham, pp. 55-71.

Mostert M. et al., 2016. Big Data in medical research and EU data protection law: challenges to the consent or anonymise approach". *Eur J Hum Genet*, Vol. 24, No. 7, pp. 956-960.

Pagallo U., 2017. The Legal Challenges of Big Data: Putting Secondary Rules First in the Field of EU Data Protection. *European Data Protection Law Review*, Vol 3, No. 1, pp 36-46.

Penasa S. et al., 2018. The EU General Data Protection Regulation: How will it impact the regulation of research biobanks? Medical Law International, Vol. 18, No. 4, pp. 1-15.

[14] Recital 13 of the GDPR.

[15] BBMRI-ERIC, Code of conduct for health research, available at: http://code-of-conduct-for-health-research.eu/

Piciocchi C. et al., 2018. Legal issues in governing genetic biobanks: The Italian framework as a case study for the implications for citizen's health through public-private initiatives. *J Community Genet*, Vol. 9, pp. 177–90.

Romio S. et al., 2013. Real-World Data From the Health Decision Maker Perspective. What Are We Talking About?, *Epidemiology, Biostatistics and Public Health*, Vol 10, No. 3, e8979.

Stendardo A. et al., 2013. Legal aspects regarding the use and integration of electronic medical records for epidemiological purposes with focus on the Italian situation. In *Epidemiology Biostatistics and Public Health,* Vol. 10, No. 3, e8971.

Thiese M. S., 2014. Observational and interventional study design types; an overview. *Biochem Med (Zagreb)*, Vol. 24, No. 2, pp. 199-210.

DATA PROCESSING APPROACHES FOR LUNG CT- IMAGE ANALYSIS UNDER RESOURCE CONSTRAINTS

Aleksandra Vatian, Artyom Lobantsev, Nikita Gorokhov, Mikhail Mirzayanov, Georgii Korneev, Natalia Gusarova and Anatoly Shalyto
ITMO University
49 Kronverksky pr., St. Petersburg, 197101, Russia

ABSTRACT

In this paper we consider the methods of the improvement the efficiency of lung tumor classification made by deep convolutional networks in conditions of limited resources. We propose two approaches of the efficiency improvenets. The first is the consideration of the specifics of the brightness distribution of CT images of lung cancer and preprocessing the data according to these specifics. Experiments with the inclusion of preprocessing proposed by this method showed an improvement in the classification efficiency on a small amount of computing resources. The second is the use of an improved network training method for the task of domain adaptation, which includes the integration of a specific loss function into the network architecture. That loss function aligns the latent feature space of the network trained on source large dataset of CT scans with the latent feature space of the target small dataset, where the network will be used for inference. The integration of this method has shown comparable results with the traditional fine-tuning method in a noticeably smaller amount of time and on the same hardware resources.

KEYWORDS

Lung Tumor, Resource Constraints, Domain Adaptation, Data Preprocessing, CT-Images

1. INTRODUCTION

In recent years, medical images have become one of the leading sources of diagnostic and therapeutic information, especially for such severe and fast diseases as cancer. Deep learning methods occupy an increasing place in the processing of medical images [Qayyum]. They are used to improve the efficiency of segmentation, classification, and abnormality detection in images generated from different clinical imaging modalities including CT, MRI, X-rays, etc. In recent years, the advantages of deep learning have been experimentally revealed in comparison with other methods of machine learning in relation to medical image processing. In particular, it is shown that features created within deep learning methods significantly outperform the predictive ability of predefined features like tumor volume and maximum diameter [Hosny].

Progress in the development of deep learning methods is associated with convolutional neural networks (CNN). In recent years, significant progress has been made in overcoming the main obstacle in applying CNN to medical images – the small amount of available medical data. UNet architecture [Ronneberger] has been developed, which allows network training on ultra-small datasets – up to several dozen of images. Generative adversarial networks (GAN), proposed in [Goodfellow], allow, in essence, to realize the synthesis of an unlimited number of images, which by statistical properties are arbitrarily close to the images from the training set. This property of GAN is already actively used in the formation of medical imaging datasets [Kazeminiaa], including for the diagnosis of lung cancer [Chuquicusma, Dakai].

Transfer learning [Weiss] also provides the ability to work with small datasets. This approach allows you to use CNN, trained on a large source dataset, to work in another, target domain, by providing additional training on a small dataset containing data from the target domain. Transfer learning has proven to be an effective to working with medical images in CNN [Shin].

But lately, negative trends have been noted in the successful development of CNN in relation to medical images [Dong Wook Kim]. As practice shows, this development follows the path of increasing the complexity of the CNN architecture, which causes an extensive increase in the required computing resources (compare, for example, [Hammack] and [Karras]). There is a tendency to reduce the threshold for entering into the problems of medical image processing by creating more and more high-level and user-friendly software products and frameworks, but their full use requires more and more expensive hardware [Koryagin]. Finally, recent published studies are increasingly reporting on the extremely high efficiency of algorithms for the diagnostic analysis of medical images. However, according to leading journals [AI diagnostics], these results are often inspired by overfitting and are weakly related to the properties of datasets themselves, and therefore cannot be applied in clinical practice.

A promising way to overcome these trends is a more complete and effective account of the properties of the images themselves and the datasets collected from them, i.e. smarter data processing. Therefore, the attention of scientists is increasingly attracted by data processing methods, which make it possible to single out the design features characteristic of datasets themselves, and to rely on them when creating reliable and robust means of clinical diagnostics.

The article discusses two approaches to improving the efficiency of classification made by CNN associated with improved data processing, which we will consider on the example of an important problem – CT diagnosis of lung cancer. It is shown that even under resource constraints, they can give a gain in the efficiency of medical image analysis comparable to the use of "heavy" methods.

2. DATA PRE-PROCESSING BASED ON ACTUAL DISTRIBUTION OF IMAGES' BRIGHTNESS

2.1 Background

Preprocessing of medical images is meaningfully defined as a sequence of actions on newly received data that is executed on on newly received data in order to achieve the best fit with existing data [Lashari]. As a rule, the preprocessing of medical data used in CNN aims to accelerate convergence when training a network.

Various algorithms have been proposed for preprocessing medical images. The authors [Öztürka], aiming to regulate brightness and contrast variations in the image and suppresses noise, use different preprocessing algorithms and their combinations, including removing the median value or background from the original image as well applying median or wiener filter and adaptive threshold. In [Perumal], at the preprocessing stage of MRI images of lung cancer, the task of noise reduction is posed, and various filtering techniques like Wiener, Median, and Gaussian are used are used for this.

However, as shown in the literature [Vatian], noises in high-tech medical images, such as CT and MRI, have complex distribution patterns (a mixture of off-center gamma distributions for CT images and Rice distribution for MRI images), and attempts at noise filtration lead not only to the loss of semantically important details, but also to the appearance of artifacts.

Attempts are being made to take into account the complex nature of the brightness distributions of medical images. In [Zhenglun Kong] the authors constructed a histogram of MRI images of brain tissue and with the help of wavelet-transformation identified 4 distinct peaks in it. Using the obtained peak values as threshold values, they constructed a procedure for tissue clustering based on CNN. Such a preprocessing provided a proven increase in the reliability of the segmentation result. In [Joris] the authors investigated the histograms of CT and MRI brain images and established their heteroscedastic nature, that is varying degrees of statistical dispersion of tissue intensity distributions. In this regard, they propose to use non-linear transformation for image preprocessing, based on current estimate of of the intensity variance for each intensity value. The proposed algorithm improves the quality of segmentation of brain tissue measured by Dice scores.

However, despite their prospects, the proposed algorithms are quite sophisticated. In addition, we were unable to find similar approaches in the available literature in relation to CT images of the lungs. In general, as the reviews [Qayyum, Parmar] show, in the real practice of using CNN for processing medical images, the simplest methods prevail. These are data zero-centering and data normalization

$$x' = x - \overline{x}, \quad x' = \frac{x - \overline{x}}{\sigma}, \quad \sigma = \frac{\sqrt{\left(x - \overline{x}\right)^2}}{N - 1}, \qquad (1)$$

where x is current pixel brightness, $\overline{x}$ and σ are mean and stddev respectively, N is number of pixels in analyzed image(s). This approach obviously assumes that the original images obey a unimodal (usually Gaussian) distribution. However, if the assumption of unimodality is not justified in practice, it will not give the desired improvement in efficiency or may even worsen the situation. In connection with the above, in our work, we posed the following question: does the consideration of the specifics of the brightness distribution of CT images of lung cancer in its preprocessing affect the effectiveness of their classification by CNN?

2.2 Materials and Method

For research, we used data from dataset LUNA16[1]. The original (total) dataset was divided into three parts, namely training, validation and test samples, with total amount of 6876, 1297 and 1622 images, 17.190.000, 3.242.500 and 4.055.000 pixels respectively. We built histograms of the brightness distributions of pictures and pixel brightness for each dataset separately and for all together, presented in figure 1, a-d. The variation of the pixel distributions of individual images is illustrated in Figure 2.

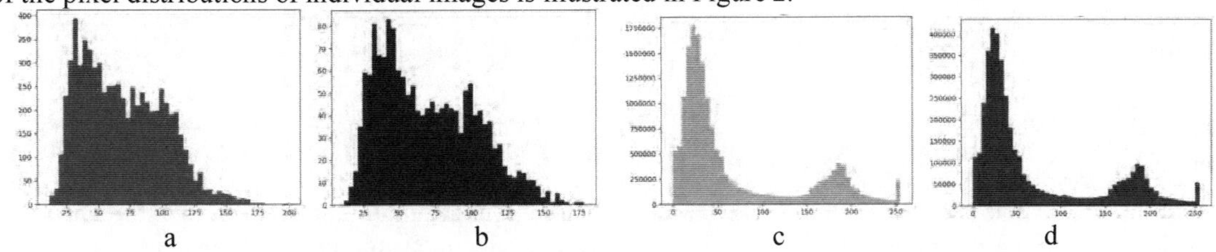

Figure 1. Images (a, b) and pixel (c, d) average brightness histograms for datasets (a,c – train, b, d – test)

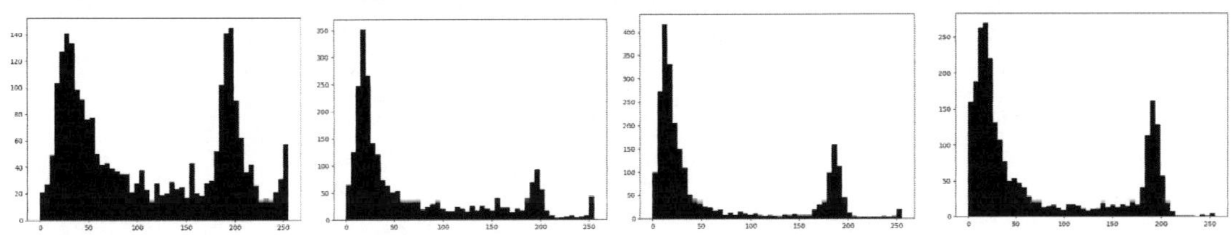

Figure 2. Examples of individual image brightness distributions

For a more subtle assessment of the brightness distributions, we divided the total pixel brightness range (0–255) into two subranges and built pixel brightness distributions for each subband. With a half division by brightness (into subbands 0–127 and 128–255), we obtained almost identical results for all datasets, which are illustrated in Figure 3, a. When the boundary is shifted, the situation is qualitatively preserved, but the ratio of the maxima changes (see example in Figure 3, b). In Figure 3, c, it is possible to see a general estimate of the ratio of the magnitudes of the maxima of the distributions to the average brightness of the images. In our experiments, we used CNN with a rather simple architecture[2] based on LeNet-5 architecture. Data preprocessing was performed in the following way. The array of pixels of each image was divided into subbands 0–127 (dark side) and 128–255 (light side). Mean and stddev values were calculated separately for the pixels of the dark side, the pixels of the light side, and for the entire image as a whole. Then we used three options:

Basic. The input of CNN is fed directly to the data from the total dataset.

Basic + mean/std. All pixels of each image are normalized to mean and stddev values according to (1) and (2).

Basic + 2 mean/std. The pixels of the dark and light sides are separately normalized to the corresponding mean value and average deviation according to (1) and (2). Assuming that the brightness distributions of pixels on the dark and light sides are approximately normal, we used the "3 sigma" rule for normalization, i.e. subtracted 3 from the dark part and added 3 to the light part.

---

[1] https://github.com/swethasubramanian/LungCancerDetection/tree/master/src/data

[2] http://datareview.info/article/kak-obuchit-glubokuyu-neyronnuyu-set-s-pomoshhyu-7-strok-koda/

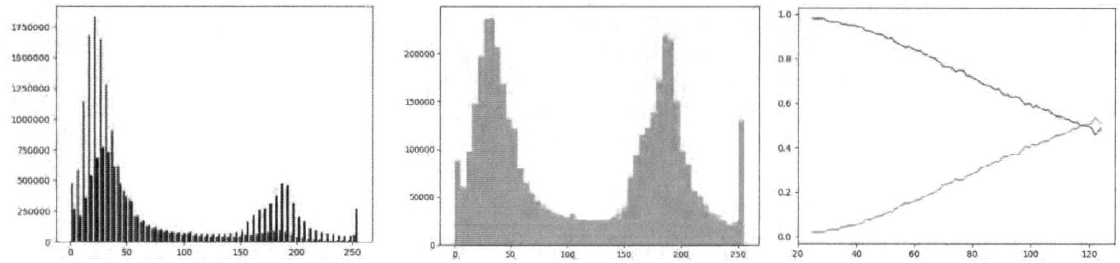

a b c

Figure 3. a – distributions of total dataset pixel brightness below average (red) and above average (blue); b – distribution of train dataset pixels with brightness above the third quartile; c – dependence of the proportion of pixels with a value of ≥128 (orange) and <128 (blue) on the average brightness of the picture

2.3 Results and Discussion

A comparative assessment of brightness distributions across all datasets constructed (Figure 1) shows that this distribution is rather far from normal. It is strongly scewed to the left, and there is some "dimple" near the mean value. The pixel brightness distribution (Figure 1) demonstrates an obvious bimodality with subsidence in the gray zone (127–128), and this distribution pattern is maintained for individual images (Figure 2). At the same time, the distribution components on the dark and light sides are quite close to normal (this assumption was successfully used in the organization of normalization in the *Basic + 2 mean/std* option).

Comparing histogrammes of individual images (Figure 2), we see that the absolute values of the maxima change, but their position remains unchanged. This regularity is confirmed on the average graphs (Figure 3, a, b). It is also worth noting the almost linear dependence of the number of "dark" (0–127) and "bright" (127–255) pixels on the average brightness of the image.

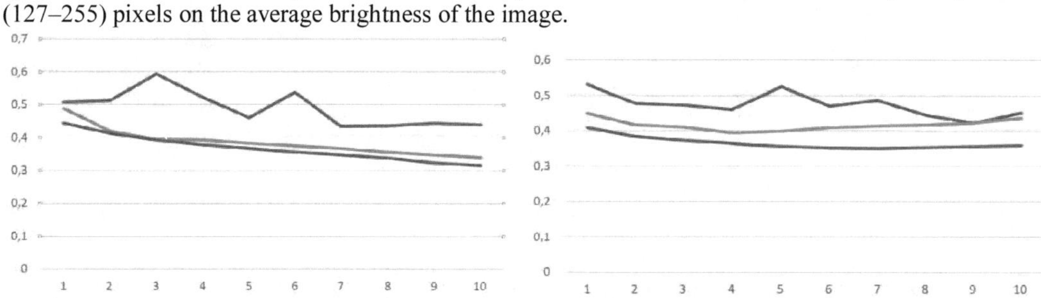

a b

Figure 4. Dependence of loss function on the number of learning epochs for train (a) and test (b) datasets: blue – *Basic* option, orange – *Basic + mean/std* option, green – *Basic + 2 mean/std* option

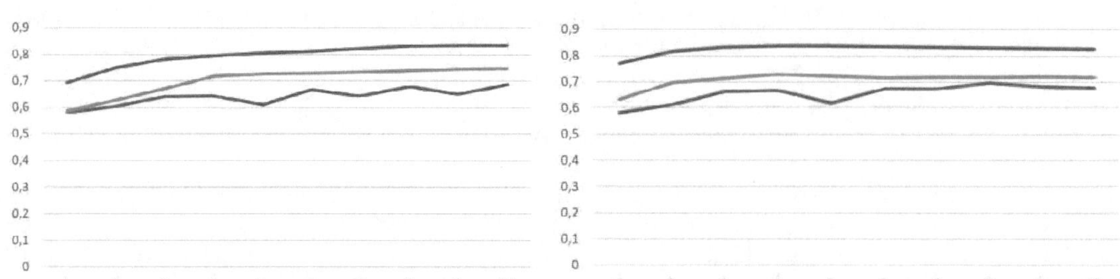

Figure 5. Dependence of roc_auc function on the number of learning epochs for train (a) and test (b) datasets: blue – *Basic* option, orange – *Basic + mean/std* option, green – *Basic + 2 mean/std* option

Graphs of the loss function and roc_auc on the training and validation samples, depending on the number of training epochs for each of the preprocessing options, are shown in Figure 4, a, b and Figure 5, a, b. Analysis of the graphs shows good convergence of results even with a small number of learning epochs, as well as the undoubted advantage of the Basic + 2 mean / std option. We emphasize that these results were obtained with very scarce resource costs, both in terms of hardware and in terms of computing time, which allows us to give a confidently positive answer to the question posed in part 2.1.

3. TRANSFER LEARNING BASED ON ACTUAL CORRELATION DEPENDENCIES IN IMAGES AND LATENT FEATURE SPACE

3.1 Background

Transfer learning is one of the popular methods of using CNN to handle small datasets. Consider CNN solving the problem T = {Y, ℓ(*)} in the domain D = {X, P(X)}. Let D_s be the source domain with the corresponding task T_s and D_t be the target domain (usually far smaller than D_s) with the task T_t. Then, according to [Weiss], transfer learning is the process of improving the function ℓ (*) for the T_t task using information from D_s. In order to do this, we first train CNN on a large dataset representing D_s. Then, leaving unchanged (freezing) the obtained weights of all layers of CNN, except the last, we replace the last layer with a new layer having the proper number of classes to solve the task T_t. Thereby we use the vector of high-level features formed by CNN on D_s to train last layer as a linear classifier for the problem T_t. In general, we can retrain not only the last layer, but also some of the previous ones, and with different learning rates. This technique is called fine tuning [Yosinski].

Transfer learning has already become the de facto standard in the processing of medical images for the diagnosis of lung cancer, as evidenced by the reviews and confirmed by experiments [Tao Tan, Nóbrega, Nishio]. However, the development of transfer learning in this area, as the analysis of the literature shows, goes mainly along the path of using more and more sofisticated CNNs and varying the parameters of layers being retrained. The statistical properties of the images themselves as a source of additional information about the relationship between D_s and D_t are practically not used. Meanwhile, within the framework of transfer learning, such opportunities exist, and one of them is provided by the DeepCoral algorithm.

DeepCORAL algorithm was proposed in [SunSaenko] as extension of the CORAL algorithm [SunFeng]. CORAL's key idea is to align second-order statistics of the source and target domain distributions. DeepCORAL extends this idea with the concept of training CNN's layers with an additional constraint on the alignment of the correlations of layer activations. For this for selected layers whose correlations are to be aligned, an additional loss function is introduced:

$$\ell_{CORAL} = \frac{1}{4d^2}\|C_S - C_T\|_F^2, \tag{2}$$

where $\|\cdot\|_F^2$ denotes the squared matrix Frobenius norm, and C_S, C_T are the covariance matrices of the source and target data respectively. The CORAL losses collected from the selected layers are then summed and added to the classification loss:

$$\ell = \ell_{CLASS} + \sum_{i=1}^{t} \lambda_i \ell_{CORAL} \tag{3}$$

The gradient is then calculated for the backward propagation step for this ensembled loss function (4). As a result, CNN forms weights in feature space, the covariance of which does not correspond to the source domain D_s, but to the target domain D_t. These additional restrictions on the search conditions, as shown by experiments [SunSaenko], provide faster network convergence.

We were unable to find literature references about the use of the DeepCORAL algorithm in detecting lung cancer from CT images.

In connection with the above, in our work, we posed the following question: does the aligning of correlations of different datasets containing CT images of lung cancer affect the effectiveness of their classification by CNN?

3.2 Materials and Method

As a source domain, we used the above mentioned dataset LUNA16. As a referent target domain, we used lung nodule dataset of SPIE[3]. As a target domain of interest, we used a new dataset, referred as Medical, formed by clinicists with strict control of CT parameters during lung cancer treatments (see Table 1 for details). As can be seen from the table 1, Medical dataset is much more statistically homogeneous than LUNA, which allows to consider them as different domains in our experiments.

Table 1. Comparison of datasets parameters

Dataset	Scans / patients in total	Types of scanners	Variants of tube peak potential	Variants of slice thicknesses	Reconstruction interval
LUNA16	1018 / 1010	6	4	11	0.45–5.0 mm
MEDICAL	20 / 20	1	1	1	1.25 mm

As a reference method of transfer learning we used fine tuning. We used a lung cancer-specific architecture[4] that is a modification of the UNet architecture for segmentation of 3D objects [Fangzhou Liao]. We implement the fine tuning method in a traditional way:

- Remove the last fully connected layers of the original NA
- Adding own fully connected layers with number of classes according to T_t, corresponding to benign and malignant nodules.
- Freezing of convolutional layers of the initial network.
- Training added layers
- Unfreeze the rest layers of the original network.
- Further training of the entire network with added layers.

To implement the DeepCORAL algorithm, we used AlexNet architecture as a base with the replacement of the activation functions ReLU with the ELU (Clevert, 2015) and the addition of BatchNorm layers (Ioffe & Szegedy, 2015), which is more relevant in modern industry practice. We conducted experiments to optimize the number of layers to be aligned and found that the best result, in contrast to the original architecture [27], gives the alignment of all five convolutional layers in AlexNet and two of the three dense layers, except the last.

3.3 Results and Discussion

The limiting values of accuracy obtained by us using the reference method of fine tuning in the retraining from LUNA16 to SPIE and MEDICAL, respectively, are presented in Table 2.

Table 2. Accuracy of lung cancer detection using the reference method of fine tuning

Dataset	w/o Fine Tuning, %	with Fine Tuning, %
SPIE	63	85
MEDICAL	61	73

We see an improvement in the accuracy up to values comparable to the level of recent achievements [Nishio], however, this improvement was achieved due to the large resource costs: the additional training of our reference network to obtain mentioned results took about two weeks on two GTX 1070 GPUs.

Figure 6 shows the results of retraining from source dataset LUNA16 to target dataset MEDICAL using DeepCORAL algorithm. Comparing figure 6, c, and figure 6, f, we see that expressive power of used AlexNet is not enough to make correct predictions on the target dataset, and ROC-AUC score oscillates around 0.5, nearly at random prediction level. In the same time, using the same CNN with the integrated DeepCORAL gives more accurate predictions for the target dataset, and even at 70th training epoch trend of ROC-AUC score still remains positive. Note that each experiment with network training for 70 epochs took only about an hour on the one GTX 1070 GPU.

[3] https://www.spiedigitallibrary.org/conference-proceedings-of-spie/7260/1/The-Lung-TIME--annotated-lung-nodule-dataset-and-nodule/10.1117/12.811645.short?SSO=1

[4] https://www.kaggle.com/c/data-science-bowl-2017

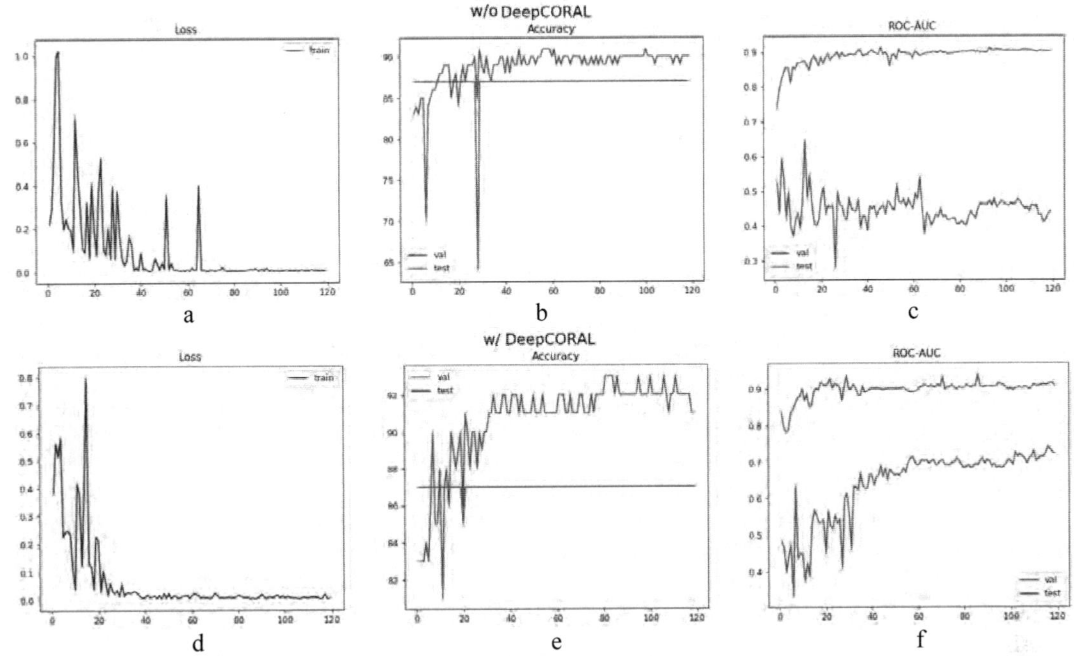

Figure 6. Dependencies of loss, accuracy and ROC-AUC on the number of training epochs not using (a–c) and using (d–f) DeepCORAL. The green line is scores for the validation subset of the source dataset, and the red line is the same for the target dataset

Thus, we have shown that using the DeepCoral method which takes into account the correlation between the images of the source and target domains for transfer learning allows us to obtain a quality of classification comparable to results of fine-tuning on a much more powerful CNN and at 120 epochs, but with a strong decrease in the required computational and time resources. It allows to give a confidently positive answer to the question posed in part 3.1.

4. CONCLUSION

The paper proposes two approaches to improving the efficiency of the classification of CT images of lung tumors, performed using CNN. These approaches are associated with improving of data processing and network training method. Data processing approach takes into account the current brightness distributions of CT scans, which showes better training results in the experiments. The network training method considers the integration of CORAL loss into network architecture, which provides the aligning of latent feature space of the CNN trained on source large dataset of CT scans with the latent feature space of the target small dataset, where the network will be used for inference. This method shows results comparable with the traditional fine-tuning method on the same hardware resources but on in a much shorter training time. In future work, we are going to use these approaches in more complex CNN architectures and we expect that this ratio should survive.

This work was financially supported by the Government of Russian Federation, Grant 08-08.

REFERENCES

AI diagnostics need attention. Nature 2018;555:285.

Chuquicusma M.J.M., Hussein S., Burt J., Bagci U. How to Fool Radiologists with Generative Adversarial Networks? A Visual Turing Test for Lung Cancer Diagnosis. posarXiv: 1710.09762v2 [cs.CV] 9 Jan 2018

Dakai Jin, Ziyue Xu, Youbao Tang, Harrison A.P., Mollura D.J. CT-Realistic Lung Nodule Simulation from 3D Conditional Generative Adversarial Networks for Robust Lung Segmentation arXiv:1806.04051v1 [cs.CV] 11 Jun 2018

Dong Wook Kim, Hye Young Jang, Kyung Won Kim, Youngbin Shin, Seong Ho Park. Design Characteristics of Studies Reporting the Performance of Artificial Intelligence Algorithms for Diagnostic Analysis of Medical Images: Results from Recently Published Papers // Korean J Radiol. 2019 Mar; 20(3)/ Pp. 405–410.

Fangzhou Liao, Ming Liang, Zhe Li, Xiaolin Hu, Sen Song. Evaluate the Malignancy of Pulmonary Nodules Using the 3D Deep Leaky Noisy-or Network // Journal Of Latex Class Files, Vol. 14, No. 8, August 2015.

Goodfellow I., Pouget-Abadie J., Mirza M., Xu B., Warde-Farley D., Ozair S., Courville A., Bengio Y. Generative adversarial nets // Advances in neural information processing systems, 2014, pp. 2672–2680.

Hammack D. Forecasting Lung Cancer Diagnoses with Deep Learning // Data Science Bowl 2017 Technical Report, April 22, 2017

Hosny A., Parmar C., Coroller T.P., Grossmann P., Zeleznik R., Kumar A., Bussink J., Gillies R.J., Mak R.H., Aerts J.W.L. Deep learning for lung cancer prognostication: A retrospective multi-cohort radiomics study // PLoS Med. 2018 Nov; 15(11): e1002711. Published online 2018 Nov 30. doi: 10.1371/journal.pmed.1002711

Joris P., Develter W., Van De Voorde W., Suetens P., Maes F., Vandermeulen D., Claes P. Preprocessing of Heteroscedastic Medical Images // IEEE Access, V. 6, 2018. Pp. 26047–26058.

Karras T., Aila T., Laine S., Lehtinen J. Progressive Growing Of Gans For Improved Quality, Stability, And Variation // ICLR 2018 Conference, Feb 2018.

Kazeminiaa S., Baurb C., Kuijperc A., van Ginnekend B., Navabb N., Albarqounib S., Mukhopadhyaya A. GANs for Medical Image Analysis. 21 Dec 2018. https://arxiv.org/abs/1809.06222

Koryagin A. Automatic lung-cancer detection on scans of Computed Tomography with RadIO. Feb. 2018. Availible: https://medium.com/data-analysis-center/automatic-lung-cancer-detecion-on-scans-of-computed-tomography-with-radio-945d781aa022

Lashari S.A., Ibrahim R. A Framework for Medical Images Classification Using Soft Set. // Procedia Technology 11 (2013) 548 – 556

Nishio M., Sugiyama O., Yakami M., Ueno S., Kubo T., Kuroda T., et al. (2018) Computer-aided diagnosis of lung nodule classification between benign nodule, primary lung cancer, and metastatic lung cancer at different image size using deep convolutional neural network with transfer learning. // PLoS ONE 13(7): e0200721. https://doi.org/10.1371/journal.pone.0200721

Nóbrega R. V., Peixoto S. A., da Silva S. P. P., Rebouças Filho P. P. Lung Nodule Classification via Deep Transfer Learning in CT Lung Images // 2018 IEEE 31st International Symposium on Computer-Based Medical Systems (CBMS). 18-21 June 2018

Öztürka Ş., Akdemir B. Effects of Histopathological Image Pre-processing on Convolutional Neural Networks // Procedia Computer Science 132 (2018) 396–403

Parmar C., Barry J.D., Hosny A., Quackenbush J., Aerts H.J.W.L. Data Analysis Strategies in Medical Imaging // Clin Cancer Res; 24(15) August 1, 2018

Perumal S. and Velmurugan T. Preprocessing by Contrast Enhancement Techniques for Medical Images // International Journal of Pure and Applied Mathematics. Volume 118. No. 18, 2018, 3681-3688

Qayyum A., Anwar S.M., Majid M., Awaisc M., Alnowamid M. Medical Image Analysis using Convolutional Neural Networks: A Review // Journal of Medical Systems. October 2018. DOI: 10.1007/s10916-018-1088-1

Ronneberger O., Fischer P., Brox T. U-Net: Convolutional Networks for Biomedical Image Segmentation. May 2015. https://arxiv.org/abs/1505.04597

Shin Hoo-Chang et al. Deep Convolutional Neural Networks for Computer-Aided Detection: CNN Architectures, Dataset Characteristics and Transfer Learning // IEEE Trans Med Imaging. 2016 May; 35(5): 1285–1298. doi:10.1109/TMI.2016.2528162.

Sun B., Feng J., Saenko, K. Return of frustratingly easy domain adaptation // AAAI. (2016)

Sun B., Saenko K. Deep CORAL: Correlation Alignment for Deep Domain Adaptation // arXiv:1607.01719v1 [cs.CV] 6 Jul 2016

Tao Tan, Zhang Li, Haixia Liu et al. Optimize transfer learning for lung diseases in bronchoscopy using a new concept: sequential fine-tuning // arXiv:1802.03617v1 [cs.CV] 10 Feb 2018

Vatian A., Gusarova N., Dobrenko N., Dudorov S., Nigmatullin N., Shalyto A., Lobantsev A. Impact of Adversarial Examples on the Efficiency of Interpretation and Use of Information from High-Tech Medical Images // FRUCT 2019

Weiss K., Khoshgoftaar T., DingDing Wang. A survey of transfer learning // Journal of Big Data (2016) 3:9 DOI 10.1186/s40537-016-0043-6

Yosinski J., Clune J., Bengio Y., Lipson H. How transferable are features in deep neural networks? // Advances in Neural Information Processing Systems 27 (NIPS '14), NIPS Foundation, 2014.

Zhenglun Kong, Ting Li, Junyi Luo, Shengpu Xu. Automatic Tissue Image Segmentation Based on Image Processing and Deep Learning // Journal of Healthcare Engineering. Volume 2019, Article ID 2912458, 10 pages

EXPLAINING AI MODELS FOR CLINICAL RESEARCH: VALIDATION THROUGH MODEL COMPARISON AND DATA SIMULATION

Qing Zeng-Treitler[1, 2, a], Yijun Shao[1, 2, a], Douglas Redd[1, 2, a], Joseph Goulet[3, 4, a], Cynthia Brandt[3, 4, b] and Bruce Bray[5, b]

[1]George Washington University, 2121 I Street, NW, Washington, DC 20052, USA
[2]DCVA Medical Center, 50 Irving St NW, Washington, DC 20422, USA
[3]Yale University Medical School, 333 Cedar Street, New Haven CT 06510, USA
[4]West Haven VA Medical Center, 950 Campbell Ave, West Haven, CT 06516, USA
[5]University of Utah Medical School , 30 N. 1900 E, Salt Lake City, Utah 84132, USA
[a] PhD
[b] MD

ABSTRACT

For clinical research to take advantage of artificial intelligence techniques such as the various types of deep neural networks, we need to be able to explain the deep neural network models to clinicians and researchers. While some explanation approaches have been developed, their validation and utilization are very limited. In this study, we evaluated a novel explainable artificial intelligence method called impact assessment by applying it to deep neural networks trained on real world and simulated data. Using real clinical data, the impact scores from deep neural networks were compared with odds ratios from logistic regression models. Using simulated data, the impact scores from deep neural networks were compared with the impact scores calculated based on the ground truth (i.e. formulas used to generate the simulated data). The correlations between impact scores and odds ratios ranged from 0.63 to 0.97. The correlations between impact scores from DNN and ground truth ranged were all above 0.99. These suggest that the impact score provide a valid explanation of the contribution of a variable in a DNN model.

KEYWORDS

Explainable AI, Validation, Clinical Research

1. INTRODUCTION

The roots of artificial intelligence (AI), with the neural network as a signature technology, span many decades. Many of the algorithms, however, could not be productively implemented with the technology available in years past. Improvements in computing power have fueled renewed enthusiasm in AI, enabling implementation of algorithms that were previously strictly theoretical. In particular, deep learning methods have resulted in very good performance in speech recognition and many computer vision tasks. It also produced stunning results in authentic game play. Deep learning approaches have advantages over more traditional statistical and machine learning methods due to their ability to model highly non-linear relationships and the avoidance of the need for extensive feature engineering. With impressive results ranging from research to consumer products, efforts have been undertaken to generalize these approaches to widespread domains. This has resulted in some inevitable exaggerated claims and unfortunate false steps.(Quach 2019) There have also been many preliminary successes in important areas.(Rajkomar, Oren et al. 2018)

AI has shown promise for many clinical research applications.(Miotto, Wang et al. 2017) With the widespread adoption of electronic health record (EHR) systems, large amounts of clinical data have become available. This has the potential to fill the need of deep learning for large volumes of labeled data. Prediction of disease risk and hospital readmission are often challenging for traditional machine learning methods, and deep neural networks (DNNs) are beginning to have an impact.(Weng, Reps et al. 2017) For example, a

DNN has been successfully used to identify diabetic retinopathy from retinal images.(Gulshan, Peng et al. 2016) Additionally, techniques such as recurrent neural networks (RNNs) are showing great promise for modeling time series data from EHRs.(Miotto, Li et al. 2016)

At the same time, deep neural network (DNN) models are viewed as black boxes, and understanding and explaining their behaviors is the subject of ongoing research. To influence clinical practice, a prediction model must have sufficient discriminating power, calibration, face validity and clinical utility. Clinicians are less inclined to act on predictions made by black-box models and researchers generally would like to investigate the significance of features (predictors) of a model. Thus, to gain clinical and research acceptance and impact, we need to provide interpretation of the DNN models.

Compared to clinical data, images of natural scenes, recordings of speech, or game board layouts are much easier for humans to understand. As a result, the majority of the DNN explanation work that has been published is intended to confirm human interpretation, e.g., if an image feature critical to the prediction of being a dog looks like the outline of a dog. With clinical prediction, human users do not have such prior knowledge and expect the model to yield information that justifies the outcome prediction.

Approaches that are being explored to explain DNN models include Attention Mechanism (Park, Hendricks et al. 2017), Local Interpretable Model-Agnostic Explanations (LIME)(Ribeiro, Singh et al. 2016) and Layer-wise Relevance Propagation (LRP).(Binder, Montavon et al. 2016) Each of these approaches has strengths and limitations: Attention mechanism provides insight into the DNN process, but can also be configured in different ways to alter the DNN process. LIME is agnostic to the DNN structure, though the manipulation of input is typically performed one variable at a time. LRP must make assumptions of the weight distribution within a neural network.

One area of ongoing research is to interpret and draw statistical inferences based on the explanations generated by these approaches.(Adadi and Berrada 2018, Kim, Rajaraman et al. 2019) Another is to validate the DNN explanations.(Lamy, Sekar et al. 2019) This paper focuses on the validation of a DNN explanation method, Impact Assessment (IA)(Shao, Cheng et al. 2019), which is developed by our team. Please note that DNN performance, though important, is not the focus of this paper.

2. METHODS

2.1 Background

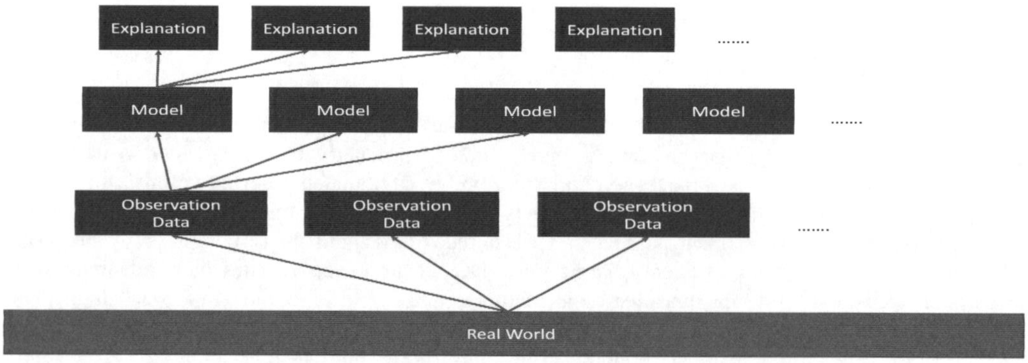

Figure 1. Relationships among real world use cases, data, model, and model explanation

Validation is an integral part of machine learning. As shown in Figure 1, multiple observation datasets may be available for one real world use case (e.g. re-hospitalization or mortality after surgery). Multiple models may be trained on a dataset and multiple explanations may be derived from a model. For the assessment of model performance and generalizability, internal and external validations are often carried out. Internal validation uses a subset of the same dataset to validate a model fitted on a different subset, while external validation uses a different dataset to validate a model fitted on a certain dataset. To understand a model's strengths and weaknesses, model comparison is also commonly performed. It typically evaluates one model using another as the reference or evaluates multiple models based on "ground truth", all on the same datasets. When ground truth is difficult to establish using data from the real world, however, validation through simulated data can be very helpful. AI models are frequently evaluated through internal and external validation. Explanation of AI models, however, cannot be easily validated against the ground truth because patterns, relationships, and pathways in the natural law that are only partially captured the observational data. In the study we describe in this paper, we compare explanations of different models trained on the same datasets. We also compare the explanation with ground truth in simulated data.

The Impact Assessment (IA)(Shao, Cheng et al. 2019) method we developed belongs to the LIME approach. For each variable, IA calculates a new measurement called impact score. Similar to the log-odds ratio from logistic regression, the Impact score represents the average rate of change of log-odds of the predicted risk.

The calculation begins by selecting a reference value. For binary variables such as diagnoses, 0 is selected as the reference as it indicates the absence of a diagnosis. For categorical variables (e.g. race), we select a category (e.g. unknown) based on convention and clinical context as the reference. For continuous variables, the mean is typically used as the reference. We then change a variable's current value to the reference value and observe change on the outcome. If the outcome does not change, this suggests that the change in value has no impact (as compared to the reference value). If the model changes, we will proceed to the following calculation.

The last layer of the DNN outputs a value p between 0 and 1 though a sigmoid function $p = \sigma(x)$. The change of prediction originates from change in x. One way to obtain x from p is the logit function: $x = \text{logit}(p) = \log\frac{p}{1-p}$ Therefore, the individual-level impact score is defined as:

$$\frac{\text{logit}(p_{\text{cur}}) - \text{logit}(p_{\text{ref}})}{(\text{current value}) - (\text{reference value})}$$

where: p_{ref} is the new value of p after changing the current value to reference value.

Each individual's impact score on a given variable varies because the DNN model can model non-linear relationships between the predictor and outcome variables. We define the impact score at the population level simply as the mean of individual scores.

2.2 Observation Data

We used data from two different clinical research studies: 1) Frailty – this study predicted the outcome of patients following major cardiovascular procedures using patient demographics and clinical characteristics including frailty status which is extracted from clinical notes; and 2) Complementary Integrative Health (CIH) – this study attempts to identify the clinical and demographic features associated with CIH use in patients with muscular skeletal disorder.

The number of features and sample size of each study is shown in Table 1.

Table 1. Characteristics of the real world datasets

	Outcome	**Features**	**Number of Features**	**Sample Size**
Frailty	Mortality	Demographics, comorbidities, prior hospitalization, prior diagnoses, vital signs, frailty status	57	21,355
CIH	Usage of Acupuncture, Biofeedback, Guided Imagery, Meditation, Tai-Chi, or Yoga	Demographics, comorbidities, vital signs, Opioid use	19	9,761

2.3 Validation through Model Comparison

Different DNN models were fitted to the 2 datasets (Table 2), while 2 logistic regression models were also fitted to each of the same datasets. IA was applied to the DNN models, with each variable's impact score as the explanation. The DNN structure was determined empirically in each case. For logistic regression, we used the odds ratio of each variable as the explanation. The Pearson and Spearman correlation was calculated between the impact scores and odds ratios for each dataset.

Table 2. DNN model description and DNN/logistic regression model performance

	DNN Model Description	DNN Performance	Logistic regression performance
Frailty	The DNN had two branches at the input end. The branch taking temporal data (represented as an image) as input was a convolutional neural network with 5 hidden layers. The branch taking non-temporal data has one fully-connected layer. The last layers from the two branches were joined using a simple concatenation to form the last hidden layer of the whole network. The initial weights in all the hidden layers and the output layer of the DNN were randomly assigned with small numbers. Then the DNN were trained on the training set with weights updated iteratively using back propagation.	AUC=0.787	AUC=0.746
CIH	The DNN was constructed with an input layer of 19 nodes, corresponding to the 19 variables, and an output layer of a single node giving the probability of CIH use. Five fully connected hidden layers were used, all using the rectified linear unit function for activation. Weights were initialized with random small numbers and updated using stochastic gradient descent with Nesterov momentum.	AUC=0.683	AUC=0.681

2.4 Simulated Data

Two sets of simulation data are generated to allow us validate the explanations against the ground truth. We selected binary outcome because many clinical risk predictions tasks have binary outcomes or categorical outcomes that can be translated into multiple binary outcomes (e.g. mortality, hospitalization, incident diagnosis and adverse event). The number of predictors in clinical studies can vary from below 50 to over 1000 and we chose n=100 for this study. The inclusion of both linear relationship and non-linear relationship is driven by the desire to represent both simple and complex datasets.

First, we use 100 variables $x_1, x_2, \ldots, x_{100}$ as predictors and a binary variable z for two outcomes represented by values 0 and 1 respectively. Among the 100 variables the first 50 are binary variables taking values 0/1, and the second 50 are continuous variables taking values ranged between 0 and 1. This setting is to resemble the real situation of patient data: some variables such as gender, diagnoses, procedures are usually treated as binary variables, while the other variables such as age, vital signs, lab results are usually treated as continuous variables. Although the continuous variables may have different numerical value ranges in their original form, they can always be normalized to range from 0 to 1.

We will experiment with two different types of relationship between the outcome variable z, or more specifically, the log odds $y = log \frac{p}{1-p}$, where $p = Prob(z = 1)$, and the predictors: 1) linear relationship, and 2) non-linear relationship.

1) the linear relationship is given by

$$y = \beta_0 + \beta_1 x_1 + \cdots + \beta_{100} x_{100}$$

The coefficients $\beta_1, \ldots, \beta_{100}$ are randomly generated from a uniform distribution with range -1 to 1. Once the coefficients are generated, they are fixed for all the following analysis. We set $\beta_0 = 1$.

Next, we generate 50,000 vectors $(x_1, x_2, \ldots, x_{100}, z)$ also through a random process. The values of the first 50 variables are sampled from a Bernoulli distribution with $p = 0.3$ (i.e., taking value 1 with probability

0.3). The values of the second 50 variables are sampled from a uniform distribution with range 0 to 1. The values of the outcome variables z is generated randomly from a Bernoulli distribution with probability p calculated using the sigmoid function

$$p = \frac{1}{1+e^{-y}}$$

2) the nonlinear relationship is given b

$$y = \beta_0 + \sum_{i=1}^{100} \beta_i x_i + \sum_{i \in I} \gamma_i x_i^2 + \sum_{j \in J, k \in K} \theta_{jk} x_j x_k$$

where I, J and K are random subsets of the index set $\{1, 2, \dots, 100\}$ of size 20 such that $J \cap K = \emptyset$. So this relationship is added with 20 square terms and 20 product terms. The coefficients β_i are generated in the same way as in 1). The coefficients γ_i and θ_{jk} are generated in the same way as β_i.

Next we generate 50,000 vectors $(x_1, x_2, \dots, x_{100}, z)$ in exactly the same way as 1) except that the non-linear relationship given in 2) is used.

2.5 Validation through Comparison with Ground Truth

After the simulation data are generated, we train a deep neural network (DNN) model on the data. The DNN has an architecture as follows. It has an input layer of 100 nodes and an output layer with 1 node. There are 5 hidden fully-connected layers between the input layer and output layer. Their numbers of nodes, from the first to the last, are 50, 70, 50, 70 and 50, respectively. Their nonlinear activation functions are all the rectified linear unit function $f(x) = max(0, x)$. The nonlinear activation function for the output layer is the sigmoid function $\sigma(x) = 1/(1+e^{-x})$ so that the DNN outputs a number between 0 and 1 representing the predicted probability, $Prob(z = 1)$. This DNN architecture is empirically designed. Since our focus is on the explanation, we did not experiment with different architectures.

For each of the two simulated datasets (linear and non-linear), we divide the dataset into 3 random subsets: training (60%), validation (20%) and testing (20%), and then we train the DNN model on the training set. The weights of the DNN is initialized with randomly generated small numbers before training, and updated using the Nesterov algorithm with a mini-batch size of 100, a learning rate of 0.001 and a momentum of 0.9 during training. After each epoch of training, the trained DNN model is tested on the validation set to measure the AUC. When the AUC reaches a peak point such that it shows no improvement over the following 10 epochs, we take the DNN model with the peak AUC as the final model.

With the final DNN model, we calculate impact scores as follows. Let $p = F(x_1, \dots, x_{100})$ be the final DNN model. By the design of DNN, we have $0 < p < 1$. First, we choose a reference value for each variable, which is zero for all variables in this case. We then proceeded to calculate the impact score as described in the Background section.

To verify the population-level impact scores calculated based on the learned DNN model, we need to calculate the true population-level impact scores based on the true underlying relationship used to generate the dataset. The calculation is similar as above, except that for the individual-level impact score formula, $logit(p_{cur})$ and $logit(p_{ref})$ are replaced by y_{cur} and y_{ref}, respectively, which are the y defined in the true relationship with x_i taking the current value/reference value. This change in definition is because when generating the data, y is calculated ahead of p using the sigmoid function.

We used several metrics to measure the difference between the true impact scores and the learned impact scores (by the DNN models), including mean absolute error (MAE), Pearson's correlation, and Spearman's (rank) correlation.

3. RESULTS

The comparison of impact score and odds ratio logistic regression using real world observation data show moderate to very strong correlations. (Table 3) This is consistent with the fact that the DNN models and logistic regressions models were trained on the same datasets. The DNN and logistic regression models are not expected to be the same, however, they are expected to have some similarities.

Table 3. Correlation coefficients between impact scores and odds ratios

	Pearson	Spearman
Frailty	0.69	0.63
CIH	0.84	0.97

Simulated evaluation results (Table 4) suggest that the DNN-derived impact score is highly correlated (Pearson's and Spearman's correlation coefficients > 0.99) with the impact score calculated based on the ground truth (i.e. the formulas used to generate the simulated score). The mean absolute errors are very small (<.07), given that the mean impact scores are around 0.5. An interesting observation is that the DNN achieved slightly higher AUC on the non-linear data and the correlations and absolute errors of impact scores are almost identical on the linear and non-linear data. Please note that when we compared the impact scores and odds ratios, the calculation of absolute errors was not possible as neither can be viewed as ground truth.

Table 4. Correlation coefficients and other performance metrics of impact score based on simulated data

	Linear	Non-linear
Training DNN Model		
Number of Epochs	30	30
Testing AUC	0.873	0.891
Impact Scores		
Mean Absolute Error (MAE)	0.058	0.065
Mean Absolute Predicted Value (MAPV)	0.466	0.514
Mean Absolute Actual Value (MAAV)	0.515	0.533
MAE / MAPV	12%	13%
MAE / MAAV	11%	12%
Pearson's Correlation	0.997	0.993
Spearman's Correlation	0.996	0.990

4. DISCUSSION

4.1 Significances

DNN models are a key AI approach. Recent studies have shown that DNN can outperform traditional machine learning (ML) methods and statistical models in clinical and non-clinical domains(Chen and Lin 2014, Belle, Thiagarajan et al. 2015, LeCun, Bengio et al. 2015, Tan, Gao et al. 2015, Pastur-Romay, Cedron et al. 2016, Erickson, Korfiatis et al. 2017, Lee, Jun et al. 2017, Miotto, Wang et al. 2017). This paper presents two approaches to validate a new explainable AI Method: model comparison and simulation. Results from both approaches suggest that the impact score provides a sensible explanation of the patterns captured by the AI models, and can be used to explain the real-world relationships. As AI is increasingly used by biomedical researchers, the demand for explanation is expected to grow. Therefore, the use of a method like IA is expected to increase as well.

4.2 Implications

Observational data is an imperfect reflection of the real world and often driven by specific clinical hypotheses and administrative requirements. Even clinical trial data have many flaws, e.g. there is often selection bias and there are always unobserved variables (which is assumed to be taken care of by the randomization process, but it is not guaranteed). Cohort studies, EHRs, insurance claims, and patient-reported datasets tend to be much larger and can have a very long follow-up, however, they also tend to be noisier and more complex to analyze.

Models are imperfect reflections of the data. As George Box said, "all models are wrong, but some are useful." Comparing models developed using real world data reflects the complexity in the real world. Different models may each represent the underlying ground truth partially, and thus offer different insights into the underlying ground truth. When we do not know the ground truth in real world data (e.g. we do not have the complete knowledge about the biological relationship between frailty and mortality), simulation provides us with ground truths. However, it builds on pre-defined rules and assumptions.

4.3 Limitations

Complex models can also be explained in different ways. There are already several approaches to explain deep learning models: some are based on an analysis of the inner structure and weights, while others are based on the change in outcome when the input is manipulated.

Currently, there is no ideal method to evaluate explanations of DNN. Model comparisons and simulations each have their strengths and weaknesses: statistical models cannot be viewed as a gold standard and simulated data tend to be simpler than real world observations. In addition, this study is limited by the model we chose for comparison, i.e. logistic regression. We chose logistic regression because the odds ratio yielded by this analysis is widely accepted by clinicians. There are of course many other modeling methods which can be compared with. There are also infinite ways to generate simulated datasets, and our simulated data can be made more complex.

4.4 Future Work

Clinical adoption requires explanations like impact scores. Our next step will be to test if impact scores can influence clinical decision making, i.e. the usefulness of impact scores. Our initial attempt at this revealed a number of challenges: e.g. when there are many variables, should we just show the ones with the highest impact? Should we display population and individual impact scores together? For continuous variables, should we provide information on change per unit or just the change? We will also need to go beyond the impact of single variables and explore multi-variable patterns.

5. CONCLUSION

We evaluated a novel explainable AI approach in two ways: comparison with the odds ratio from logistic regression and comparison with ground truth from simulated data. The results suggest the impact scores provide valid explanations of the contribution/importance of a feature in a DNN model. The impact scores can help facilitate the utilization of DNN in clinical research and practice.

ACKNOWLEDGEMENT

The work is funded in part by several US NIH and VA grants: 1UL1TR001876-01, I01 HX001145-02, 1I01HX002308-01A1, and R56 AG052536-01A1.

REFERENCES

Adadi, A. and M. Berrada (2018). "Peeking Inside the Black-Box: A Survey on Explainable Artificial Intelligence (XAI)." IEEE Access 6.

Belle, A., R. Thiagarajan, S. M. Soroushmehr, F. Navidi, D. A. Beard and K. Najarian (2015). "Big Data Analytics in Healthcare." Biomed Res Int 2015: 370194.

Binder, A., G. Montavon, S. Lapuschkin, K.-R. Müller and W. Samek (2016). Layer-wise relevance propagation for neural networks with local renormalization layers. International Conference on Artificial Neural Networks, Springer.

Chen, X. W. and X. T. Lin (2014). "Big Data Deep Learning: Challenges and Perspectives." Ieee Access 2: 514-525.

Erickson, B. J., P. Korfiatis, Z. Akkus and T. L. Kline (2017). "Machine Learning for Medical Imaging." Radiographics 37(2): 505-515.

Gulshan, V., L. Peng, M. Coram, M. C. Stumpe, D. Wu, A. Narayanaswamy, S. Venugopalan, K. Widner, T. Madams, J. Cuadros, R. Kim, R. Raman, P. C. Nelson, J. L. Mega and D. R. Webster (2016). "Development and Validation of a Deep Learning Algorithm for Detection of Diabetic Retinopathy in Retinal Fundus Photographs." JAMA 316(22): 2402-2410.

Kim, I., S. Rajaraman and A. S (2019). "Visual Interpretation of Convolutional Neural Network Predictions in Classifying Medical Image Modalities." Diagnostics 9(2): 38.

Lamy, J., B. Sekar, G. Guezennec, J. Bouaud and S. B (2019). "Explainable artificial intelligence for breast cancer: A visual case-based reasoning approach." Artif Intell Med(94): 42-53.

LeCun, Y., Y. Bengio and G. Hinton (2015). "Deep learning." Nature 521(7553): 436-444.

Lee, J. G., S. Jun, Y. W. Cho, H. Lee, G. B. Kim, J. B. Seo and N. Kim (2017). "Deep Learning in Medical Imaging: General Overview." Korean J Radiol 18(4): 570-584.

Miotto, R., L. Li, B. A. Kidd and J. T. Dudley (2016). "Deep patient: an unsupervised representation to predict the future of patients from the electronic health records." Scientific reports 6: 26094.

Miotto, R., F. Wang, S. Wang, X. Jiang and J. T. Dudley (2017). "Deep learning for healthcare: review, opportunities and challenges." Brief Bioinform.

Miotto, R., F. Wang, S. Wang, X. Jiang and J. T. Dudley (2017). "Deep learning for healthcare: review, opportunities and challenges." Briefings in Bioinformatics 19(6): 1236-1246.

Park, D. H., L. A. Hendricks, Z. Akata, A. Rohrbach, B. Schiele, T. Darrell and M. Rohrbach (2017). Attentive Explanations: Justifying Decisions and Pointing to the Evidence. Interpretable ML Symposium (NIPS 2017). Long Beach, CA, USA.

Pastur-Romay, L. A., F. Cedron, A. Pazos and A. B. Porto-Pazos (2016). "Deep Artificial Neural Networks and Neuromorphic Chips for Big Data Analysis: Pharmaceutical and Bioinformatics Applications." Int J Mol Sci 17(8).

Quach, K. (2019, 18 Apr 2019). "IBM Watson Health cuts back Drug Discovery 'artificial intelligence' after lackluster sales." Retrieved 3 May 2019, from https://www.theregister.co.uk/2019/04/18/ibm_watson_health.

Rajkomar, A., E. Oren, K. Chen, A. M. Dai, N. Hajaj, M. Hardt, P. J. Liu, X. Liu, J. Marcus, M. Sun, P. Sundberg, H. Yee, K. Zhang, Y. Zhang, G. Flores, G. E. Duggan, J. Irvine, Q. Le, K. Litsch, A. Mossin, J. Tansuwan, D. Wang, J. Wexler, J. Wilson, D. Ludwig, S. L. Volchenboum, K. Chou, M. Pearson, S. Madabushi, N. H. Shah, A. J. Butte, M. D. Howell, C. Cui, G. S. Corrado and J. Dean (2018). "Scalable and accurate deep learning with electronic health records." npj Digital Medicine 1(1): 18.

Ribeiro, M. T., S. Singh and C. Guestrin (2016). Why should i trust you?: Explaining the predictions of any classifier. Proceedings of the 22nd ACM SIGKDD international conference on knowledge discovery and data mining, ACM.

Shao, Y., Y. Cheng, R. Shah, C. Weir, B. Bray and Q. Zeng-Treitler (2019). Shedding Light on the Black Box: Explaining Deep Neural Network Prediction of Clinical Outcome. ICHI 2019 : International Conference on Health Informatics. Rome, Italy.

Tan, S. S., G. Gao and S. Koch (2015). "Big Data and Analytics in Healthcare." Methods Inf Med 54(6): 546-547.

Weng, S. F., J. Reps, J. Kai, J. M. Garibaldi and N. Qureshi (2017). "Can machine-learning improve cardiovascular risk prediction using routine clinical data?" PloS one 12(4): e0174944.

THE PERFORMANCE OF SOME MACHINE LEARNING APPROACHES IN HUMAN MOVEMENT ASSESSMENT

Johan Hagelbäck[1], Pavlo Liapota[2], Alisa Lincke[1] and Welf Löwe[1]
[1]*DISA, Linnaeus University, Växjö/Kalmar, Sweden*
[2]*Softwerk AB, Växjö, Sweden*

ABSTRACT

The advent of commodity 3D sensor technology enabled, amongst other things, the efficient and effective assessment of human movements. Statistical and machine learning approaches map recorded movement instances to expert scores to train models for the automated assessment of new movements. However, there are many variations in selecting the approaches and setting the parameters for achieving high performance, i.e., high accuracy and low response time. The present paper researches the design space and the impact of approaches of statistical and machine learning on accuracy and response time in human movement assessment. Results show that a random forest regression approach outperforms linear regression, support vector regression and neuronal network approaches. Since the results do not rely on the movement specifics, they can help improving the performance of automated human movement assessment, in general.

KEYWORDS

Statistical, Machine Learning, Human Movement Assessment

1. INTRODUCTION

Healthcare is becoming more and more value centered optimizing patient value (best possible health achieved) and system value (effective treatments at efficient costs). Until recently, measuring value, e.g., clinical outcomes, was not a capability of hospital or health IT (Elton and O'Riordan 2016). With inexpensive sensor technologies and data analytics becoming increasingly available, it is nowadays possible to collect data on the clinical activities of healthcare, the health status of a patient and the change of this status after treatment. We contribute with an automated objective method for the assessment of physical health of the human musculoskeletal system to help diagnose, predict or prevent related pain, injuries and long-lasting diseases. We use commodity 3D camera technology, such as the Microsoft Kinect.[1] Unlike similar tools for physical therapists[2,3] not aiming to support the caring or nursing process, our approach localizes potential issues and quantifies their severity.

Regardless of the actual physical movement, their assessment in existing studies is usually done by comparing a patient's performance in the movement to the performance of healthy reference participants. Participants repeat the same movement several times recorded by sensors, video cameras or other devices. Then features are defined and extracted from the recordings. Finally, different machine learning (ML) algorithms apply supervised learning using the features and expert scores as labels. This process makes it is quite efficient to add new movements and to assess them accurately (Dressler et al. 2019a).

As in many data driven applications, selecting the right ML approach is still a tedious, partially manual process characterized by prototyping, trial and error. This is because there are quite a few variants to select from spanning a huge search space. At the same time the variant selection has a large impact on the performance of the final application. Variants include: feature selection, the selection of preprocessing steps, movement alignment (cutting and matching), ML model selection, and the selection of training and testing data. The present paper focuses on selecting variants of ML approaches. More specifically, we assess

[1] https://en.wikipedia.org/wiki/Kinect
[2] https://www.qinematic.com
[3] https://kinetisense.com

accuracy and response time of scoring and time for training the following models: linear regression, random forest, support vector regression, and neuronal networks.

The paper contributes with (i) a general human movement assessment framework, (ii) a component of this framework for substituting and assessing ML models, and (iii) an experimental evaluation of accuracy and response and training time of four different ML models.

It is organized as follows: Section 2 discusses related work in human movement assessment. Section 3 gives basic definitions. Section 4 introduces the human movement assessment framework (i). Section 5 explains the component for human movement quality scoring for different ML models (ii). Section 6 experimentally evaluates the approaches wrt. assessment accuracy and response time (iii). Section 7 concludes the paper and mentions future work.

2. RELATED WORK

Features necessary for human movement assessment extracted with commodity 3D sensor technology, such as the Kinect, seem to agree with the same features extracted with high-end motion capturing systems (Fern'ndez-Baena et al. 2012, Müller et al. 2017, Ma et al. 2018) or with other sensor technology (Baldewijns et al. 2014, Lopez et al. 2018). Only few studies found differences in, e.g., foot placement locations and step lengths (Geerse et al. 2017).

Many studies research the *recognition* of human movements using commodity 3D sensors (Wang et al. 2012, Wang and Zheng 2012, Xia et al. 2012, Jiang et al. 2015, Pazhoumand-Dar et al. 2015). While these approaches are similar to ours in feature extraction and preprocessing, their goal is the classification of movements not their quality assessment.

The same technology has also been used in movement *quality assessment*, e.g., for identifying anterior cruciate ligament (ACL), for injury risks (Gray et al. 2014), for human abduction-adduction motion analysis (Kim et al. 2016), for ergonomic assessment in real workplace conditions (Plantard et al. 2017), for the measurement of lower extremity jump landing and squatting kinematics, for sit-to-stand analysis (Cippitelli et al. 2015), and for static and dynamic balance assessment (Eltoukhy et al. 2018). Similar studies include (Wang et al. 2013, Paiement et al. 2014, Parmar and Morris 2016, Tao et al. 2016). However, these studies aim at *qualitative* assessments—checking, e.g., whether movements are correctly or incorrectly, well or badly performed—or at identifying different abnormal movement patterns rather than *quantitatively* scoring the quality of a movement and localizing impairments. Some of the studies (Paiement et al. 2014, Tao et al. 2016) use preprocessing transformations such as dimensionality reduction based on manifold learning (Elgammal and Lee 2008), which enables a fast online assessment but even prohibits impairment localization.

Pirsiavash et al. (2014) suggest a learning framework for training models able to quantitatively assess the quality of human movements from 2D videos. Their approach trains a regression model from spatiotemporal pose features to scores obtained from expert judges. Features are extracted using unsupervised feature learning directly from 2D video data. Therefore, a localization of issues is not possible.

Dressler et al. (2019b), (2019a) suggest an automated, quantitative assessment of human movements based on the Kinect camera that scores the overall movement quality and localizes and scores issue. It provides the baseline of our assessment framework and performance comparisons. They use linear regression mapping movement features to expert scores.

Other ML approaches for estimating movement quality classes and scores include support vector machines (Taylor et al. 2010, Parmar and Morris 2016, Bao et al. 2019), Adaboost (Taylor et al. 2010, Parmar and Morris 2016), random forest (Leightley et al. 2017), multi-layer neural network (Parmar and Morris 2016), and distance functions (Houmanfar et al. 2016, Parmar and Morris 2016, Bao et al. 2019).

Liao et al. (2019) use deep neural networks, such as CNN, RNN, and hierarchical neural networks, for assessing the quality of ten different movements with high accuracy. Movements include deep squat, hurdle step, and sit to stand. Our study assesses linear regression, random forest, support vector regression, and neuronal networks to using the deep squat movement.

3. BASIC DEFINITIONS

The Kinect identifies 24 "joints". Their *x*, *y*, and *z* coordinates are given in a 3D coordinate system with the camera in its origin and a horizontal *x*, *z* plane with *z* the depth coordinate. Because of low reliability of the other joints, we only use the following: head, neck, spine shoulder, spine mid, spine base, left/right shoulder, left/right elbow, left/right wrist, left/right hip, left/right knee, and left/right ankle.

A *sequence* is a list of frames from a recording of a human movement, e.g., a deep squat. It is a *master* sequence if the movement is assessed and scored by a human expert and used for building a (statistical) model of the movement. Any sequence that is to be scored automatically is called a *user* sequence. A *frame* is a record of features. It is a part of a sequence and describes the body posture at a specific point in time during the recorded movement. A *feature* is a part of a frame and describes an aspect of the body posture at specific point in time. A feature is called *direct* if it is directly measured by the 3D camera or *indirect* if it is computed from the direct or other indirect features. The direct features include the *x*, *y*, and *z* coordinates of skeleton joints. Indirect features include the angles between limbs and angles between limbs and the axes of the 3D coordinate system. An *aggregated sequence* is a list of aggregated frames aggregating one or more sequences into one. An *aggregated frame* aggregates two or more frames into one. It is a vector of sample distributions of the feature values of each feature. For scoring, we compare a user sequence with an aggregated master sequence.

4. A FRAMEWORK FOR MOVEMENT ASSESSMENT

All variants of our movement assessment algorithm perform five steps: (1) *Building an aggregated master sequence* is only performed once while the following steps are performed for each user sequence. (2) *Preparing the sequences* mitigates noisy feature values caused by random camera and skeleton recognition errors. (3) *Cutting and Matching* cuts the leading and trailing frames that do not belong to the movement and aligns a user sequence with the aggregated master sequence. (4) *Normalizing* applies linear transformations to the sequences and adds indirect features. (5) *Scoring* computes indicators for individual features and an overall movement score. If not explicitly mentioned otherwise, all steps are performed automatically. Details are introduced in the following paragraphs and precisely defined in (Dressler et al. 2019b).

Building an Aggregated Master Sequence builds an initial aggregated master sequence and aggregates it with the other master sequences: (1.1) *selects* the best master sequence *S* manually. It should have a constant movement speed without any delays and stops. (1.2) *cuts* off manually leading and trailing frames of *S* containing postures that do not belong to the movement. (1.3) *prepares S*, i.e., applies Step (2). (1.4) *aggregates subsequent frames* of *S*. We aggregate frames into 15 groups each containing the same number of frames (+/-1). Each group contributes to a separated aggregated frame. (1.5) *normalizes S*, i.e., applies all but the first sub-steps of Step (4). (1.6) *computes a numerical sample distribution* of the feature values, separately for each group and for each feature.

For any of the remaining master sequences *S'*, (1.7) *prepares S'*, i.e., performs Step (2). (1.8) *matches S'* with current aggregated master sequence, i.e., performs Step (3). This maps each frame of *S'* to an aggregated frame of the aggregated master sequence. (1.9) *normalizes* the master sequence, i.e., performs Step (4). (1.10) *aggregates subsequent frames* of *S'*, i.e., for each frame and each feature of *S'*, it adds the feature value to the sample distribution of the respective feature of the mapped aggregated frame.

Preparing the Sequences: (2.1) *aligns the floor clip plane*, i.e., rotates the joint position vectors for each frame such that the floor clip plane is parallel to the *x*, *z* plane. (2.2) *smoothening* computes a sliding average of feature values for all *direct* features. (2.3) *interpolates* positions if a joint was not visible for $<k$ consecutive frames. If a joint was not visible for $\geq k$ consecutive frames, the joint is considered not tracked and an error is reported.

Cutting and Matching: (3.1) *cutting* removes leading and trailing frames of postures not belonging to the movement. (3.2) *matching* aligns a sequence with the aggregated master sequence.

Normalizing: (4.1) *computes a scaling* transformation that moves each joint of the first frame of a sequence to the corresponding mean joint position of the first aggregated frame of the aggregated frame sequence. It applies this scaling to all frames of the sequence. (4.2) *computes a hip rotation* transformation for the first frame of a sequence that lets skeletons "look" towards the camera. It applies this rotation to all frames of the sequence. (4.3) *computes the indirect features* from the direct features for each frame.

Scoring: (5.1) *calculates frame weights* for the aggregated master sequence frames based on the extremeness of the poses in these frames, i.e., the deviation of the features from a normal null pose. This is done automatically but offline, once for the aggregated master sequence. (5.2) *calculates the deviations* of user sequences feature values v as the z-score of v in the corresponding sample distributions from sub-step (1.10) and multiplies them with the respective weight calculated in the previous sub-step. (5.3) *computes for each feature* $v_{avg\text{-}max}$ as the average of the three largest weighted feature deviation from the previous sub-step. (5.4.) *calculates a feature score* for each $v_{avg\text{-}max}$ as the sample probability that a lower (better) maximum weighted deviation ever exists for a feature, i.e., that the previous sub-step ever gives a smaller (better) value than $v_{avg\text{-}max}$. Feature scores are between 0 (best) and 1 (worst) and locate movement issues and their severity. The next and final step (5.5) uses these feature scores in order to compute an overall movement quality score using an ML approach, cf. Section 5.

5. STATISTICAL AND MACHINE LEARNING COMPONENT

Various ML approaches apply to estimate an overall quality score. We assess linear, random forest, and support vector regression, and feed forward deep neural network.

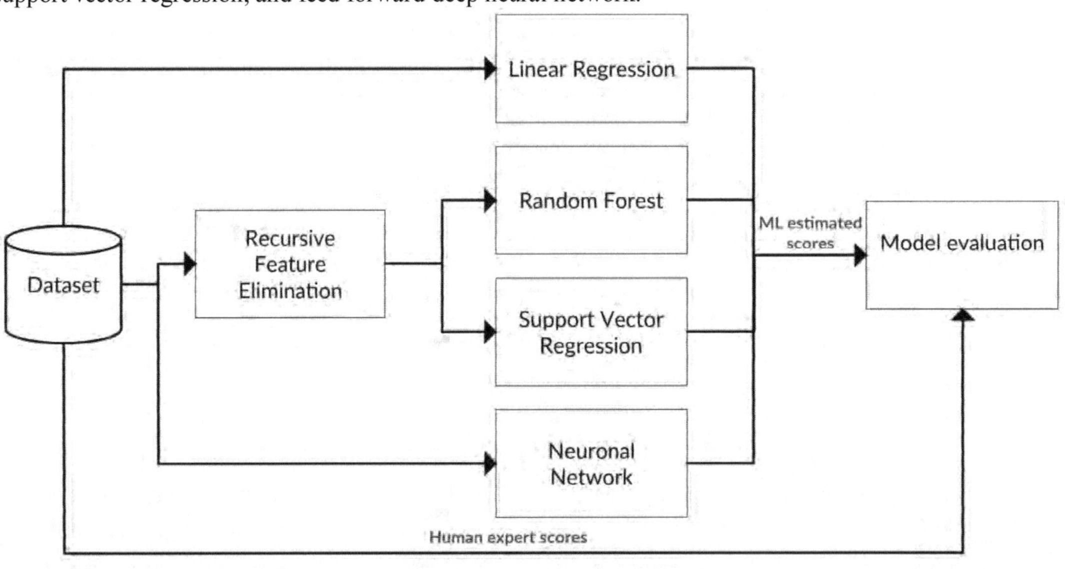

Figure 1. ML Component

The ML component depicted in Figure *1* uses recursive feature elimination for the random forest and the support vector regression approaches to select the subset of features with the best estimation results (Guyon and Elisseeff 2003). For linear regression and neural network, recursive feature elimination is not applied, because the importance of features is inherently regarded. Parameter tuning is applied in all models. The dataset contains the following 40 features scores calculated in step 5: 13 angles deviation between limbs adjacent to the same joint, e.g., elbows, hips, head, neck, spine, 24 weak links features, e.g., heels rise, knees more inwards or outwards, arms fall forward, etc., two time-deviation features relative to the start and end frames, and the mean of these scores. Time-coherence is not considered relevant. Hence, we use aggregated feature scores, not sequences of features. The output is an estimated overall quality movement score. The estimated scores are passed to a model evaluation step where they are compared with human experts scores.

To avoid overfitting the training data, regression is constrained such that pairs of symmetric features, e.g., deviations in the left/right knees, get the same weights. This is achieved by regressing on the *sum* of these instead of the individual feature scores.

6. EXPERIMENTAL ASSESSMENT OF PERFORMANCES

Method: We use an overhead deep squat as the evaluation movement, an exercise standardized by the National Academy of Sports Medicine (NASM)[4] that comes with movement execution and scoring specifications.

We assess the accuracy of the different approaches as follows. The baseline approach is the digitalization of the NASM overhead deep squat assessment and scoring, i.e., it implements the NASM scoring specification. Manual human-expert-based NASM overhead deep squat assessment and its digitalization show a high agreement (Dressler et al. 2019b) so that the digitalization serves as our ground truth. We assess the agreement of the scores for the four ML learning approaches introduced in Section 5 with the scores of the baseline approach.

A high agreement of is considered a high accuracy. We therefore calculate the Pearson coefficient r for the correlation of scores coming from different approaches with baseline. Altman and Bland argue that a high correlation is necessary but not sufficient for a good agreement between two approaches and suggest an additional analysis of differences (Altman and Bland 1983). Therefore, we additionally assess the differences of the baseline score and score of the approach with the highest correlation to baseline. The accuracy results depend on the sample in training and test data. We therefore repeat the experiments and report the statistics of 100 runs with fair random splitting in training and test data sets each containing 50% of the sample data for a run. We randomly select one run for discussing Bland-Altman statistics.

For time evaluations, preprocessing, cutting, matching, normalization, and feature scoring are excluded from the assessment as these steps are common to all ML variants. Note that we are interested in the relative response time between the variants not in minimizing the absolute response time.

Data: The assessment is based on a sample of 2094 user sequences recorded in 2018, the bulk during so-called company health days in Germany where AIMO[5] recorded deep squats of health interested employees. The sample data contains scans of female and male persons, 25-65 years old, working in office jobs. No further effort was made to get a representative and fair sample of the full population.

Implementation: All steps of Section 4 use a *Matlab* implementation and so does scoring with linear regression. The machine learning based scoring uses the *Sklearn* (random forest and support vector regression) and the *Keras* libraries based on *TesorFlow* (neuronal network) for *Python.* The baseline approach uses AIMO's production code.

There are 40 features in total. Random forest was eventually trained and tested with 33 features after recursive feature elimination and a maximum depth of 10 (the deeper the tree the slower are training and testing). Support vector regression was trained with 23 features after recursive feature elimination. Linear regression and neuronal network use all features. Different neural network layer architectures were tested. The following layers were found empirically by maximizing the correlation: input layer with 40 neurons activation function *relu*, one hidden layer with 40 neurons and activation function *relu*, and one output layer with 1 neuron and activation function *sigmoid*, optimizer *adam* and loss *binary_crossentropy*.

Response time experiments are run on a MacBook Pro with a 2,3 GHz Intel Core i7 processor.

Results: Figure 2 (left) summarizes the accuracy results. It shows the minimum, first quartile, median, third quartile, and maximum (five-number summary) of the correlation coefficients of the 100 runs assessing the four ML approaches. The mean correlation coefficients (standard errors) are 0.78 (0.9×10^{-3}) for linear regression, 0.83 (1.0×10^{-3}) for random forest, 0.78 (0.9×10^{-3}) for support vector regression, and 0,81 (1.3×10^{-3}) for the neuronal network.

Figure 2 (right) summarizes the response time results. For a movement instance, training (scoring) takes 3 (3) *msec* for linear regression, 1320 (17) *msec* for random forest, 30 (8) *msec* for support vector regression, and 20575 (756) *msec* for neuronal network.

---

[4] https://en.wikipedia.org/wiki/National_Academy_of_Sports_Medicine

[5] Our industry partner AIMO AB, https://www.aimo-health.com

Figure 3 shows the results of Bland-Altman statistics for linear regression (top, left), random forest (top, right), support vector regression (bottom, left) and neuronal network (bottom, right): 95% of the differences between the baseline and the respective approach are in the range of -0.02±0.26 score points for linear regression, -0.01±0.21 score points for random forest, -0.03±0.25 score points for support vector regression, and -0.02±0.23 score points for neuronal network, i.e., between 23% for random forest and neuronal network and 25.5% for linear and support vector regression of the maximum possible range of differences.

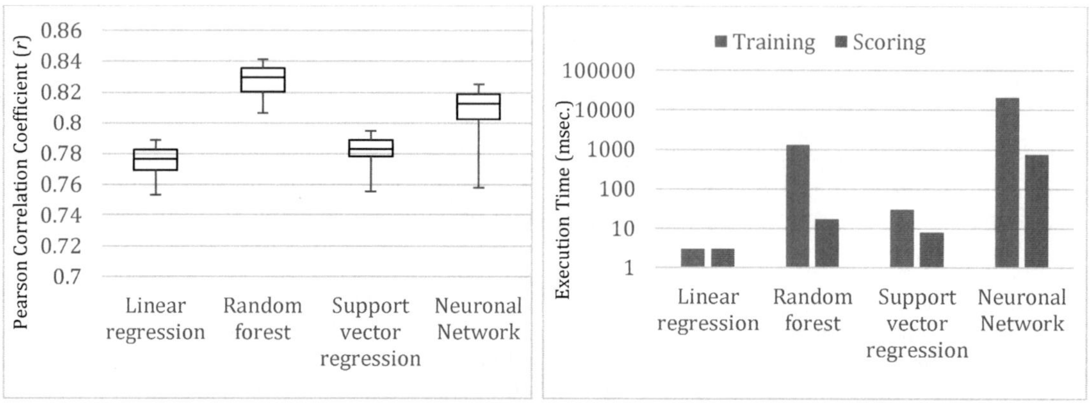

Figure 2. Pearson coefficient of correlation r of the ML approaches with the baseline approach (left) and time of training and scoring per movement instance (in msec, log scale) of the approaches (right)

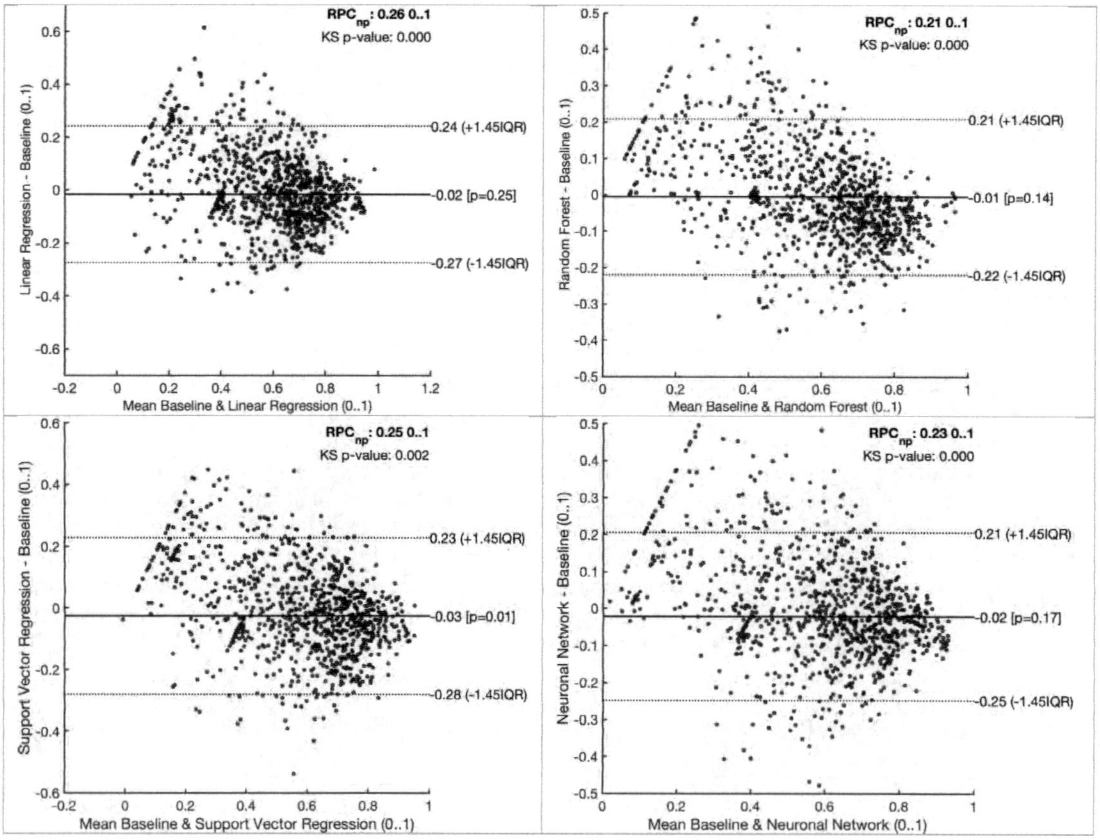

Figure 3. Differences of the ML approaches and the baseline approach

Discussion: The *random forest* approach shows the highest correlation (*r=83*) and the lowest range of differences with the scores of the digitalized expert opinions. It responds quite fast in scoring (*t=17 msec.*) and is sufficiently fast in training new models. It is therefore our champion approach. The neural network performs slightly better than linear and support vector regression. It would require more training data in order to improve further. All approaches show a high correlation and an acceptably low range of differences with the scores of the digitalization of expert opinions. However, the neuronal network approach is quite expensive in training as the *Keras* library runs on CPU and not on a *Tensorflow* GPU.

7. CONCLUSION

The paper presented (i) a general human movement assessment framework allowing the assessments of movements based on data from commodity 3D sensor technology. It reduces the noise of the input sensor and extracts relevant features of movements automatically. These features are input to (ii) a component assessing different ML models that score the quality of movements. We (iii) experimentally evaluated the accuracy and response and training times of four such models. Random forest turned out to be our champion approach.

The selected features, the preprocessing, the matching, and the scoring are agnostic w.r.t. the specific type of movement. None of the steps would be different if another movement, e.g., sit-to-stand, was assessed instead, only the data. Thus, the introduced technology contributes to value-centered healthcare with an improved objective assessment of musculoskeletal insufficiencies of patients. However, for stable conclusions, we need to assess other types of movements on larger and more representative samples of the population, which is matter of future work.

Moreover, the ML approaches outperform the baseline approach (linear regression) only by ca. 6%, which leaves room for improvements. Future work could add new sequential features and use CNN and RNN models trained on a bigger dataset or apply other machine learning models such as probabilistic neural networks.

Finally, we currently develop a mobile client based movement assessment using online algorithms allowing for pipelined preprocessing, cutting, matching, normalization, and scoring with the champion approach. Future work also aims at substituting the 3D with 2D camera technology in the full movement assessment framework allowing for more efficient data collection.

REFERENCES

Altman, D. G. and J. M. Bland, 1983. Measurement in medicine: the analysis of method comparison studies. *The Statistician* 32: 307-317.

Baldewijns, G., G. Verheyden, B. Vanrumste and T. Croonenborghs, 2014, Validation of the kinect for gait analysis using the GAITRite walkway. *36th Annual Inter. Conf. of the IEEE Engineering in Medicine and Biology Society, EMBC 2014, Chicago, IL, USA, August 26-30, 2014.*

Bao, T., B. N. Klatt, S. L. Whitney, K. H. Sienko and J. Wiens, 2019. Automatically evaluating balance: a machine learning approach. *IEEE transactions on neural systems and rehabilitation engineering* 27(2): 179-186.

Cippitelli, E., S. Gasparrini, S. Spinsante, E. Gambi, F. Verdini, L. Burattini, F. Di Nardo and S. Fioretti, 2015, Validation of an optimized algorithm to use Kinect in a non-structured environment for Sit-to-Stand analysis. *2015 37th Annual International Conference of the IEEE Engineering in Medicine and Biology Society (EMBC).*

Dressler, D., P. Liapota and W. Löwe, 2019a, Data Driven Human Movement Assessment. *11th International Conference on Innovation in Knowledge Based and Intelligent Engineering Systems, (KES), Invited Session on Digital Health, Distance Learning and decision support for eHealth.*

Dressler, D., P. Liapota and W. Löwe, 2019b, Towards an Automated Assessment of Musculoskeletal Insufficiencies. *11th International Conference on Innovation in Knowledge Based and Intelligent Engineering Systems, (KES), Invited Session on Data Selection in Machine Learning.*

Elgammal, A. M. and C.-S. Lee, 2008, The Role of Manifold Learning in Human Motion Analysis. *Human Motion – Understanding, Modelling, Capture, and Animation*, Springer.

Elton, J. and A. O'Riordan, 2016. *Healthcare Disrupted: Next Generation Business Models and Strategies*, Wiley.

Eltoukhy, M. A., C. Kuenze, J. Oh and J. F. Signorile, 2018. Validation of Static and Dynamic Balance Assessment Using Microsoft Kinect for Young and Elderly Populations. *Jour. of Biomedical and Health Info.* 22(1): 147-153.

Fern'ndez-Baena, A., A. Susín and X. Lligadas, 2012, Biomechanical Validation of Upper-Body and Lower-Body Joint Movements of Kinect Motion Capture Data for Rehabilitation Treatments. *2012 Fourth International Conference on Intelligent Networking and Collaborative Systems*.

Geerse, D., B. Coolen, D. Kolijn and M. Roerdink, 2017. Validation of Foot Placement Locations from Ankle Data of a Kinect v2 Sensor. *Sensors* 17(10): 2301.

Gray, A. D., J. M. Marks, E. E. Stone, M. C. Butler, M. Skubic and S. L. Sherman, 2014. Validation of the Microsoft Kinect as a Portable and Inexpensive Screening Tool for Identifying ACL Injury Risk. *Orthopaedic Journal of Sports Medicine* 2.

Guyon, I. and A. Elisseeff, 2003. An introduction to variable and feature selection. *Journal of machine learning research* 3(Mar): 1157-1182.

Houmanfar, R., M. Karg and D. Kulić, 2016. Movement analysis of rehabilitation exercises: Distance metrics for measuring patient progress. *IEEE Systems Journal* 10(3): 1014-1025.

Jiang, M., J. Kong, G. Bebis and H. Huo, 2015. Informative joints based human action recognition using skeleton contexts. *Signal Processing: Image Communication* 33: 29-40.

Kim, C. Y., J. S. Hong and K. J. Chun, 2016. Validation of feasibility of two depth sensor-based Microsoft Kinect cameras for human abduction-adduction motion analysis. *International Journal of Precision Engineering and Manufacturing* 17(9): 1209-1214.

Leightley, D., J. S. McPhee and M. H. Yap, 2017. Automated analysis and quantification of human mobility using a depth sensor. *IEEE journal of biomedical and health informatics* 21(4): 939-948.

Liao, Y., A. Vakanski and M. Xian, 2019. A deep learning framework for assessment of quality of rehabilitation exercises. *arXiv preprint arXiv:1901.10435*.

Lopez, N., E. Perez, E. Tello, A. Rodrigo and M. E. Valentinuzzi, 2018. Statistical Validation for Clinical Measures: Repeatability and Agreement of Kinect™-Based Software. *BioMed Research International*.

Ma, M., R. Proffitt and M. Skubic, 2018. Validation of a Kinect V2 based rehabilitation game. *PLOS ONE* 13(8): 1-15.

Müller, B., W. Ilg, M. A. Giese and N. Ludolph, 2017. Validation of enhanced kinect sensor based motion capturing for gait assessment. *PLOS ONE* 12(4): 1-18.

Paiement, A., L. Tao, M. Camplani, S. Hannuna, D. Damen and M. Mirmehdi, 2014, Online quality assessment of human motion from skeleton data. *Proceedings of the British Machine Vision Conference*, BMVA Press.

Parmar, P. and B. T. Morris, 2016, Measuring the quality of exercises. *2016 38th Annual International Conference of the IEEE Engineering in Medicine and Biology Society (EMBC)*, IEEE.

Pazhoumand-Dar, H., C.-P. Lam and M. Masek, 2015. Joint movement similarities for robust 3D action recognition using skeletal data. *Journal of Visual Communication and Image Representation* 30: 10-21.

Pirsiavash, H., C. Vondrick and A. Torralba, 2014, Assessing the Quality of Actions. *Computer Vision – ECCV 2014*, Cham, Springer International Publishing.

Plantard, P., H. P. H. Shum, A.-S. L. Pierres and F. Multon, 2017. Validation of an ergonomic assessment method using Kinect data in real workplace conditions. *Applied Ergonomics* 65: 562-569.

Tao, L., A. Paiement, D. Damen, M. Mirmehdi, S. Hannuna, M. Camplani, T. Burghardt and I. Craddock, 2016. A comparative study of pose representation and dynamics modelling for online motion quality assessment. *Computer Vision and Image Understanding* 148: 136-152.

Taylor, P. E., G. J. M. Almeida, T. Kanade and J. K. Hodgins, 2010, Classifying human motion quality for knee osteoarthritis using accelerometers. *2010 Annual Inter. Conf.of the IEEE engineering in medicine and biology*, IEEE.

Wang, J., Z. Liu, Y. Wu and J. Yuan, 2012, Mining actionlet ensemble for action recognition with depth cameras. *IEEE Conference on Computer Vision and Pattern Recognition*, IEEE Computer Society.

Wang, J. and H. Zheng, 2012, View-robust action recognition based on temporal self-similarities and dynamic time warping. *2012 IEEE International Conference on Computer Science and Automation Engineering (CSAE)*.

Wang, R., G. Medioni, C. J. Winstein and C. Blanco, 2013, Home Monitoring Musculo-skeletal Disorders with a Single 3D Sensor. *2013 IEEE Conference on Computer Vision and Pattern Recognition Workshops*.

Xia, L., C.-C. Chen and J. K. Aggarwal, 2012, View invariant human action recognition using histograms of 3D joints. *CVPR Workshops*, IEEE Computer Society.

SEMANTIC INTEGRATION OF MULTIPLE HEALTH DATA FOR TREATMENT DECISION-MAKING IN LOW-RESOURCE SETTINGS

Olawande Daramola[1] and Thomas Moser[2]
[1]Cape Peninsula University of Technology
PO Box 652 Cape Town 8000 | Corner of Keizergracht and Tennant Street, Zonnebloem
South Africa
[2]St. Pölten University of Applied Sciences, Matthias Corvinus-Straße 15, 3100 St. Pölten
Austria

ABSTRACT

Accurate treatment decision-making for disease treatment is important, particularly if it is done in a way that helps to overcome the challenges associated with low resource setting. This include shortage of qualified personnel, infrastructure, ready access to devices, and healthcare by common people. Although real-time gait analysis has been used for the diagnosis of diseases that are associated with gait impairments, the need to improve on the performance of gait-based prediction by augmenting it with other sources of knowledge in a complementary way has been acknowledged. In contrast, to existing approaches, this paper presents the design of a conceptual framework that will enable the semantic integration and use of information from multiple health data sources in order to ensure efficient treatment decision making for gait-related diseases in a low resource setting such as South Africa. The analytical evaluation of the proposed framework suggests that it has sufficient merit and potential to be usable and useful in low-resource settings.

KEYWORDS

Low Resource Settings, Semantic Integration, Health Data Integration, Electronic Health Records, Disease Treatment, Decision Making

1. INTRODUCTION

Accurate and prompt treatment decision-making by medical practitioners in dealing with patients is critical to ensuring the health and well-being of patients. It is particularly important when this is done in low resource settings where there are myriad of challenges that impedes good healthcare conditions. Some of the challenges of low resource settings, which any viable healthcare solution must overcome include shortage of qualified personnel, lack of good infrastructure, lack of ready access to mobile devices, and high cost of healthcare for common people (Fritz et al., 2015; Zargaran et al., 2014). The advent of new digital technologies such as cloud computing, mobile computing, wearable sensors, and linked open data has created new opportunities for expert decision-making in terms of disease diagnosis, disease treatment, and patient's rehabilitation - irrespective of the physical distance between the patient and the medical practitioner.

Gait analysis (GA), which is also known as locomotive analysis entails collecting quantitative data on the pattern of physical movements of a human in order to understand the etiology of gait defects and the formulation of an appropriate treatment plan. GA has been used as a basis for disease diagnosis for patients with walking impairment, and the treatment of many neurological disorders and cardiovascular diseases (Tang and Su, 2013). Currently, it is possible to obtain real-time information on a patient's gait pattern while the patient is walking both within and outside the hospital environment. Critical gait parameters such as gait speed, cadence, step length, stride length and many more can be measured by using wearable body sensors that are attached to either knees, using sensor-insoles for shoes, or sensors for legs and limbs (Horak et al., 2015; Horsak et al., 2016). These gait parameters have been associated with the symptoms of some specific diseases, which make them suitable as a basis for disease diagnoses, disease treatment, gait retraining, and gait rehabilitation of patients that are being treated for gait impairment. However, real-time gait analysis is still confronted with several challenges, which include the need to i) improve the precision of diseases diagnosis based on gait factors; ii) adapt real-time gait analysis for application in resource-constrained settings; iii) identify specific gait parameters that are more associated with specific types of diseases; and iv) engender gait improvement during gait retraining and gait rehabilitation by using multisensory feedback mechanism (Tang and Su, 2013; Zhou et al., 2012; Zillner and Sonntag, 2012). Hence, there is the need to augment real-time gait analysis predictions with knowledge from other sources in order to enhance its efficiency (Horak et al., 2015; Zhou et al., 2012; Zillner and Sonntag, 2012). It is particularly more challenging to attain efficient treatment decision-making for gait-related diseases in low resource settings.

This paper, by focusing on gait-related diseases, presents the conceptual design of a framework (SemTreat) that leverages the semantic integration of data from multiple health data sources such as real-time gait analysis data, linked electronic health records (EHR), epidemiological data, and other open data sources in order to facilitate decision making for disease diagnosis and disease treatment in low resource settings . It is expected that the provision of our type of proposed framework will enable i) medical practitioners to make more informed decisions on the treatment of patients based on their medical information; ii) patients to have continuation of their treatment if they happen to change their location/city of residence; and iii) patients to identify the type of diseases that they might have, and make a decision on the appropriate self-management strategies that could be adopted.

The rest of this paper is organized as follows. Section 2 presents the background and related work, while Section 3 gives an overview of the methodology and design adopted for the SemTreat framework. Section 4 describes the approaches that will be used for the evaluation of the proposed framework. The paper is concluded in Section 5 with a brief note and overview of future work.

2. BACKGROUND AND RELATED WORK

This section presents an overview of related work in the area of real-time gait analysis, and background on electronic health records (EHR), and semantic integration.

So far, many applications of gait analysis have been reported in the literature, particularly to aid the treatment of diseases that are associated gait impairment such as neurological disorders and cardiovascular diseases (Horak et al., 2015; Tang and Su, 2013). Also, it has been used in conjunction with sonification for the purpose of gait training and gait rehabilitation (Effenberg et al., 2011; Rodger et al., 2014). However, most of these systems have been deployed within the laboratory setting, which makes them unable to support everyday use outside the hospital environment. A few of the exceptions to this are reported in (Noehren et al., 2011; Redd and Bamberg, 2012), but even then, these systems were not been designed or deployed in low resource settings, hence the ability of these real-time gait analysis systems to function in low resource settings has not be verified. Low cost and affordable sensor technologies for gait-data acquisition are required in order to facilitate real-time gait analysis. Another issue is the need to improve the accuracy of gait measures as a basis for disease diagnosis because gait factors are dependent on other external factors such as emotions and environment. Hence, the need to augment gait prediction with other sources of health data such as EHR and open linked data.

Electronic health records (EHR) are repositories of personal medical information of individuals. EHR provide a rich source of information on a patient including pathology data, family history, medical history, primary care data, and secondary care data (Lobo et al., 2017; Zhao and Weng, 2011). If accessible, EHR can be integrated with real-time gait data as a basis for accurate disease diagnosis, and formulation of a suitable treatment plan for a patient. Recently, the instances of the use of EHR for research purposes have been on the increase. Examples of these include CALIBER (Morley et al., 2014), HealthID[1], epSOS (Fonseca et al., 2015), openEHR[2], HEOR (Asaria et al., 2016), semantichealthnet (Martinez-Costa et al., 2014). Many of the EHR approaches focussed on the interoperability of EHR (Adebesin et al., 2013; Adenuga et al., 2015; Fonseca et al., 2015; Martinez-Costa et al., 2014), and the use of EHR for the purpose of informed decision

[1] HealthID: the technology that puts your patients' health records in your hands. available at: https://www.discovery.co.za/medical-aid/health-id

[2] OpenEHR. An open domain-driven platform for developing flexible e-health systems. http://www.openehr.org

making on patient's treatment (Asaria et al., 2016; Lobo et al., 2017; Morley et al., 2014; Zhao and Weng, 2011).

Semantic integration enables meaningful communication between different terms that have been used for similar domains concepts. It enables diverse systems to understand one another and engage in meaningful conversation (Aldred et al., 2006; Doan and Halevy, 2005; Hohpe, 2006; Moser and Biffl, 2012; Noy, 2004). In contrast to existing approaches, this research will explore the semantic integration of real-time gait data, EHR, epidemiological data, open linked data, and other open data sources for treatment decision making in low resource settings for some diseases that are associated with gait impairment.

3. METHODOLOGY AND DESIGN

This section starts with describing the requirements of the proposed SemTreat framework as well as the included stakeholders, then continues with defining the proposed architecture of the SemTreat framework, and finally identifies potential use case scenarios for the usage of the SemTreat framework.

3.1 SemTreat Requirements Elicitation

The requirements were gathered in order to understand both the technical as well as the user requirements for the semantic integration of linked EHR, epidemiological data, and real-time gait data. This was done by the help of a requirements analyst, medical experts, and epidemiologists in order to elicit knowledge from the domain experts and relevant stakeholders. During the requirements elicitation process, the following stakeholders of the SemTreat framework were identified: medical practitioners, patients, and governmental organizations.

The requirements of the SemTreat framework are structured into the following categories:

- practicability of the SemTreat framework
    - centralized overview of distributed health data of a single patient: should be usable both by the patient as well as by medical practitioners that have been granted such authorized access
- functionality in low-resource settings
    - limited connectivity regarding both signal strength as well as bandwidth
    - electrical power availability;
    - Accessibility of technology to patients in terms of cost of devices and internet data;
- usability of the SemTreat frameworks application (mobile)
    - smartphone vs. regular mobile phone version
    - usage should also be possible (in a limited way) with low literacy as well as low technical affinity (e.g. elderly persons)
- privacy and data protection
    - personal health data is very sensitive information and needs to be well protected
    - each patient should be able to define for which other parties he/she wants to share the personal health data
    - aggregated but anonymized health data should be available for usage by governmental organizations, e.g. for statistical analysis, data analytics, and strategic planning

3.2 Description of the SemTreat Architectural Design

The SemTreat architecture is conceptualized in way that makes it suitable to facilitate the envisioned capability to aid treatment decision making on gait-related diseases in low resource settings.

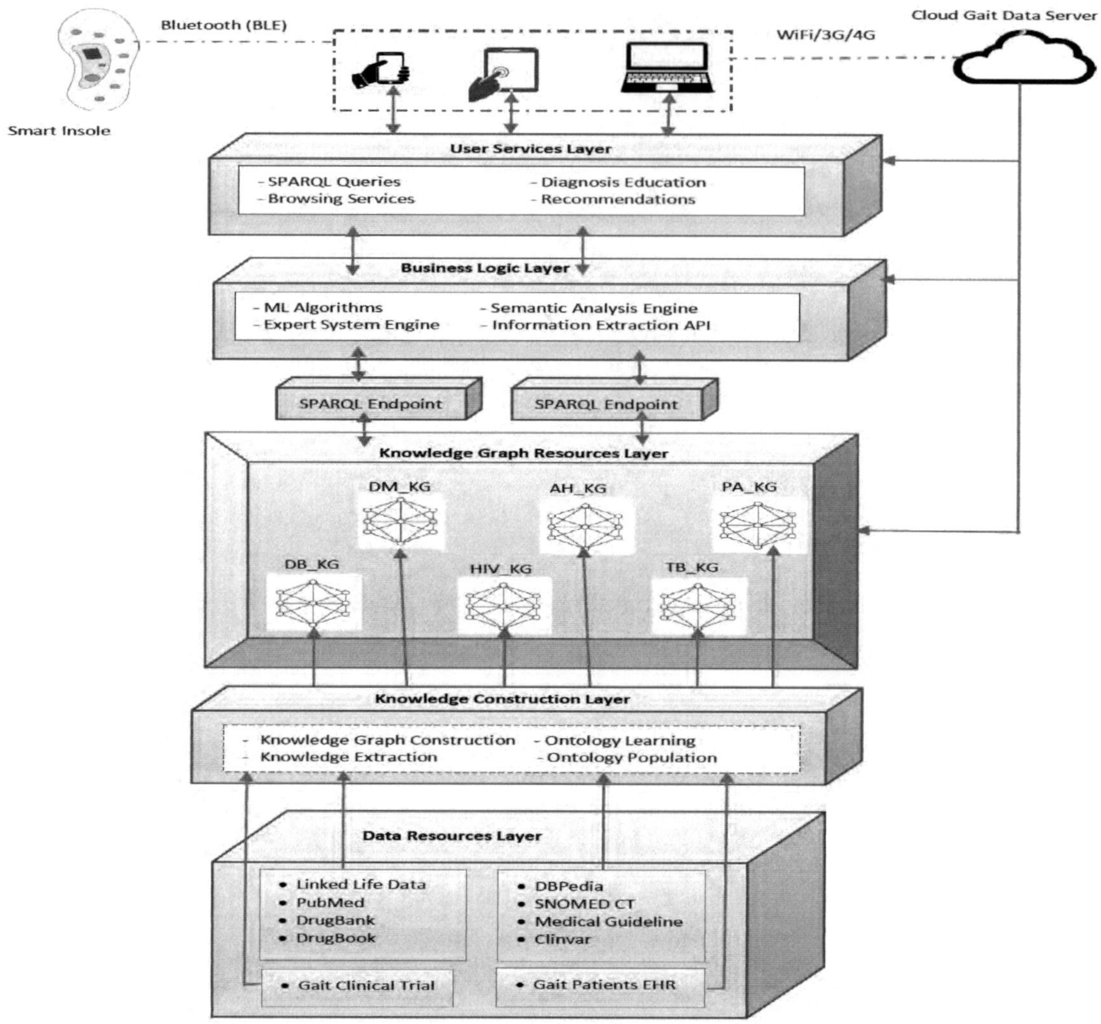

Figure 1. SemTreat Architecture Diagram

The SemTreat architecture (see Figure 1) is composed of four layers, which are described as follows:

User services Layer: this layer presents a user with opportunities to access the various services that the SemTreat architecture affords. This includes search and query services, recommendations, health education, and diagnosis.

Business Logic Layer: this layer enables the SemTreat architecture to construct an appropriate response to a user's query in a timely manner. The various algorithms, and computational modules that performs machine-learning-based prediction based on data, expert systems diagnosis, recommendations on health and wellness requests from users, and semantic analysis are contained in this layer. The components of this layer function based on raw data and semantic web data that are retrieved by using SPARQL endpoints in form of RDF triples from specific knowledge graphs (KG) in the lower level of the SemTreat architecture.

Knowledge Graph Resources Layer: this layer is composed of different knowledge graphs that have been pre-constructed through an offline process for specific types of diseases such as tuberculosis (TB_KG), HIV AIDS (HIV_KG), Diabetes (DB_KG), Dementia (DM_KG), Alzheimer's Disease, Parkinson's Disease.

Knowledge Graph Construction Layer: this layer subsumes the myriad of offline activities that enable the construction of relevant knowledge graphs (KG) from multiple health data sources. Key operations that are deployed at this level include ontology learning, ontology population, and information extraction from raw health data sources. The knowledge graphs are constructed from aggregating data from diverse relevant open sources in order to build a rich data pool for specific diseases. In order to build/construct the KG for a particular disease, the clinical gait data, and electronic health records of gait patients of that particular disease, and relevant epidemiological data are used to compose the data pool for that specific disease.

Data Resources Layer: this is a suite of open health data resources, and sources of textual medical knowledge (TMK) that provide good basis for generating a credible knowledgebase for specific types of diseases. The knowledge graph construction entails extracting relevant knowledge from these sources, ensuring their semantic integration in form of knowledge graphs in a way that can be used by web services and computational modules at run-time for decision making. A description of some exemplary medical knowledge sources from which data could be extracted is shown in Table 1.

Table 1. Medical Knowledge Sources to be integrated into the SemTreat framework

S/n	Sources	Description	
1	Linked Life Data (LLD)	LLD as a data-as-a-service platform enables 25 public biomedical databases to be accessed through a single access point	http://linkedlifedata.com/
2	PubMed	PubMed contains over 29 million open access articles from biomedical literature.	https://www.ncbi.nlm.nih.gov/pubmed/
3	DrugBank	This database contains information on drug data and drug targets. It can be accessed using an API.	https://www.drugbank.ca/
4	DrugBook	This is an e-book that offers rich textual knowledge on drugs	
5	DBPedia	This exposes data in Wikipedia in form of a database for open access	https://wiki.dbpedia.org/
6	SNOMED CT	This is a comprehensive vocabulary of clinical healthcare terminology.	http://www.snomed.org/
7	Clinvar	this provides access to information about the relationships asserted between human variation and observed health status, and the history of that interpretation.	https://www.ncbi.nlm.nih.gov/clinvar/

3.3 SemTreat Framework Use Case Scenarios

In this section, we present few use cases of the SemTreat framework as relevant to a patient, or a doctor by using exemplar narratives.

Case 1. Mr X, is a 70 years old male with gait impairment due to Tuberculosis (TB) ailment. X wants to receive recommendation on suitable recreational activities as a form self-management that can aid his wellness. As a result of wearing a smart insole for a period of 3 months, gait data such as his cadence, gait speed, step length, and stride length was constantly monitored and recorded. Since X is domiciled in a low-resource setting, gait data were temporarily stored in his mobile device from time to time before they are subsequently transferred to the cloud via WiFi or mobile (3G/4G) connectivity. When X seeks recommendation to be advised on the type of recreational activities that could help to improve his personal fitness, the recommendation for Mr X, will based on the level of severity of TB ailment as inferred from the stored gait patterns that are inherent in his gait data, and the diverse forms of knowledge about TB that are stored in TB KG. This will include using the documentation of TB treatment cases, clinical trials, epidemiological data, and TB knowledge from diverse sources as contained in the TB knowledge graph, and the EHR of Mr X in order to give him an appropriate recommendation.

Case 2. Dr Y is an expert in the treatment of Type II Diabetes and needs to respond to a request for advice from Patient B, female, who is Type II Diabetes Patient that is personally know by him, but who is currently in a remote location. The severity of Patient B's ailment has affected her gait, and such as worn the smart insole for the purpose of real-time gait analysis for 2 months. Dr Y wants to use the expert diagnosis service of the SemTreat framework as decision support tool in order to attend to Patient's B query. The request from Patient B is that she has witnessed some symptoms lately that appear strange and needs to know what is going on. The request from Patient B as any other request of the same nature is uniquely identified by the unique medical record number/medical insurance number of Patient B. Dr Y needs to access the electronic health record of Patient B in order to know her medical history. An encrypted message from Dr Y asking Patient B to give consent/permission to Dr Y to access her medical records is sent to Patient B. Patient B must give an approval/ consent before access is granted to Dr Y. Once the permission to access personal EHR is granted by Patient B, her pre-stored gait data also becomes available for the SemTreat algorithms to use in order to construct a medical advice on Patient B for the consideration of Dr Y. By using the combination of relevant extracted knowledge from the Diabetes KG, Gait data, and EHR of Patient B, Dr Y will receive machine learning-based decision support on how to respond to the request from Patient B.

4. EVALUATION CONCEPT

The evaluation concept for the SemTreat framework will be guided by two perspectives. The first perspective of the evaluation will focus on performance and usability. For this, the emphasis will be on the atomic/individual user-based services that the SemTreat framework affords to assess their performance and usability. The services, such as expert diagnosis, health recommender system, and information search shall be experimentally evaluated by using standard performance metrics such as FI, recall, precision, and accuracy. Also, the usability of the user-based services shall be assessed both within and outside the clinic environment to determine users' perceptions of the services. The usability evaluation shall be done mainly through qualitative evaluation by user experiments feedbacks, questionnaire surveys, interviews and case studies. The second perspective of the evaluation is to assess the practicability and effectiveness of the framework in terms of its support for treatment decision-making for selected gait-related diseases from the viewpoints of patients/users, and medical practitioners. The emphasis of this particular evaluation is to determine whether the framework can indeed attain its envisioned goals for low-resource settings and can be integrated effectively into the operational workflow of medical practitioners for use, and also the daily lifestyle of an average user/patient. To do this, the Goal Question Metric (GQM) paradigm (Basili et al., 1994) will be used to conduct a holistic evaluation of the framework in order to assess the extent to which it can attain the goals and objectives that are associated with providing support for treatment decision making in low resource settings.

5. CONCLUSION AND FUTURE WORK

In this paper, we have presented the description of a conceptual framework ((SemTreat) that can enable the sematic integration of multiple health data sources for treatment decision-making for gait-related diseases in low-resource settings. The framework affords a suite of user-based services that aid the decision-making process of medical practitioners and help users/patients in the area of self-management for healthy living and wellness. The SemTreat framework is presented as capable of leveraging aggregated sematic knowledge from diverse sources to deliver intelligent services such as search and browsing, health recommendations, and expert disease diagnosis, while providing support for low cost and affordable technologies for real-time gait analysis.

However, the eventual realization of the SemTreat framework depends on giving attention to some additional non-functions requirements that will ensure the overall success of the framework. This includes the following:

- **Security:** dealing with health information require that all data must be treated with security by ensuring that that personal data of patients are protected, and the data privacy is not compromised
- **Accessibility**: the framework must provide easy ways for people to assess his resources
- **Cost**: there must be emphasis on the use for low-cost technologies that are affordable for poor people. This include the use of ensuring that the design and product development is suitable for.

In future work, we will prototypically implement the components/subsystems that make up the SemTreat framework, so that we can proceed with the evaluation.

ACKNOWLEDGEMENT

This work was supported by the Centre for International Cooperation & Mobility (ICM) of the Austrian Agency for International Cooperation in Education and Research (OeAD-GmbH) under Grant No. ZA 10/2019. It is also supported by the National Research Foundation of South Africa under Grant No. STGR 180414320796.

REFERENCES

Adebesin, F., Foster, R., Kotzé, P., Van Greunen, D., 2013. A Review of Interoperability Standards in E-health and Imperatives for their Adoption in Africa. South African Computer Journal 50. https://doi.org/10.18489/sacj.v50i1.176

Adenuga, O.A., Kekwaletswe, R.M., Coleman, A., 2015. eHealth integration and interoperability issues: towards a solution through enterprise architecture. Health Information Science and Systems 3. https://doi.org/10.1186/s13755-015-0009-7

Aldred, L., Aalst, W.M.P., Dumas, M., Ter Hofstede, A., 2006. Understanding the challenges in getting together: The semantics of decoupling in middleware.

Asaria, M., Grasic, K., Walker, S., 2016. Using Linked Electronic Health Records to Estimate Healthcare Costs: Key Challenges and Opportunities. PharmacoEconomics 34, 155–160. https://doi.org/10.1007/s40273-015-0358-8

Basili, V.R., Caldiera, G., Rombach, H.D., 1994. The Goal Question Metric Approach, in: Encyclopedia of Software Engineering. Wiley.

Doan, A., Halevy, A.Y., 2005. Semantic-integration research in the database community. AI Mag. 26, 83–94.

Effenberg, A., Fehse, U., Weber, A., 2011. Movement Sonification: Audiovisual benefits on motor learning. BIO Web of Conferences 1, 00022. https://doi.org/10.1051/bioconf/20110100022

Fonseca, M., Karkaletsis, K., Cruz, I.A., Berler,. A., Oliveira, I.C., 2015. OpenNCP: a novel framework to foster cross-border e-Health services. Studies in Health Technology and Informatics 617–621. https://doi.org/10.3233/978-1-61499-512-8-617

Fritz, F., Tilahun, B., Dugas, M., 2015. Success criteria for electronic medical record implementations in low-resource settings: a systematic review. Journal of the American Medical Informatics Association 22, 479–488. https://doi.org/10.1093/jamia/ocu038

Hohpe, G., 2006. Conversation Patterns: Workshop Report. Dagstuhl Seminar 7.

Horak, F., King, L., Mancini, M., 2015. Role of Body-Worn Movement Monitor Technology for Balance and Gait Rehabilitation. Physical Therapy 95, 461–470. https://doi.org/10.2522/ptj.20140253

Horsak, B., Dlapka, R., Iber, M., Gorgas, A.-M., Kiselka, A., Gradl, C., Siragy, T., Doppler, J., 2016. SONIGait: a wireless instrumented insole device for real-time sonification of gait. Journal on Multimodal User Interfaces 10, 195–206. https://doi.org/10.1007/s12193-016-0216-9

Lobo, J.M., Denton, B.T., Wilson, J.R., Shah, N.D., Smith, S.A., 2017. Using claims data linked with electronic health records to monitor and improve adherence to medication. IISE Transactions on Healthcare Systems Engineering 7, 194–214. https://doi.org/10.1080/24725579.2017.1346728

Martinez-Costa, C., Legaz-Garcia, M.C., Schulz, S., Fernandez-Breis, J.T., 2014. Ontology-based infrastructure for a meaningful EHR representation and use, in: IEEE-EMBS International Conference on Biomedical and Health Informatics (BHI). Presented at the 2014 IEEE-EMBS International Conference on Biomedical and Health Informatics (BHI), IEEE, Valencia, Spain, pp. 535–538. https://doi.org/10.1109/BHI.2014.6864420

Morley, K.I., Wallace, J., Denaxas, S.C., Hunter, R.J., Patel, R.S., Perel, P., Shah, A.D., Timmis, A.D., Schilling, R.J., Hemingway, H., 2014. Defining Disease Phenotypes Using National Linked Electronic Health Records: A Case Study of Atrial Fibrillation. PLoS ONE 9, e110900. https://doi.org/10.1371/journal.pone.0110900

Moser, T., Biffl, S., 2012. Semantic Integration of Software and Systems Engineering Environments. IEEE Transactions on Systems, Man, and Cybernetics, Part C (Applications and Reviews) 42, 38–50. https://doi.org/10.1109/TSMCC.2011.2136377

Noehren, B., Scholz, J., Davis, I., 2011. The effect of real-time gait retraining on hip kinematics, pain and function in subjects with patellofemoral pain syndrome. British Journal of Sports Medicine 45, 691–696. https://doi.org/10.1136/bjsm.2009.069112

Noy, N.F., 2004. Semantic integration: a survey of ontology-based approaches. ACM SIGMOD Record 33, 65. https://doi.org/10.1145/1041410.1041421

Redd, C.B., Bamberg, S.J.M., 2012. A Wireless Sensory Feedback Device for Real-Time Gait Feedback and Training. IEEE/ASME Transactions on Mechatronics 17, 425–433. https://doi.org/10.1109/TMECH.2012.2189014

Rodger, M.W.M., Young, W.R., Craig, C.M., 2014. Synthesis of Walking Sounds for Alleviating Gait Disturbances in Parkinson's Disease. IEEE Transactions on Neural Systems and Rehabilitation Engineering 22, 543–548. https://doi.org/10.1109/TNSRE.2013.2285410

Tang, W., Su, D., 2013. Locomotion analysis and its applications in neurological disorders detection: state-of-art review. Network Modeling Analysis in Health Informatics and Bioinformatics 2, 1–12. https://doi.org/10.1007/s13721-012-0020-8

Zargaran, E., Schuurman, N., Nicol, A.J., Matzopoulos, R., Cinnamon, J., Taulu, T., Ricker, B., Garbutt Brown, D.R., Navsaria, P., Hameed, S.M., 2014. The Electronic Trauma Health Record: Design and Usability of a Novel Tablet-Based Tool for Trauma Care and Injury Surveillance in Low Resource Settings. Journal of the American College of Surgeons 218, 41–50. https://doi.org/10.1016/j.jamcollsurg.2013.10.001

Zhao, D., Weng, C., 2011. Combining PubMed knowledge and EHR data to develop a weighted bayesian network for pancreatic cancer prediction. Journal of Biomedical Informatics 44, 859–868. https://doi.org/10.1016/j.jbi.2011.05.004

Zhou, Y., Qureshi, R., Sacan, A., 2012. Data simulation and regulatory network reconstruction from time-series microarray data using stepwise multiple linear regression. Network Modeling Analysis in Health Informatics and Bioinformatics 1, 3–17. https://doi.org/10.1007/s13721-012-0008-4

Zillner, S., Sonntag, D., 2012. Image metadata reasoning for improved clinical decision support. Network Modeling Analysis in Health Informatics and Bioinformatics 1, 37–46. https://doi.org/10.1007/s13721-012-0003-9

ACCEPTANCE AND USABILITY OF MEDICAL INFORMATICS AMONG PHYSICIANS IN BAHRAIN

Mishleen M. Aqleh, Mukhtar S. AL-Hashimi, Allam Hamdan and Anjum Razzaque
College of Business and Finance, Ahlia University
Manama, Bahrain

ABSTRACT

Purpose: the main objective is to explore the usability of medical mobile applications and to study the factors affecting the acceptance of the Electronic Medical Record Systems (EMRs) among physicians in Bahrain. Methodology: The study followed a quantitative design employing a questionnaire based on the Extended Technology acceptance Model (TAM2) to test the relationship between the acceptance of EMRs and the factors affecting it; System Perceived Usefulness (PU), Perceived Ease of Use (PEOU), Job Relevance and Output Quality in addition to descriptive questions to understand the extent of use of the mobile medical Apps among the physicians. The study covered 53 physicians from various hospitals across Bahrain. SPSS was used to analyze data. Findings: Acceptance of EMR is positively affected by US, EOU, Job Relevance and Output Quality and thus accepts all the intended hypothesis. Implication: the research emphasizes on the role of hospitals' management to introduce the importance, usefulness and impact of such technological advancements on the physicians work flow and quality of patient care and to provide effective training and support on how to use the implemented EMR so end-users may own the system.

KEYWORDS

Medical Information Systems, Electronic Medical Records, Technology Acceptance, Physicians, Mobile Health, Smartphones, Medical Applications, Usability

1. INTRODUCTION AND BACKGROUND

Advancements medical practice and introducing Medical Informatics in the diagnosis, treatment and processing of the information has helped the health organizations profoundly improve its functioning. Medical Informatics, in the context, is a huge technological advancement that has interacted with the medical professionals in different domains of work and were described by J C Wyatt and J L Y Liu as "*Mechanical computing in medicine" and they defined it as: "the study and application of methods to improve the management of patient data, clinical knowledge, population data, and other information relevant to patient care and community health" (Wyatt & Liu, 2002). There* is a huge variety of applications and software available that range from simple (e.g. transaction processing systems) to complex systems such as clinical decision support systems (CDSS) and used according to the need and interactivity of the health professionals. Likewise, Smartphones have captured the attention of the physicians worldwide as the medical and patient oriented applications available on these devices enable the doctors at the point of care to get the information needed for diagnosing, treating and controlling different patient health conditions and diseases (Okazaki et al., 2015; Wyatt et al., 2013). One of the most supported medical informatics systems is the patient centered type, Electronic Medical Records (EMR) through which the analyses of patients' data helps reaching the best diagnosis of any health problem, whether it was an acute or a chronic disease, faced by the profession, and deciding the most appropriate treatment needed for the patient, in addition to enhancing the quality of service delivery (Marques et al., 2011). EMR is nowadays available and implemented in hospitals, clinics and medical centers in many countries, and Bahrain is one of them, thus understanding the acceptance of EMR by physicians and the factors affecting it is very crucial to make sure that the EMR systems are finely integrated and that the anticipated benefits will be eventually demonstrable (Gagnon et al., 2014). In addition, with the increase reliability of doctors on the medical apps and their ability to carry textbooks and access different data resources (e.g. medical calculator, and medications prescriptions) on their smartphones (Payne et al.,

2012), it's important to study the usability of smartphones in the medical practice and the kinds of apps that are most widely used among doctors in Bahrain. Nowadays, there are more than 10,000 easily accessed health care applications stored in iOS (Apple operating system) serving all areas of medicine such as patient management, educational, research, and referencing.

The health industry in some countries is utilizing these devices and apps for the management of medical data, retrieval the patients' electronic medical records, clinical communication between physicians and patients (such as HealthJump App), trackers eye and skin diagnosis, blood pressure measurement and many more (Gan et al, 2016) (Gan et al., 2016). Using such improvements as EMR, and mobile-health (m-health) and likewise have been extensively recommended for their roles in enhancing the patient care quality in medical area (Moeinedin, 2011) together with the frequent and dramatic changes in the medical domain, which mandates the full acceptance of these systems by physicians and healthcare providers (Pai and Huang, 2011). The term "*Acceptance*" was defined in most of the literature referring to the original definition that had been constructed by Davis et al., described it as "*users' decision about how and when they will use technology*" (Davis et al., 1989). Research was conducted in UK in 2015 by R.K Patel et al, , evaluating surgical doctors' utilization of smartphones, medical apps popularity and use of internet activities (Illiger et al., 2014). Through a questionnaire approach to pinpoint the percentage of surgical doctors of all levels utilizing smartphones in their clinical practice (ibid). They assessed the usage of medical applications, and online medical resources, and evaluated if smartphones users were likely to utilize their own smartphone for clinical context. The results indicated that 54.2% of the doctors downloaded medical applications and 86% used smartphones for accessing information/online: especially juniors more than the seniors, (ibid) which contradicts with Illiger et al. (2014) who stated that using the tablets does not depend on age. Although most of Bahrain local and private hospitals have implemented the EMR systems and have started using them, there are no studies on the acceptance and usability of these systems among the physicians, and the factors affecting their intention to use them. Additionally, the availability of specialized medical apps on the Smartphones with the increasing usage of such devices makes it important to quantify their usability in the medical practice and the frequency of usage among physicians in Bahrain. The physicians are the ones who take the decision if they would use a new technology implemented in their workplace or not it's very crucial to find out the key factors affecting their decision to use such technology. So, the objectives of this study are to give a brief image about the EMR acceptance and the factors affecting it, in addition to the usability of Smartphones in clinical practice.

2. RESEARCH METHODOLOGY

2.1 Research Model

Extended Technology Acceptance Model (TAM 2) theoretical model that has been introduced by Venkatesh and Davis in 2000 to scientifically study how such factors affect technology acceptance by physicians was adopted and employed in this research to be able to answer the research questions. Many studies such as have found that TAM 2 also known as (extended TAM) was one of the most useful methodologies and models to describe the intentions of health healthcare professionals to use technologies and can be adequately applied to improve the way they understand technology care (Sezgin & Özkan, 2014; Chang, 2015). Each variable's indicators were determined referring to literature studies that have been reviewed.

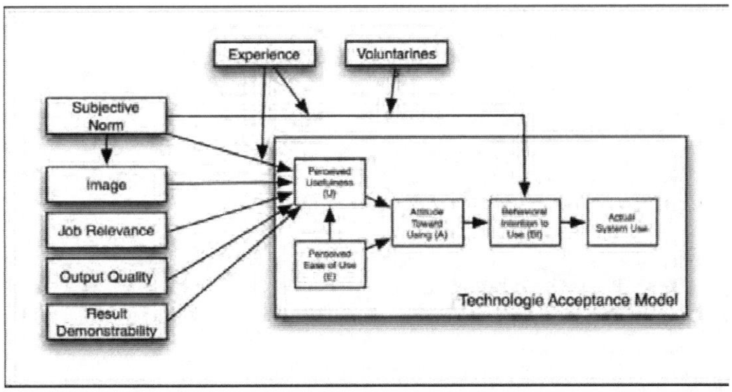

Figure 1. TAM 2

Table 1. TAM2 Constructs

Construct	Definition
Behavioral Intention (BI)	Person's intent to perform activities
Attitude	User's positive or negative reaction towards an activity
Perceived Usefulness (PU)	User's belief that utilizing a system would enhance his/her performance.
Subjective Norm	The perceived social effect on an individual to participate or not in an activity.
Image	The extent to which a person understands the technology use to strengthen his or her position among people.
Voluntariness	The degree to which a person decides to adopt and accept technology willingly.
Job Relevance	Awareness of how relevant the new technology is to a person's job.
Output Quality	The extent to which a system complete tasks related to a person's job.
Result Demonstrability	The tangibility of the aftereffects of utilizing the technology.
Experience	Knowledge or practical wisdom gained from what one has observed, encountered, or undergone.

2.2 Research Hypotheses

For the EMR acceptance, the research hypotheses were adopted from previous literature and studies that were conducted to study the EMR acceptance among doctors and the factors affecting them using the theoretical models of Technology acceptance (Gagnon et al., 2014; Davis et al., 1989). For this study purposes, 4 variables will be hypothesized and discussed. H1: Physicians' "*perceived usefulness*" of EMR has positive influence on their intention to use the system. H2: Physicians' "*perceived ease of use*" of EMR has positive influence on their intention to use the system. H3: Job relevance is positively associated with physicians' "*perceived usefulness*" H4: Quality of output is positively associated with physicians' "*perceived usefulness*"

2.3 Research Design

A quantitative cross- sectional approach using a survey design was chosen to serve the purpose of the study. The questionnaire of 34 questions was adopted from (Payne et al., 2012; Patel et al., 2015; Sweis et al., 2014; Hamidfar et al., 2008): divided in 3 sections; A - participant demographic and job-related details, B measures participant smartphone medical applications usability, and C measured acceptance of EMRs and factors affecting it. Section C (from 11 to 34) employ "a five-point Likert scale" where 1 being "*strongly disagree*" and 5 being "*strongly agree*". Target population was physician using EMRs. Sample selection was thru non-probability convenience sampling technique. The questionnaire was distributed to physicians from different hospitals, clinics and health centers in Bahrain from different specialties and job positions by using the reliable website Survey Monkey® link, and for the limitations that faced the researcher which are mentioned in the introduction chapter, 53 valid fully answered questionnaires were used for the analysis out of only 70 collected. SPSS version 20 software for statistical analysis and data manipulations. Descriptive statistics were conducted to show the percentages of the demographical variables and the percentage of the smartphone applications usage among doctors in Bahrain. Moreover, to test the relationship between the dependent variable, intention to use the EMRs, and the independent variables; PEOU and PU, and the interrelationships among variables such as the relationship between the job relevance, and quality of output and PU of the EMR, the researcher employed multiple regression analysis, which is a statistical test used to describe the effect and the relationships between the dependent and the independent variables. F statistics with P—significance (significance level of 0.01 was chose) besides Beta coefficients and Adjusted R2 values are used to analyze and study the results of regression tests.

3. RESULT AND DISCUSSION

3.1 Respondent's Demography

Table 2 represents the participants' demographics and percentages. As can be seen, most of the respondents were females 64.2% and the age range was almost around 30-39-year-old 45%. A percentage of 34% were of junior doctors and 30.2% of seniors with 35.8% from medicine specialty and 22.6% of Obstetrics and Gynecology. The years of experience of the respondents ranged between less than 1 year and more than 10 years.

Table 2. Demographic Profile of the Participants

		Frequency	Valid %
Gender	Male	19	35.8
	Female	34	64.2
age	20 – 29	21	39.6
	30 – 39	24	45.3
	40 - 49	8	15.1
Seniority level	Intern	2	3.8
	Junior	18	34
	Senior	16	30.2
	Chief	9	17
	Consultant	8	15.1
Specialization	GP	4	7.5
	ENT	1	1.9
	Medicine	19	35.8
	Obs & Gyne	12	22.6
	Pediatrics	5	9.4
	Surgery	6	11.3
	ER	3	5.7
	Nephrocyte	2	3.8
	Family	1	1.9
Work experience	< 1 year	15	28.3
	1 – 3 years	8	15.1
	4 - 6 years	11	20.8
	7 – 10 years	9	17
	> 10 years	10	18.9

3.2 Smartphone Apps Usability

This section of the analysis is projected at the usability of Smartphones among doctors in Bahrain, showing the important role that this device plays nowadays in the clinical settings, and what are the most medical applications used by the doctors. Table 3 shows that 100% of the physicians own Smartphones with almost equal percentages between the iPhone and the Android phones (52.8% and 47.2% respectively). The usability of medical related app among the doctors in Bahrain seems to be high, as shown in the table, where 94.3% of the participants use medical applications in their clinical practice. Among these medical apps, the most commonly used application with the highest usage percentage (34%) was "Medscape" which is "the leading medical resource for physicians, medical students, nurses, and other healthcare professionals that offers fast and accurate clinical answers that help in decision making at the point-of-care" (Medscape.com, 2018). It gives meticulous information about diseases including its prevalence, diagnoses and the treatment options (drugs, procedures and follow-up) (ibid). The second most used application is "Uptodate" with a usage percentage among the respondents of 20.8%. Uptodate is an application that can answer its subscribers' clinical questions anytime (UpToDate, 2018). It consists of a search tab that the physicians can write the keywords of any dilemma that they incur, and it gives them the answer directly and updated day by day (ibid). Then comes Omnio and BNF, and many other medical apps that were mentioned by the doctors beside the "other" choice of this question such as, Pedi-stat, instant anatomy, gestation age for pregnant women, Skyscape apps, etc. Most of the doctors of the sample use their medical applications during the working hours between 30 minutes 52.8% and an hour 30.2%. Only 9.4% of the respondents do not use the medical apps they own. About the frequency of usage, the table shows that 28.3% of the participants use their applications 2 to 3 times a week, and 26.4% use them many times each day, while 3.8% do not use them and 13.2% use them infrequently, which suggests that the medical apps are being frequently used among physicians in Bahrain. Moreover, the usage percentages of the most prevalent types of medical applications are mentioned. As shown, the most used types of applications that seemingly have a distinctive role in the decision making among doctors were drug formulary apps 69.8% of respondents use them), then come disease diagnoses and management apps with a usage percentage of 62.3%, then come medical calculators and journals access with percentages of usage of 56.6% and 47.2% respectively. Internet access and calendars are not widely used among doctors in Bahrain. Other types of medical applications that were mentioned by the doctors in the open-ended part of this question "others" were "research es", "iBooks" and "EHR" with low percentage of usage.

Table 3. Smartphone Usability

		Frequency	Valid %
Have smart phone?	Yes, i-Phone	28	52.8 %
	Yes, android	25	47.2 %
I use medical app on the smart phone.	No	3	5.7 %
	Yes	50	94.3 %
What medical apps do you use	Medscape	8	34 %
	Omnio	16	11.3 %
	Uptodate	11	20.8 %
	BNF	2	3.8 %
	Others	16	30.2 %
What approximate time you spend using these apps?	None	5	9.4 %
	Up to 30 mins	28	52.8 %
	Up to 1 hour	16	30.2 %
	More than 1 hour	4	7.5 %
How frequently you use medical apps?	I do not know them	2	3.8 %
	Infrequently	7	13.2 %
	Once a week	5	9.4 %
	2 – 3 times per week	15	28.3 %
	1 – 2 times per day	10	18.9 %
	Many times, a day	14	26.4 %
Forming drugs using the medical apps	No	16	30.2 %
	Yes	37	69.8 %
Diagnosing diseases using the medical apps	No	20	37.7 %
	Yes	33	62.3 %
Medical calendar using the medical apps	No	23	43.4 %
	Yes	30	56.6 %
Access Journal/Database	No	28	52.8 %
	Yes	25	47.2 %
Internet/E-mail access using the medical apps	No	29	54.7 %
	Yes	24	45.3 %
Calendar access using the medical apps	No	38	71.7 %
	Yes	15	28.3 %
Others	No	48	90.6 %
	Yes	5	9.4 %

3.3 EMR Acceptance

A descriptive analysis was executed to know the percentages of agreement on each statement among the participants, in addition to regression analysis to test the linear relationship among the hypothesized variables. Table 4 shows the percentages of the usage of EMR in the clinical practice among participants. As resulted from the analysis, 62% of the respondents do use the EMR in their practice while 37.7% still do not use them.

Table 4. EMR in Practice

		Frequency	Valid Percent
Do you use EMRs in your practice?	No	20	37.7
	Yes	33	62.3
	Total	53	100

3.4 Regression Analysis and Hypotheses Testing

3.4.1 Usefulness of EMR and the Intention to use it Among Doctors

Table 5 shows that the PU has a significant positive relationship to Intention to Use of EMR, with a Beta coefficient value of (B = 0.438) suggesting a strong relationship with a significance level of < 0.01 which gives an evidence that the first hypothesis (H1) is correct and can be accepted.

Table 5. Regression for PU Effect on Acceptance of EMR

Dependent Variable: Acceptance (Intention to Use EMRs)					
F= 10.195 (p= 0.003 < 0.01), $R2 = 0.192$, adjusted $R2 = 0.173$					
Model	Unstandardized Coefficients		Standardized Coefficients	t	Sig.
	B	Std. Error	Beta		
(Constant)	1.803	0.597		3.020	0.004
PU	0.503	0.158	0.438	3.193	0.003

3.4.2 Ease of Use of EMR and the Intention to use it among Doctors

Table 6 shows that the PEOU has a significant positive relationship to Intention to Use of EMR, with a Beta coefficient value of (B = 0.464) suggesting also strong relationship with a significance level of < 0.01 which gives an evidence that the first hypothesis (H2) is correct and can be accepted.

Table 6. Regression for PEOU Effect on Acceptance of EMR

Dependent Variable: Acceptance (Intention to Use EMRs)					
F= 11.779 (p= 0.001 < 0.01), $R2 = 0.215$, adjusted $R2 = 0.197$					
Model	Unstandardized Coefficients		Standardized Coefficients	t	Sig.
	B	Std. Error	Beta		
(Constant)	1.810	0.555		3.260	0.002
EOU	0.503	0.147	0.464	3.432	0.001

3.4.3 Job Relevance and Quality of Output of EMR and the Usefulness of EMR

Table 7 shows that the Job Relevance and Output Quality have significant positive relationships to PU of EMR, with Beta coefficients value of (B = 0.415 and 0.478 respectively) suggesting strong relationship with significance levels of < 0.01 which gives an evidence that the first hypothesis (H3 and H4 respectively) are correct and can be accepted.

Table 7. Regression for Job Relevance and Output Quality Effect on PU of EMR

Model	Dependent Variable: PU F= 50.647 (p= 0.000 < 0.01), R2 = 0.707, adjusted R2 = 0.693				
	Unstandardized Coefficients		Standardized Coefficients		
	B	Std. Error	Beta	t	Sig.
(Constant)	0.883	0.291		3.035	.004
RELEVENCE	0.352	0.111	0.415	3.156	.003
OUTPUT QUALITY	0.424	0.116	0.478	3.640	.001

4. CONCLUSION

Bahrain is a developing and advanced country that is found to have extensive modernization utilization of the medical information over several years. The Kingdom has shown a tremendous growth in integrating the information technology in the functioning of the hospitals in the country which helps the healthcare professionals and patients to have quality healthcare with cost and time efficiency starting as early as 1993 with ALCARE system at the BDF hospital until now with many more improvements and internet based EMRs and Hospital Information Systems across all governmental hospitals. Also, it was quite clear that people have access to high tech gadgets in the Kingdom which increases their interface interaction with the technology and help them learn new techniques and technologies in a better manner. Therefore, effective strategies that are adopted in a systemic manner will help to achieve success and growth in the healthcare. Medical Informatics have found to have a positive effect on the work that provides faster and ease access to information and thus lets the healthcare professional make final decisions. Productivity, speed and efficiency at work is found to be increased which is like that of who claims that EMR helps to perform people in a better way that is helpful for the patients which lets the researcher accepts the four hypotheses. The hypotheses used in the research helped to reach the objective of the study which is to determine the acceptance and usability of the end-users for the EMR and mobile Medical Apps. It was found that people have hesitations for implementation of a new system which is in accordance to (Illiger et al., 2014) is a major factor of accepting a system. Any new change, nevertheless, has different resistances from people as they don't like to change their normal and well-settled routines but effective leadership, training and understanding of the system can help people own a change and let them implement it in a successful manner and this apply on physicians. Thus, it is important for organizations to understand the core concepts of the change management while implementing EMRs in the hospital and effectively use these techniques to let people own the change. In Hospital organizations as patients are at the centric core of the model thus both internal and external stakeholders should be taken into confidence while implementing the new systems which is installed in a way that change process can be smooth. The analysis regarding the usability of Smartphones in the clinical practice by the doctors in Bahrain showed that the doctors in Bahrain have medical related apps on their own Smartphones and they tend to use these applications frequently in their medical practice to help them take decisions at the point of care and ease up the process of diagnoses. Different apps with different functions are downloaded on their phones and used daily. On the other hand, regarding the Acceptance of EMRs, it was concluded from this study that the perceived ease of use and perceived usefulness of EMRs play an important role in determining the intention of the doctors in Bahrain to use and adopt such system in their clinical practice and this lead to the acceptance of H1 and H2 in this research, agreeing with the previous studies that were done in regards of the EMRs among physicians in many other countries and mentioned earlier in the research background. Moreover, the job relevance and the quality of output of the system also proved to have a significance relationship with the Perceived Usefulness (PU) affecting the intention to use the EMR indirectly, and this also agree with the previous literature.

REFERENCES

Chang, C. P. (2015) The Technology Acceptance Model and Its Application in a Telehealth Program for the Elderly with Chronic Illnesses. *Hu Li Za Zhi*, 62(3), 11.

Davis, F. D., Bagozzi, R. P., & Warshaw, P. R. (1989), User Acceptance of computer technology: A comparison of two theoretical models. *Management Science*, 35, 982–1003, doi:10.1287/mnsc.35.8.982

Gagnon, M. P., Ghandour, E. K., Talla, P. K., Simonyan, D., Godin, G., Labrecque, M., Ouimet, M., & Rousseau, M. (2014). Electronic health record acceptance by physicians: testing an integrated theoretical model. *Journal of biomedical informatics*, 48, 17-27.

Gan, S. K. E., Koshy, C., Nguyen, P. V. & Haw, Y. X. (2016) An overview of clinically and healthcare related apps in Google and Apple app stores: connecting patients, drugs, and clinicians. *Scientific phone apps and mobile devices*, 2(1), 8.

Hamidfar, M., Limayem, M., & Zegordi, S. H. (2008) Using the UTAUT model to explore Iranian physicians and nurses' intention to adopt electronic patient record. Proceedings of the International Conference on E-Learning, E-Business, Enterprise Information Systems, and E-Government, Las Vegas, Nevada, USA.

Illiger, K., Hupka, M., von Jan, U., Wichelhaus, D., & Albrecht, U. V. (2014) Mobile Technologies: Expectancy, Usage, and Acceptance of Clinical Staff and Patients at a University Medical Center. *JMIR mHealth and uHealth*, 2(4).

Ketikidis, P., Dimitrovski, T., Lazuras, L., & Bath, P. A. (2012). Acceptance of health information technology in health professionals: an application of the revised technology Acceptance model. Health informatics journal, 18(2), 124-134.

Lakbala, P., & Dindarloo, K. (2014) Physicians' perception and attitude toward electronic medical record. *SpringerPlus*, 3(1), 63.

Marques, A., Oliveira, T., & Martins, M. O. (2011). Medical Records System Adoption in European Hospitals. *The Electronic Journal Information Systems Evaluation*, 14(1), 89-99.

Medscape.com. (2018). Medscape App. [online] Available at: https://www.medscape.com/public/medscapeapp [Accessed 8 May 2018].

Moeinedin, F. (2011) Assessing Primary Care Physicians' Attitudes Towards Adoption of an Electronic Tool to Support Cancer Diagnosis. MS. University of Toronto.

Okazaki S., Blas S. S., & Castañeda J. A. (2015). Physician Adoption of Mobile Health Monitoring Systems in Spain: Competing Models and Impact of Prior Experience. *Journal of Electronic Commerce Research*, 16(3), 194.

Pai, F. Y., & Huang, K. I. (2011) Applying the Technology Acceptance Model to the introduction of healthcare information systems. *Technological Forecasting and Social Change*, 78(4), 650-660.

Patel, R. K., Sayers, A. E., Patrick, N. L., Hughes, K., Armitage, J., & Hunter, I. A. A (2015) UK perspective on smartphone use amongst doctors within the surgical profession. *Annuals of Medicine and Surgery*, 4(2), 107-112.

Payne K. F. B., Wharrad, H., & Watts, K. (2012) Smartphone and medical related App use among medical students and junior doctors in the United Kingdom (UK): a regional survey. *BMC medical informatics and decision making*, 12(1), 121.

Saunders, M., Lewis, P. & Thornhill, A. (2012) Research Methods for Business Students. (6th Ed.). England: Pearson.

Sclafani, J., Tirrell, T. F., & Franko, O. I. (2013) Mobile tablet use among academic physicians and trainees. *Journal of medical systems*, 37(1), 1-6.

Sezgin, E., & Özkan, S. (2014) Assessment of Information Technology Use in Small and Medium-Sized Enterprises: Empirical Investigation in Five Cases. In Information Systems for Small and Medium-sized Enterprises. *Springer Berlin Heidelberg*. 97-121

Sweis, R., Isa, A., Azzeh, H., Shtyh, B., Musa, E., & AlBtoush, R. (2014) Nurses' resistance to the adoption of information technology in Jordanian hospitals. *Life Science Journal*, 11(4s), 8–18.

UpToDate. (2018). UpToDate Individual Subscriptions for Medical Professionals. [online] Available at: https://www.uptodate.com/home/uptodate-subscription-options-clinicians [Accessed 8 May 2018].

Venkatesh V., & Davis F. D. A (2000) Theoretical extension of the technology acceptance model: Four longitudinal field studies. *Management science*, 46(2), 186-204.

Wyatt G., Sikorskii A., & Wills C. E. (2013). Development and Initial Validation of a Complementary and Alternative Medicine (CAM) Knowledge Instrument. *Journal of nursing measurement* 21(1), 55-63.

Wyatt J., & Liu J. (2002). Basic concepts in medical informatics. *Journal of Epidemiology & Community Health*; 56, 808-812. doi 10.1136/jech.56.11.808

DEPLOYMENT OF ELECTRONIC MEDICAL PRESCRIPTIONS FOR AMBULATORY CARE IN BELGIUM TRACING AND ELIMINATING TECHNICAL ERRORS

Marc Nyssen[1] and Yelina Piedra Salomón[2]
[1]*Independent Author, JB. Vandercammenstraat 25, Brussels, Belgium*
[2] *University of Havana, Havana, Cuba*

ABSTRACT

Electronic prescriptions in ambulatory care are considered to be the second most important e-health goal, after the electronic medical record in many countries. Belgium is no exception: electronic prescription are part of the successive national e-health plans since more than 10 years. This paper reviews the scientific production on the topic of electronic prescriptions (via the Dimensions database). In the second part, we then explains the fundamental principles that were adopted and faithfully realized, the main functional characteristics and the technical realization and deployment of the ambulatory electronic prescription system "Recip-e"in Belgium. Then we highlight the evolution of the main parameters and technical problems that were encountered during the roll-out phase of the Recip-e project and the remaining challenges.

KEYWORDS

Electronic Prescription, e-Prescription, Ambulatory Care, e-Health, Deployment

1. INTRODUCTION

Since 2005, the World Health Organization [WHO 2005] encourages its member states to define the national e-health plans and even provides a "toolkit" to facilitate these exercises. Many countries have elaborated such plans and usually, electronic medical records and electronic prescriptions (intra-muros as well as ambulatory) are cornerstones of such plans. In the EU, the EPSOS project developed the tools for trans-border exchange of summary records and electronic prescriptions, this was evaluated in the field [Moharra 2015] and will most probably be followed by an effective deployment during the following years.

In this paper we first make a literature survey, on the subject of the electronic prescriptions in ambulatory care, to show the importance of the related scientific production, the evolution in time of relevant scientific publications and the international relationships that link these publications to one another.

The second part of the paper explains the architecture of the ambulatory electronic prescription system in Belgium and shows the results of the deployment since 2011, in three different stages.

The purpose of this paper is to give insight in the field of ambulatory electronic prescription in general and to focus on a specific country-wide implementation that is now mature enough so that lessons can be learned which will help others in optimizing their implementations. These deployments will have to take into account the common requirements that underly electronic prescribing but will also adapt to the specific characteristics of the health care system of each country.

2. ELECTRONIC AMBULATORY PRESCRIPTIONS

After making a comprehensive literature search, we will focus on the implementation and deployment in Belgium.

2.1 Literature Research

For the systematic collection of scientific production on the subject, we used the Dimensions database[1]. This is a linked research knowledge system that offers novel approaches to discover relevant publications and access to research. Developed by Digital Science in collaboration with over 100 leading research organizations around the world, Dimensions encompasses grants, publications, citations, alternative metrics, clinical trials, patents and policy documents to deliver a platform that enables users to find and access the most relevant information faster, analyze the academic and broader outcomes of research, and gather insights to inform future strategies (Dimensions, 2019).

By searching "electronic prescription" OR "electronic prescriptions" OR "e-prescription", a total of 582 documents were collected. We found 428 articles, 68 chapters, 50 books, 33 proceedings and 3 monographs. A total of 194 patents and 34 policy documents were also part of the sample. We limited further work to articles, books and proceedings because an organizational subscription is requested by the database to access the other types, which seemed also less relevant to us for this study. Microsoft Excel and Tableau Public 2018.3 were used to generate the frequency tables. For the bibliometric mapping of titles, co-word analysis and countries collaboration networks the data was exported to a *.cvs* file. To construct and visualize bibliometric networks we used VoSviewer 1.6.10.

The most productive authors on the topic, based on the number of published articles are Michelle Anne Chui (13) from University of Wisconsin–Madison, United States; Olufunmilola K Odukoya (10) from University of Pittsburgh, United States and Johanna Timonen (10) from University of Eastern Finland, Finland. The first two authors form an invisible group because they collaborate in publishing. Their most relevant researches according database quotation counts are focused on exploring the consequences and contributing factors of e-prescribing errors in community pharmacies [Odukoya, Stone & Chui, 2014], this article obtained 30 quotations. The second one obtained 26 quotations and focuses in exploring pharmacy staff perceptions of the strengths and weaknesses of electronic prescribing (e-prescribing) design in retail pharmacies using the socio-technical systems framework [Odukoya & Stone, 2011].

The first contribution (with 6 quotations) is a semistructured interview among primary health care physicians in Finland to find out about how medication and the usability of electronic prescribing is perceived by physicians [Kauppinen, Ahonen, Mäntyselkä & Timonen, 2017]. The second [Kauppinen, Ahonen & Timonen, 2017] and the third [Lämsä et al, 2017] most relevant works of Timonen obtained 5 quotations each. These publications mainly focus on the impact of electronic prescriptions on medication safety in Finnish community pharmacies, using a survey of pharmacists and in the pharmacy customer´s experiences with My Kanta, the national onl-ine service for viewing electronic prescriptions in Finland.

The most cited document of the whole sample is authored by Fisher *et al* (2010). This article is focused on evaluating primary care non-adherence in community-based practices and identifies predictors of non-adherence. The authors concludes that many e-prescriptions were not handed-in to pharmacies and previous studies of medication non-adherence failed to capture these prescriptions. Efforts to increase primary adherence could dramatically improve the effectiveness of medication therapy.

Interventions that target specific medication classes may be most effective (citation-wise). According to the database, this publication in *Journal of General Internal Medicine* has been cited 260 times. 32% of its citations have been received in the past two years, which is higher than you might expect, suggesting that it is currently receiving a lot of interest. Compared to other publications in the same field, this publication is scores very high regarding citations as it has received approximately 28 times more citations than average. This research output has an Altmetric Attention Score of 409, as was calculated when the research output was last mentioned on 16 October 2017. This is the Dimension´s high-level measure of the quality and quantity of on-line attention that it has received. It's in the top 5% of all research outputs ever tracked by Altmetric[2]. This publication was found relevant by 18 Dimension´s categories[3] according to different research fields.

---

[1] For more information about the product visit https://dimensions.ai.

[2] https://dimensions.altmetric.com/details/1373293/citations

[3] Public Health and Health Services, Information Systems, Clinical Sciences, Policy and Administration, Data Format, Applied Economics, Artificial Intelligence and Image Processing, Computer Software, Law, Business and Management, Psychology, Cardiorespiratory Medicine and Hematology, Pharmacology and Pharmaceutical Sciences, Sociology, Computation Theory and Mathematics, Specialist Studies in Education, Communications Technologies, Nursing, Political Science, Physical Chemistry (incl. Structural), Veterinary Sciences, Curriculum and Pedagogy, Banking, Finance and Investment, Linguistics, Philosophy

In the period between 1989 and 2019, the scientific production collected was as depicted in Figure 1:

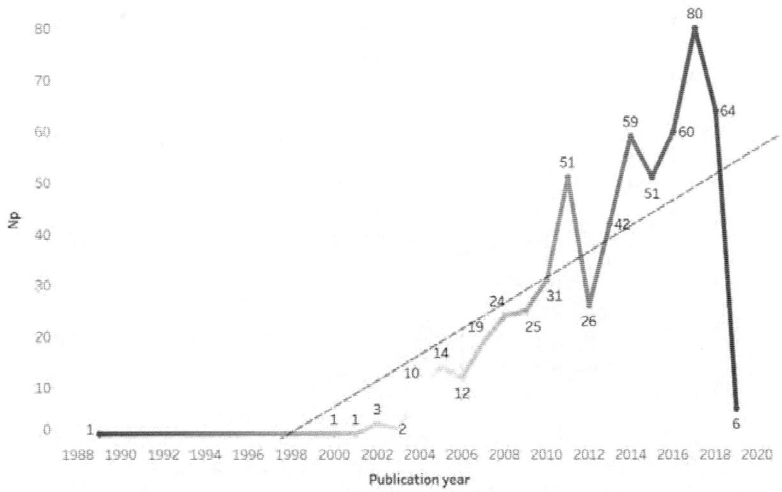

Figure 1. Chronological distribution of electronic-prescriptions scientific production in Dimensions

As shown, in general, there is a growing trend line mainly starting in 2005, indeed the period in which electronic prescription systems became operational in several regions of the world. Also in Belgium the first designs and proposals for a national electronic prescription system in ambulatory care date from that period.

2.2 Deployment of Electronic Prescriptions in Belgium

2.2.1 General Principles

Ambulatory prescriptions in Belgium originate from (independent) physicians, dentists, midwives and hospitals (upon release of patients), while the medication delivery is performed in independent pharmacies (4840 country-wide). Patients can freely choose any pharmacy. Prescribers make use of electronic medical record systems, which all have prescription modules (either in private practice, in polyclinics, hospitals), provided by 25 different software vendors. All pharmacies have pharmacy information systems, provided by 8 vendors.

It was decided to make maximal use of the national "eHealth-platform", guaranteeing safe and secured data transmission and a user-and-access management system comprising all health care workers. Patients are identified by their national identifier (not only Belgians, but also given to all inhabitants residing at least three months in the country).

2.2.2 Implementation Methodology

In order to keep the prescriptions, submitted by prescribers, available to the pharmacist, freely chosen by the patient, we developed a secure database application: "Recip-e engine". Enabled and authentified prescribers submit prescriptions uniquely identified prescriptions in KMEHR-XML format to this database. The unique identifiers (Recip-e Identifier or RID) are attributed by the central service upon creation.

The pharmacists, chosen by the patient can deliver the medication and thereby realize a status change in the server, prohibiting thereby multiple deliveries of a prescription; finally archiving by the pharmacist enables the Recip-e server to delete the prescription from the database, thereby releasing space.

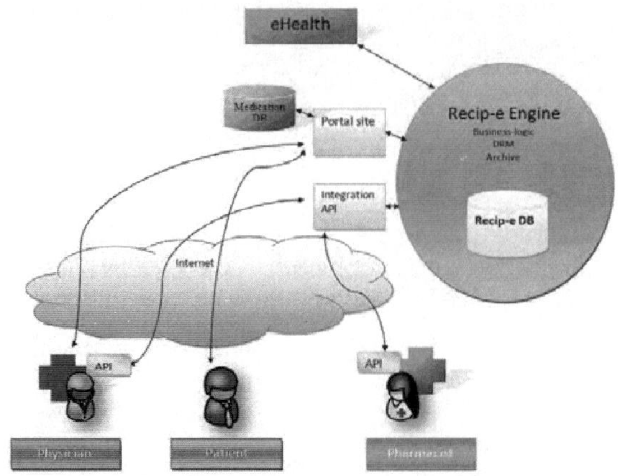

Figure 2. General Recip-e Architecture

The Recip-e system was realized by a not for profit organization: Recip-e, managed by representatives of all medical professions. The financing came from the Belgian Federal Social Security organization RIZIV/INAMI.

Three implementations were realized: “test”, “acceptance” and “production”. The “test” environment allows developers of the central software to develop and test new server releases. Once OK, these are transferred to “acceptance” which is a complete eHealth-platform environment, allowing third parties (software vendors for prescribers and pharmacies) to test their end-user applications by calling services such as Recip-e, but also others: sharing of medical records, requesting information on patient’s insurance statuses and sending in or consulting summary records, as typical examples. The “production” environment is the real thing: severely controlled and updated only a few times per year with enforcement of strict regulations regarding access and identity controls.

Triple encryption end-to-end assures security, but also limits the controls than can be provided centrally. We can just rely on “header information” to assure good technical functioning; content-wise we are blind: the prescriber and the executors are supposed to use the same XML formats, medication databases and codes as well as structured posologies in their prescription messages.

2.2.3 Deployment Time-Line

The deployment in production (since 2011) was accompanied by regulations by the state: first phase 2011-2016 paper prescriptions simultaneous with electronic prescriptions identified by a RID (unique identifier per prescription) bar-code on top of the paper print-out; second phase: 2017–2020 electronic prescriptions with printout of a “proof of electronic prescription” containing the RID and medication print-out but without any legal value of the paper, just assisting recovery of the electronic prescription (usually via the bar-code readers at the pharmacies); third phase: starting January 2020 – obligatory usage of electronic prescriptions, possibly without printout of a “proof of electronic prescription” and recovery at the pharmacy via the identity of the patient.

Deployment was progressive, as presented in the graph below, showing the monthly figures of three key indicators: the number of prescriptions received from prescribers (top blue line), the number of prescriptions, delivered by pharmacists (center red line) and the number of prescriptions archived by pharmacists (lower green line).

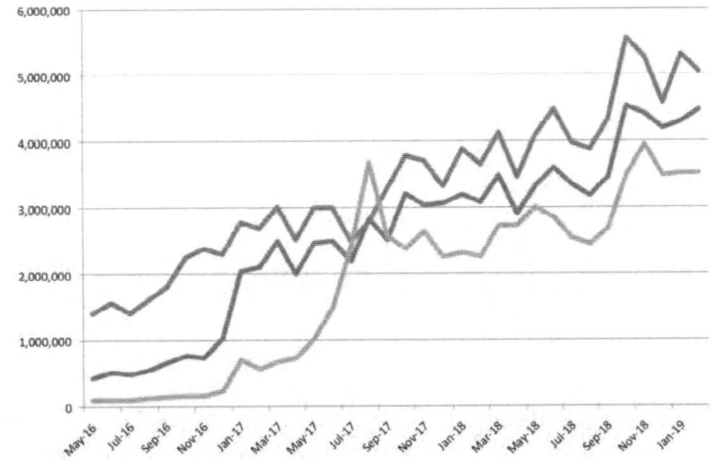

Figure 3. Key Deployment Indicators (May 2016 – February 2019)

In January 2017, a sharp rise in volume was due to the introduction of the "proof of electronic prescription" (with no legal value) that needs to be accompanied by the electronic prescription in the pharmacy, as the only legal document if no paper based prescription is present.

The graph shows that at the beginning of 2019, we reached the 5 million incoming electronic prescriptions per month milestone. Top days are Mondays, with just over 300.000 incoming prescriptions.

2.2.4 State Diagram

The Recip-e system was formally described as a "state machine", defining the states in which every electronic prescription is at any moment in time. State modifications are activated by the corresponding webservice calls, emanating directly from the systems of the users (prescribers and pharmacists). When a prescriber creates a new prescription on his/her electronic medical record system, the electronic version of this prescription is then transmitted to the Recip-e server with the Create_Prescription webservice. Each new prescription, identified by its unique Recip-e ID obtains the status: "Not Delivered"; meaning that any pharmacy can now pick it up via a "Get_Prescription" webservice call, emanating directly from the pharmacy.

This is illustrated in Figure 3. Suppose a pharmacy gets the prescription by activating the webservice "Get_Prescription RID", at this point the status is changed into "In Process", meaning that no other pharmacy can obtain it: the prescription is "locked" at the first pharmacy, where at this point any of three transition possibilities can take place: either the pharmacist delivers the medication, present in the prescription, his system must then communicate "Mark_as_delivered" and the status of the prescription moves to: "Delivered"; otherwise, the prescription is "handed back to the patient", no delivery takes place and the patient can go to any other pharmacy or come back later: this is indicated by the "Mark_as_undelivered" webservice and results again in the status: "Not Delivered"; finally, the pharmacist can "revoke" the prescription, as it is no longer useful by "Mark_as_revoked": the status becomes "Revoked": the prescription can no longer be used, its contents can be deleted.

After delivery, the pharmacist will make sure the prescriptions which he delivered are properly archived, either by himself or by an external instance (tarification/archiving service), this must be confirmed to the Recip-e server via a "Mark_as_archived" webservice call, resulting in the status: "Archived".

On daily basis, the Recip-e server deletes the prescription with status "Archived" from the database, to recover storage space.

Webservices play an eminent role in the whole e-health environment, as they are well defined and allow strict syntax and user access controls.

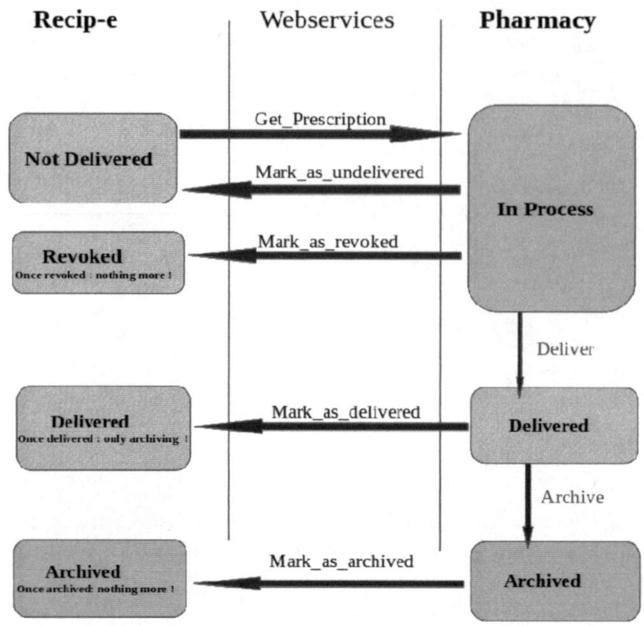

Figure 4. State Diagram of a Prescription and Corresponding Web-Service Calls

To make the system consistent, third party software at prescribres' locations and pharmacies should keep trace of the statuses of all the prescriptions they manage, following-up on the pharmacist's actions immediately so that there is coherence between the status in the Recip-e database and the effective status "in the field".

Note that prescribers can also revoke prescriptions (for example during the consultation when they realize they made an error) by performing Mark_as_revoked, but of course only for the prescriptions they had made.

Patients have access to the system via the "MyHealth" national portal: they can list, visualize the contents of their prescriptions and delete those they don't need anymore by issuing revoke requests.

Currently, about 130 patients consult their prescriptions every day, but only 3 prescriptions are revoked by them.

2.2.5 Problem Follow-up and Remedies

The Recip-e system has an audit trail with associated web interface, allowing to follow up on every individual electronic prescription, perform searches in time windows (up to the minute), indicate conditions on prescriber, executor and other conditions such as error codes.

This audit trail is used for follow-up purposes, to discover erroneous operations and to pinpoint their origins.

Table 1 shows the observed error rates, associated with the different end-user operations, at two Mondays, separated by 1 year: March 5th 2018 and March 6th 2019.

On March 5th 2018, 9777 prescribers (8630 general practitioners, 431 independent specialists, 961 dentists and 25 hospitals sent in 207804 prescriptions.

On March 4th 2019, 10409 prescribers (mainly GP's: 8517, but also 714 independent specialists, 1055 dentists, 1 midwife and 122 hospitals) sent in 246027 prescriptions.

The increase in total users is mainly due to the hospitals (there are a total of 132 hospitals in the country).

Table 1. Error Rates (Comparison between March 2018 and March 2019)

Function	Prescribers	Pharmacies	Patients	Error rates 5 March 2018	Error rate 4 March 2019
Create prescription	*			0.27%	0.011%
Revoke (delete) prescription	*	*	*	2.6%	6.3%
List open prescriptions	*		*	0	0
Get prescription	*	*	*	1.0%	2.0%
Mark as delivered		*		.40%	1.0%
Mark as undelivered		*		9.4%	7.0%
Archive prescription		*		1.6%	0.98%
Announce prescription	*			0%	0%
Create feedback message		*		0%	0%
List feedback message	*			0%	0%

Technical Errors due to Prescribers

At the input side, we observe tree types of technical errors, leading to non-admission of an electronic prescription into the system:

- identity problems: the identity of the physician who started the session (SAML token, which we consider as the signature, validated by the prescriber's electronic certificate and electronic signature with his e-ID) is not the same as the identity of the prescriber in the header of the prescription;
- wrong prescription format (name or header badly formatted by the prescriber software)
- technical error: time-stamping service not available or other server errors

Each of these accounts for about equal portions of the 2018 measurement: 0.09% of incoming prescriptions.

Improving the bandwidth of the main bus, capacity of the time-stamping server at the eHealth-Platform and follow-up towards the vendors of prescribing softwares, virtually eliminated the last two reasons for technical errors; currently only the first (prescriber taking over a session of a colleague) one remains.

We try to eliminate these errors by strongly recommending vendors of prescribing packages to re-check the correspondence of the identity of the prescriber versus the one in the current SAML token, before sending in a prescription and thus avoiding useless transmissions, error messages and possible user frustration.

Prescribers can revoke prescriptions they have made, surprisingly, the absolute numbers of successful revokes by prescribers (from 771 to 1488) have almost doubled and the error rate for this operation went up from 4.4% to 6.3%.

Technical Errors due to Pharmacists

Here, the error rates are higher: the main origin seems to be the difficulty pharmacy packages have to keep track of the status of each Recip-e prescription that is handled in the pharmacy (in particular when pharmacies operate via multiple terminals) and in the interpretation of error messages.

The rather high rate of errors with the "get-Prescription" operation is due to two causes: re-doing the operation in the same pharmacy, after delivery took place and repeating the operation, not taking into account the previous error messages that were sent by the server.

Also: here we observe a few cases of "fraud attempts" by patients who try to obtain multiple deliveries, based on a single prescription.

Thanks to insisting on "good practices", the error rates regarding archiving gave gone down significantly and we hope that in 2019, the number of "archived" prescriptions will effectively equal the number of "delivered" ones.

3. CONCLUSION

The literature exploration learns us that the scientific production on the topic of electronic prescriptions is still growing, that international cooperations exist and that the main topics are situated in the fields of adoption and error tracing.

Our experience with the roll-out of a large scale electronic prescription system learns that it is possible to operate with many vendors, provided that clear specifications are given and that the system's characteristics are close to the daily practice of the health care professionals who have to use it.

A comprehensive and informative audit trial system is essential in tracing technical errors and subsequent actions can be made very efficient, provided not encrypted header information of the messages contains the source (health care worker) and the tools that are used (program vendor, program name and version that is in use by the prescriber) to generate the prescription.

Precise feedback to software vendors must be given by the central system, as end-users are not capable of discerning where problems exactly originate.

ACKNOWLEDGEMENT

Thanks to the Belgian RIZIV-INAMI, sponsor of the Recip-e system; to D. Broeckx for formatting Figure 3.

REFERENCES

Fischer M A et al, 2010, "*Primary Medication Non-Adherence: Analysis of 195,930 Electronic Prescriptions*", Journal of General Internal Medicine, 195 p 930.

Kauppinen H et al, 2017, "*Medication safety and the usability of electronic prescribing as perceived by physicians—A semistructured interview among primary health care physicians in Finland*", Journal of Evaluation in Clinical Practice, 23(6), 1187-1194, 2017.

Kauppinen H et al, 2017, "*The impact of electronic prescriptions on medication safety in Finnish community pharmacies: A survey of pharmacists*", l Journal of Medical Informatics.

Lämsä E et al, 2017, "*Pharmacy customers' experiences with the national on-line service for viewing electronic prescriptions in Finland*", International Journal of Medical Informatics.

Moharra, M et al, 2015, "Implementation of a cross-border health service: physician and pharmacists' opinions from the epSOS project", Fam. Pract, 2015, Oct, 32(5) pp. 564-567.

Odukoya O et al, 2014, "*E-prescribing errors in community pharmacies: Exploring consequences and contributing factors*", International Journal of Medical Informatics, 83(6), 427-437.

Odukoya O, Chui M A, 2012, "*Retail pharmacy staff perceptions of design strengths and weaknesses of electronic prescribing*", Journal of the American Medical Informatics Association, 19(6), 1059-1065.

Papshev D et al, 2001, "*Electronic Prescribing in Ambulatory Practice: Promises, Pitfalls, and Potential Solutions*", Am. J. Manag. Care, Jul 7 (7), pp. 725-736.

Theilgaard J, 1997, "*Message Implementation Guide for Prescription*", version 1.0, 21.03.97 for the CoCo project, on behalf of Medcom, Denmark. Zipped file (120 K) of a Word-document of 117 pages.

WHO, 2005, "*World Health Assembly Resolution 58.28*" and "*ITU World Telecomunication Development Conference Resolution 65 (2010)*".

EVALUATING A REMOTE HEALTH MONITORING APPLICATION POWERED BY BLUETOOTH

Maryam Vahabi, Hossein Fotouhi, Mats Björkman and Maria Lindén
Mälardalen University; School of Innovation, Design and Technology, Sweden

ABSTRACT

It has become widely accepted that the Internet of Things (IoT) devices and technologies are the key enablers for many emerging applications including remote health monitoring. Various physiological sensing devices have been designed and equipped with different radio technologies. The choice of radio hardware plays an important role on the overall performance of the system since it imposes some limitations on the delivered quality of service. Hence, it is critical to properly evaluate the embedded radio technology based on the application requirements. In this paper, we perform extensive experiments on Shimmer physiological sensors that is one of the leading providers of wearable wireless sensor products powered by Bluetooth classic radio. Shimmer sensors are designed and used for monitoring various human health information such as temperature, heart rate, movement, etc. We review and investigate different scenarios in which Shimmer devices are used by medical practitioners to monitor the ECG signal and the movement of a human. This study shows that the Shimmer device can provide reliable data delivery by using a specific configuration. For instance, employing a maximum number of seven Shimmer devices attached on a body at home environment within the range of at most 5 m and with the sampling rate of 512 Hz would result in a reasonable quality of service, while varying these parameters may degrade the overall performance. Mobility of human body, noisy environment, and higher packet transmission rates are some examples that will reduce the system quality.

KEYWORDS

Remote Health Monitoring, Internet of Things, Wireless Communication, Bluetooth, Interference

1. INTRODUCTION

The advancement in the Internet of Things (IoT), wireless communication and hardware miniaturization is playing a key role in the development of remote health monitoring applications, making them more available and affordable [Doukas & Maglogiannis, 2012]. The number of elderly people (over 60 years of age) is expected to grow more than double, from 841 million individuals in 2013 to more than 2 billion in 2050 [United Nations, 2013]. This means that the health sector will require more support from digitalization through remote health monitoring in order to reduce the cost and increase health services to elderly people. Currently, there are a lot of effort in all over the world to provide digital healthcare. For instance, in Sweden, the vision for eHealth is that in 2025, Sweden will be best in the world at using the opportunities offered by digitization and eHealth to make it easier for people to achieve good and equal health and welfare, and to develop and strengthen their own resources for increased independence and participation in the life of society [Ministry of Health and Social Affairs, 2016].

IoT is a growing community that initially was focusing only on connectivity issues, where the main idea was to connect all devices (things) to the Internet [Reiter, 2014]. Thus, IoT was an extension to the traditional wireless sensor networks (WSNs), which supports remote access to sensing devices. Though, the new IoT term covers wider area that integrates more aspects of physical devices rather than mere connectivity aspect, such as hardware design, computing power, software properties, communication protocols, data collection, data storage, data processing, data management and control loop. This way, IoT has become a multidisciplinary research area where both academia and industry are putting a lot of attention [Ray, 2018]. However, still in many IoT applications, such as remote health monitoring, one of the major research areas is providing reliable data communication over wireless medium with high QoS demands, while tackling various challenges, such as limited resources in terms of channel bandwidth, processing, battery power and memory [Premsankar et al., 2018], [Gardašević et al., 2017], [Fotouhi et al., 2016].

Due to the growing number of IoT devices, it is highly desired to devise algorithms and protocols in order to cope with channel bandwidth limitation, especially when devices are sharing same frequency band [Chen & Kunz, 2016]. Most of indoor-used IoT devices operate at 2.4 GHz frequency, which is shared among various wireless technologies, such as IEEE 802.15.4, Bluetooth, WiFi, etc. [Natarajan et al., 2016]. There exist various technologies that is used for wearable wireless network as in remote health care applications such as IEEE 802.11a/b/g/n (WiFi), IEEE 802.15.1 (Bluetooth), IEEE 802.15.1 (Bluetooth-LE), IEEE 802.15.4 (Zigbee), IEEE 802.15.4a (UWB), IEEE 802.15.4j (MBAN), IEEE 802.15.4j (MBAN) [Alam et al., 2016]. IoT devices with low-power transmission (e.g. 802.15.4 and Bluetooth) are more prone to packet losses when operating in the vicinity of higher power transmission devices such as WiFi and microwave. Thus, it is necessary to set up strategies to achieve better performance while dealing with the interference and noisy environment.

Bluetooth is one of the technologies that is widely used in medical devices. As stated earlier, it operates in 2.4 GHz frequency band and supports short range of communication (10 to 100 m) with 1 Mbps data rate for a network of at most 8 nodes [Baker, 2005]. No sleep cycle is implemented, which permits data streaming through different sensors, such as Electromyogram (EMG), Electroencephalogram (EEG) and Electrocardiogram (ECG). These features make Bluetooth a suitable choice for remote health monitoring services that requires data streaming. Shimmer device is an example of a physiological sensing device equipped with classical Bluetooth radio, which is widely used in research and development [Gardašević et al., 2017], [Ahmed, 2017], [Santos, 2016], [Sarkar & Misra, 2016]. It can monitor the status of various health elements by measuring physiological parameters such as ECG, EMG, Electrodermal Activity (EDA), heart rate and Galvanic Skin Response [Tomasic et al., 2017], [Shimmer, 2019].

Contributions. This paper focuses on evaluating the Bluetooth classic radio of Shimmer sensor devices in various network and environment setup. Main parameters for the evaluation are number of nodes, sampling rate, mobility of nodes, and different sources of interference including the presence of human body. The results of experiments are shown in terms of packet reception rate (PRR) where together with the received signal strength indicator (RSSI), more insight into the experimental environment can be obtained.

Organization of the paper. Section 2 describes the system architecture, followed by the results and discussions of various experimental scenarios in Section 3. Finally, Section 4 concludes the paper.

2. SYSTEM ARCHITECTURE

This section explains the system architecture of the considered remote health monitoring ecosystem. Figure 1 depicts a schematic view of all the components used in our system design. In such systems, patient health related information is measured by a set of wearable sensors. The measured value is then transmitted through a Master node to a local data base. The details of the main components used in the system are as follows:

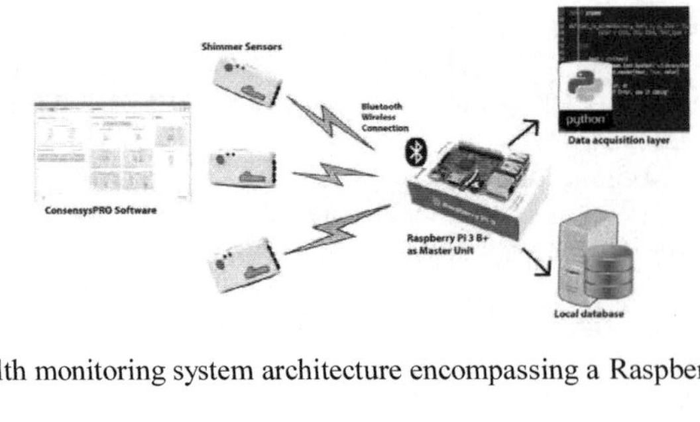

Figure 1. Remote health monitoring system architecture encompassing a Raspberry Pi and Shimmer sensors

Shimmer devices (medical sensors). As previously stated, Shimmer is a wearable physiological sensing device used to monitor the status of various health parameters. Shimmer wireless sensing technology offers two modes for data acquisition. Measured data can be either stored locally on the sensing unit or streamed through Bluetooth to the Master node. The Shimmer3 device is equipped with a microcontroller, a built in microSD Flash socket for additional storage and Bluetooth v2.1 + Enhanced Data Rate (EDR) radio module [MICROCHIP, 2015]. In Bluetooth Classic, there are 79 channels, each with a channel bandwidth of 1 MHz and a raw symbol rate of 1 M symbol/s. The modulation scheme could be Gaussian frequency shift keying (GFSK), quadrature phase shift keying (4PSK), or 8PSK [Chang, 2014]. The maximum sampling rate in the Shimmer3 device is 2048 Hz that can run up to 10 hours at a single battery charge [Gardašević et al., 2017].

Master node and the local database. The preliminary role of a Master node is to create a piconets topology with Shimmer3 sensing devices. It can also store streamed data of physiological devices locally. The Raspberry Pi B+ is the latest model of the Raspberry Pi series (released 2018) that were used as a Master node. The main reason for using Raspberry Pi is the low cost of employing this device at home environment for remote health monitoring. The optimal operating system for use in the Raspberry Pi is Raspbian[1], which is based on Debian/Linux operating system. The free operating system "Raspbian stretch with desktop" comes with a graphical user interface (GUI) and a pre-installed version of the Python language. The most remarkable hardware update related to the Raspberry Pi B+ is its chipset for wireless connectivity that supports Bluetooth 4.2 Low Energy (BLE) [MagPi, 2018].

Consensys. Consensys is a free license software that is used to configure Shimmer devices. It is compatible with Microsoft operating system. There are several features that are provided by Consensys-Basic[2] including, changing the sampling rate, managing the type of data collection and real-time data streaming.

Data acquisition layer. The streamed data by Shimmer3 devices is sent to an open source Python script. The part of the Python script that is responsible for creating communication with the Shimmer3 devices was developed by the Shimmers Research Group [Shimmer RG., 2018]. We developed the second part of the script that allows storing the data locally on the Master unit (Raspberry Pi in this setup). The Python script comes with some standard libraries that can be executed on many different operating systems such as Windows, macOS and most distributions of Linux.

3. EVALUATIONS

To evaluate the robustness of the designed architecture, four experiments were carried out by using different configurations of the control system. This section explains the experimental setup and the results from real world testbed.

Experimental setup. We conducted a set of experiments to evaluate the system performance in terms of packet received ratio (PRR) in different environments: (i) at home with low level of interference, and (ii) at work with higher level of interference. We selected university library to simulate the noisy working environment since more WiFi access points and larger number of users are generally exist. A spectrum analyzer was employed to discover the surrounding WiFi signals that could potentially create interference in our experiments.

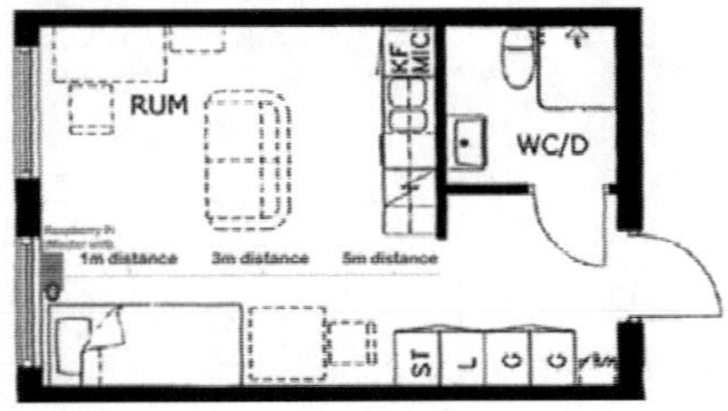

Figure 2. Position of the Master unit and the Shimmer devices from different distances

[1] https://www.raspberrypi.org/downloads/raspbian/
[2] http://www.shimmersensing.com/products/consensys

Figure 2 shows the positions of the Master node and the Shimmer devices at home environment. As it is shown, the Shimmer devices have been located in different proximity of the Master node, 1m, 3m and 5m. Besides the distance, we have also evaluated the network performance based on the number of shimmer devices in the piconet, sensors' sampling rates, and also the effect of human body presence and the mobility.

Impact of number of nodes. As each health-monitoring application may need a specific number of physiological sensors, varying the number of sensors in the experiment is one of the first steps towards system evaluation. We configured the Shimmer devices by using Consensys-Basic. The sampling rate was set to 512 Hz. The 18-byte Bluetooth frame encompasses the data of low-noise accelerometer, wide-range accelerometer, and battery voltage. This frame size format is also appropriate for sending ECG, EMG or respiration measurements. We conducted experiments with different number of nodes, while nodes were locating in different distances from the Master node. Each experiment was run for a duration of 10 min. the stored data of sensing measurements and RSSI values were stored in files in the Master node.

Figure 3 to 8 shows the performance of the network in terms of PRR together with the relative RSSI value. The three numbers in the RSSI value graph represents the maximum, average and minimum reading of nodes' RSSI values for each scenario. The results from Figure 4 shows a high value of RSSI reading from all Shimmer devices in 1m distance. The high PRR values (nearly 100% packet deliver) reveals that there is a high correlation between the RSSI and PRR when the RSSI is strong. Thus, keeping the Master node closer to physiological devices regardless on number of nodes would result in a reliable system. It is also important to note that we considered 512 Hz sampling rate, which is for example, the minimum required frequency for collecting ECG parameter in order to monitor heart condition.

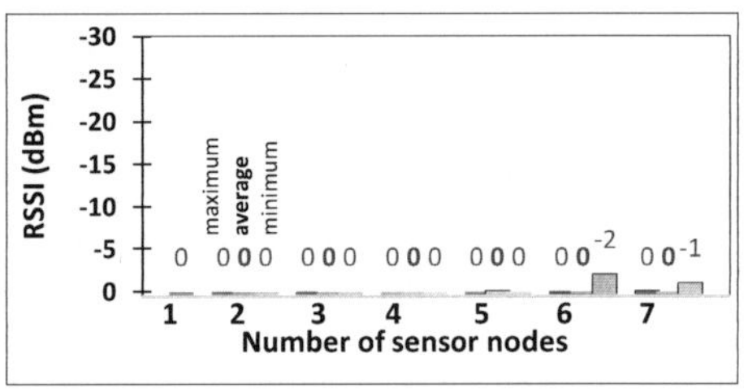

Figure 3. Measured RSSI values for different number of sensors located in 1m distance from the Master node

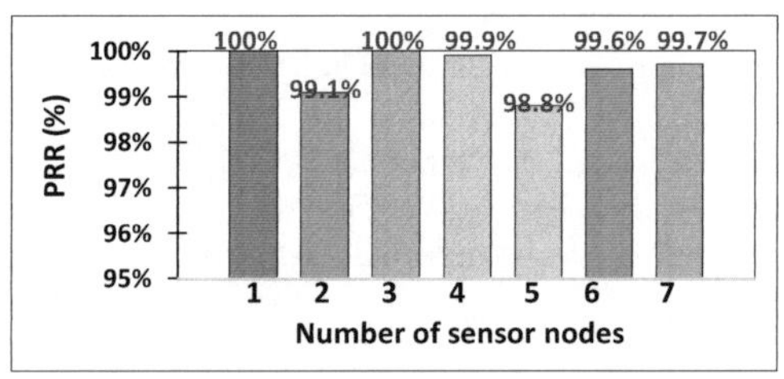

Figure 4. Packet reception rate for different number of sensors located in 1m distance from the Master node

Figure 5 and 6 illustrate the results of RSSI and PRR when sensor nodes are located 3m far from the destination. In this scenario we observe some fluctuation in the RSSI value as well as a sudden drop in signal strength. Moreover, the higher the number of nodes the lower the PRR, meaning that by increasing the number of physiological sensors there is a meaningful reduction in successful data transmission.

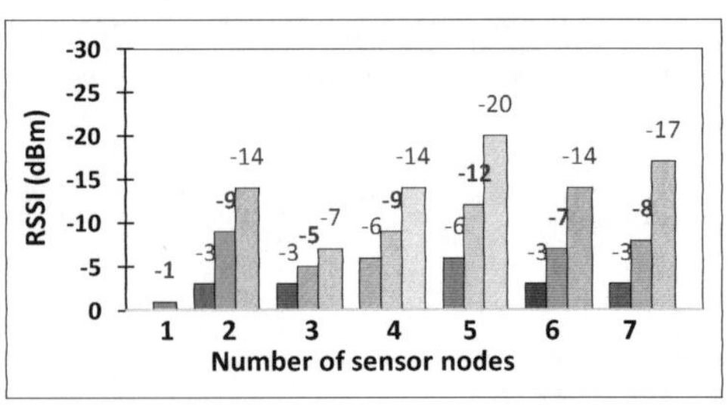

Figure 5. Measured RSSI values for different number of sensors located in 3m distance from the Master node

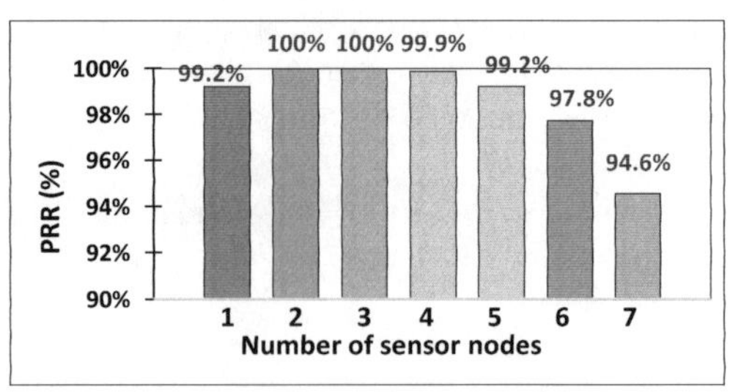

Figure 6. Packet reception rate for different number of sensors located in 3m distance from the master node

Further increase in the distance between sensor nodes and the Master node degrades system performance as illustrated in Figure 7 and 8. However, the system is still able to provide high reliability with low number of nodes (less than 4 nodes). This justifies that the transmission power of the Bluetooth radio is strong enough to provide reliable links up to 5m distance. We have limited our experiments up to 5m distance as there were space limitation in the experiment area.

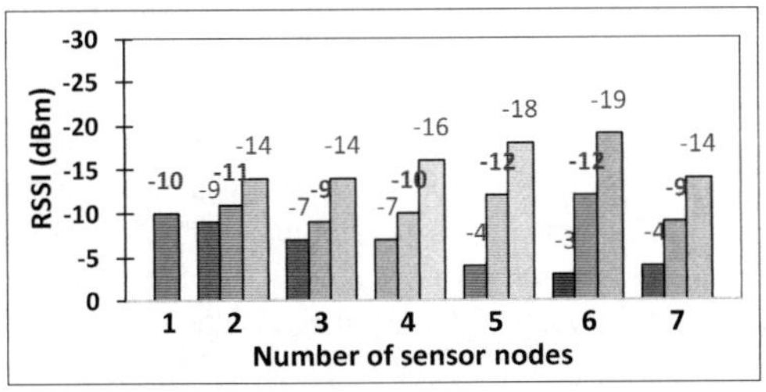

Figure 7. Measured RSSI values for different number of sensors located in 5m distance from the Master node

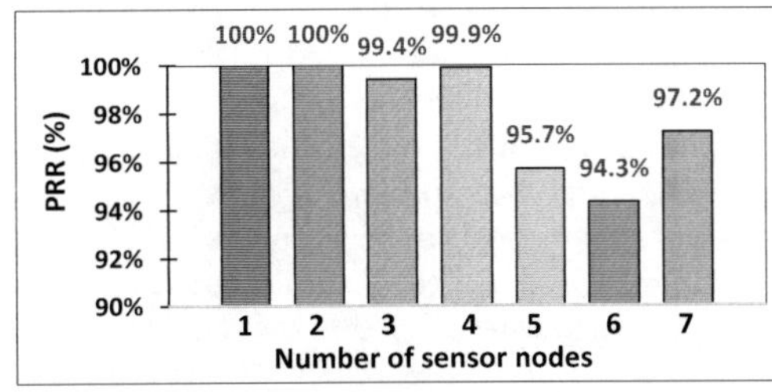

Figure 8. Packet reception rate for different number of sensors located in 5m distance from the Master node

Impact of sampling rate. Sampling rate is one of the major parameters that could have high impact on the overall network performance. We run a set of experiment to evaluate the network performance with different sampling rate of 256 Hz, 512 Hz and 1024 Hz. We used the maximum possible Shimmer devices for this set of experiments (7 Shimmers) while varying the distance from 1m to 5m.

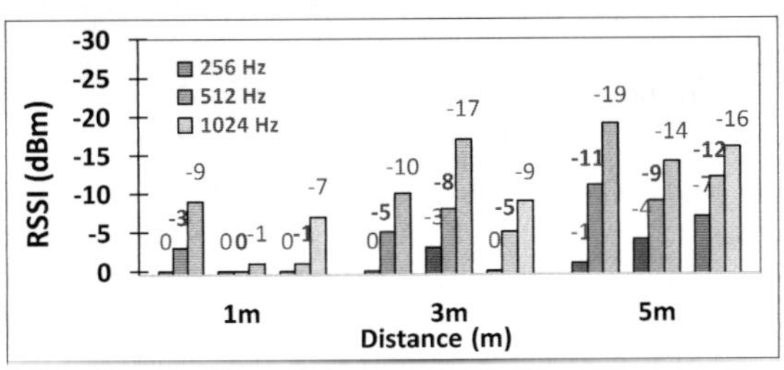

Figure 9. Measured RSSI values with different sampling rates of 256 Hz, 512 Hz and 1024 Hz when varying the distance to Master node from 1 to 5 m

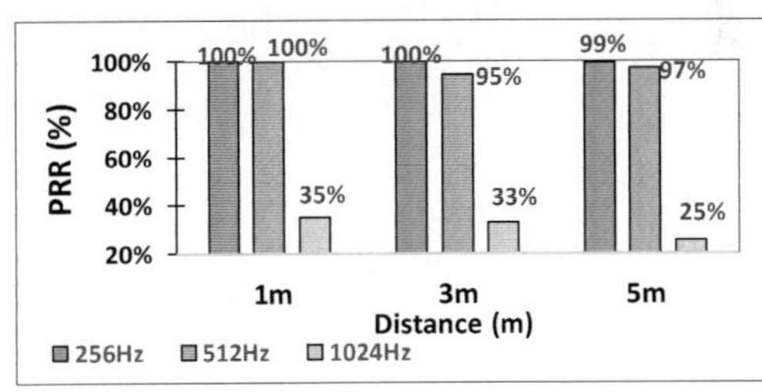

Figure 10. Packet reception rate with different sampling rates of 256 Hz, 512 Hz and 1024 Hz when varying the distance to Master node from 1 to 5 m

Figure 9 shows that regardless of the chosen sampling rate, the RSSI has more fluctuations and lower strength in longer distances. However, as shown in Figure 10 there is a high correlation between the PRR and the sampling rate. As expected increasing the sampling rate to 1024 Hz would significantly degrade the network performance in all distances. This experiment further proves that sampling rate is one of the crucial parameters that must be tune properly in order to maintain a reliable system.

Impact of human body presence. Human body has been known as one of the major sources of interference for sensor devices. To investigate the impact of human body, we attached seven Shimmer devices on a person and ask the person to stand at the selected distances from the Master node during the experiment while wearing the sensors.

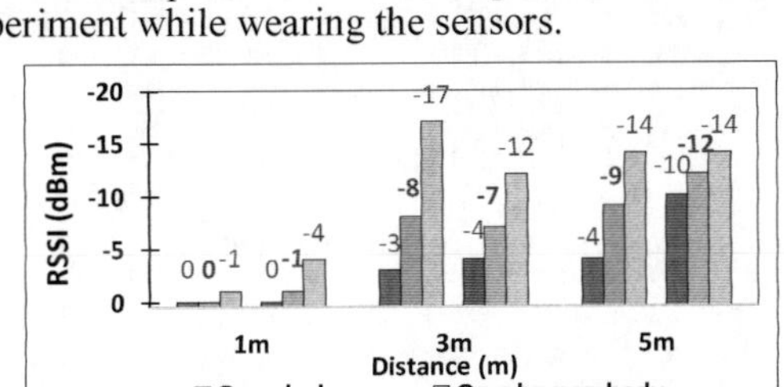

Figure 11. Measured RSSI with 512 Hz sampling rate when locating either on a desk or attached on human body

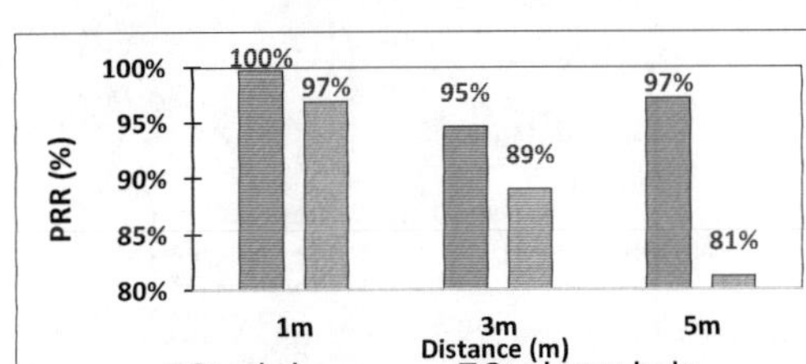

Figure 12. Packet reception rate with 512 Hz sampling rate when locating either on a desk or attached on human body

Results from Figure 11 and 12 show the RSSI and PRR values for scenarios where Shimmer devices are either located on a desk or attached to a human body. Unlike our expectation, we see quite similar results for measured RSSI value in both scenarios, meaning that the human body has no significant impact on the RSSI value. Thanks to the strong transmission power that covers the weak interference caused by the human body. However, we observe a reduction in the network reliability by having less PRR when nodes are attached to human body. This observation makes it clear that although the RSSI value remained fairly unchanged but the higher level of noise floor caused by the presence of a human body would degrade the system performance when the RSSI value is relatively low (ie, at 5m distance).

Impact of mobility. We evaluated the performance of the system in a more realistic scenario, where the Shimmer devices are worn by a person. We considered two cases: (i) the person with attached sensors is standing at a fixed position in a 5 m distance from the Master node, and (ii) the person walks randomly with a normal human walk speed in a room, up to 5 m distance from the Master node. We compare these two cases with the scenario where the Shimmer devices were located on a desk locating 5 m far from the Master node.

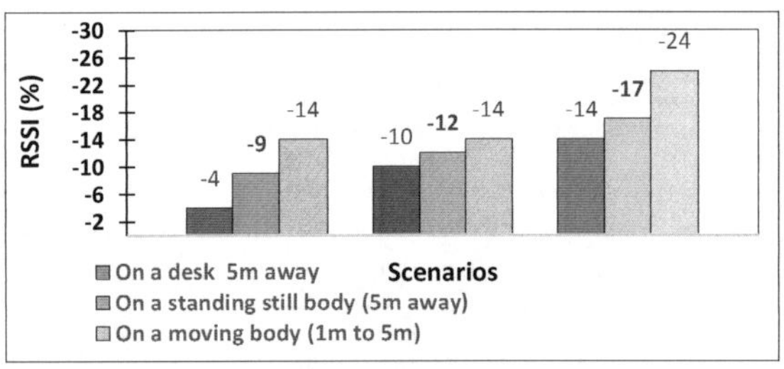

Figure 13. Measured RSSI values with 512 Hz sampling rate for the static and mobile scenarios

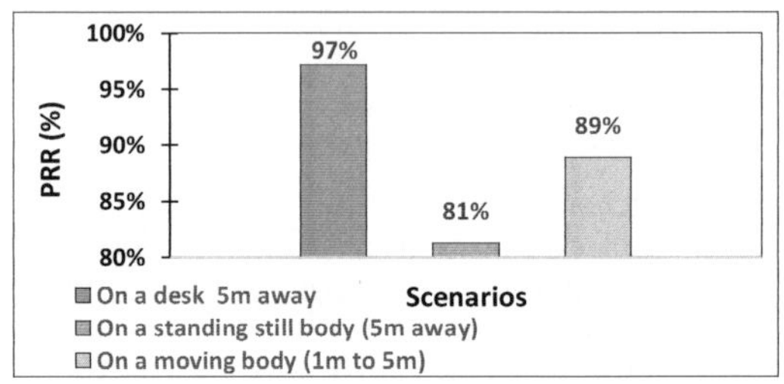

Figure 14. Packet reception rate with 512 Hz sampling rate for the static and mobile scenarios

Figure 13 once more shows that the RSSI drops gradually when sensors are attached on the body, but it degrades more when the person moves randomly in the room. The minimum RSSI value in the mobile scenario reaches to -24 dBm, which is a meaningful drop comparing to its counterpart in the static scenario. The same pattern is seen as before in the PRR. Figure 14 reveals that the system reliability degrades when sensors are attached to the human body. However, we noticed more packet drops for the case where the person is standing at a fixed position. The reason is that we have selected the worst-case scenario, where the person is standing 5 m away from the Master node, while in a mobile scenario, the person is moving randomly in the room with the maximum length of 5m.

Impact of interference. As stated earlier, wireless networks that operate in 2.4 GHz are the main sources of interference for the Bluetooth-enabled network. In this work, we conducted another set of experiments at a library environment at MDH university. We conducted the tests during a busy period of the day at the library, in order to provide similar noise condition as in a hospital. Besides people using their mobile phones, at the hospital, patients' medical information and images are transferred through WiFi access points.

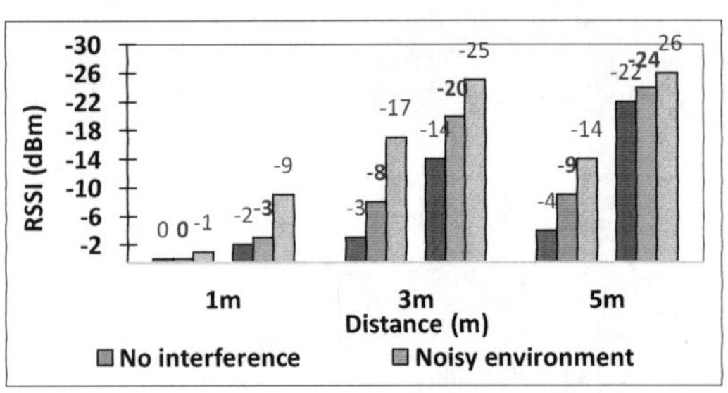

Figure 15. Measured RSSI in a non-noisy scenario with a noisy scenario at library in a static network when locating Shimmer devices in a 1 to 5 m distance

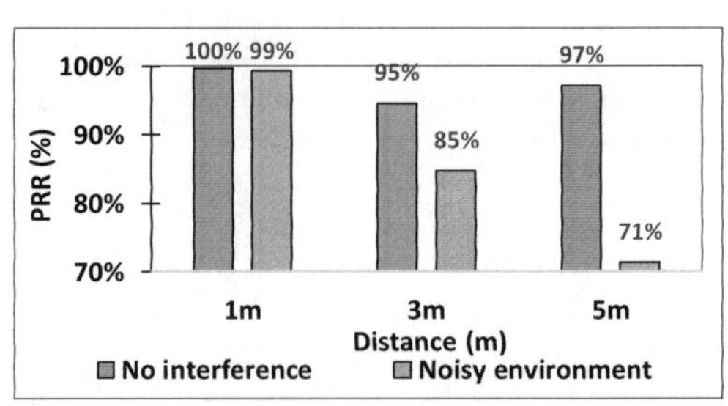

Figure 16. Packet reception rate in a non-noisy scenario with a noisy scenario at library in a static network when locating Shimmer devices in a 1 to 5 m distance

Figure 15 shows that the RSSI value reduces remarkably in a library environment as the distance between the sensors and the Master node grows. RSSI reduction is mainly due to the presence of people and the furniture that attenuate the signal. Figure 16 shows that the system can be still reliable in a noisy environment if the Shimmer devices and the Master node are at most 1m apart. Otherwise, up to 30% of data can be lost which is not acceptable in the critical health monitoring applications.

4. CONCLUSIONS

In this work, we have extensively evaluated the performance of Shimmer devices in terms of wireless connectivity. Shimmer devices are equipped with Bluetooth classic radio that operates in 2.4 GHz frequency, sharing the same frequency band with WiFi. The main goal of the experiment was to study the impact of different sampling rate, presence of human body, mobility and interference on the overall system performance. We found that the Bluetooth radio is reliable at home environment where the distance between the sensor devices and the Master node does not exceed 5m. The maximum sampling rate to achieve high PRR is 512 Hz, which is sufficient for EXG signal monitoring with streaming requirement. The presence of human body can degrade the system performance. Mobility of the person is another source of performance degradation, where we recorded about 9% more packet losses when a person with attached sensors moves randomly. Finally, the extreme noisy environment, such as in hospital and library, degrades the network performance drastically. We observed 10% more packet losses in 3 m distance and 25.8% more packet losses in 5 m distance.

ACKNOWLEDGEMENT

The work presented in this paper is supported by the Swedish Research Council (Vetenskapsrådet) through starting grant no. 2018-04582 via the MobiFog project, and by the Swedish Knowledge Foundation (KKS), through Prospekt grant via the FlexiHealth project and research profile ESS-H plus, and the EU Celtic_Plus/Vinnova project, Health5G (Future eHealth powered by 5G).

REFERENCES

Ahmed, M. U., Fotouhi, H., Köckemann, U., Lindén, M., Tomasic, I., Tsiftes, N., & Voigt, T. (2017). Run-time assurance for the e-care@ home system. *International Conference on IoT Technologies for HealthCare,* 107-110. Springer.

Alam, M. M., Hamida, E. B., Rehmani, M. H., & Pathan, A. S. K. (2016). Wearable wireless sensor networks: Applications, standards, and research trends. *Emerging Communication Technologies Based on Wireless Sensor Networks: Current Research and Future Applications,* 59-88. CRC Press.

Baker, N. (2005). ZigBee and bluetooth: Strengths and weaknesses for industrial applications. *Computing and Control Engineering, 16*(2), 20-25.

Chang, K. H. (2014). Bluetooth: A viable solution for IoT? [Industry Perspectives]. *Wireless Communications, 21*(6), 6-7. IEEE.

Chen, Y., & Kunz, T. (2016). Performance evaluation of IoT protocols under a constrained wireless access network. *International Conference on Selected Topics in Mobile & Wireless Networking,* 1-7. IEEE.

Doukas, C., & Maglogiannis, I. (2012). Bringing IoT and cloud computing towards pervasive healthcare. *International Conference on Innovative Mobile and Internet Services in Ubiquitous Computing,* 922-926. IEEE.

Fotouhi, H., Vahabi, M., Ray, A., & Björkman, M. (2016). Reliable communication in health monitoring applications. *International Conference on IoT Technologies for HealthCare,* 64-70. Springer.

Gardašević, G., Fotouhi, H., Tomasic, I., Vahabi, M., Björkman, M., & Lindén, M. (2017). A heterogeneous IoT-based architecture for remote monitoring of physiological and environmental parameters. *International Conference on IoT Technologies for HealthCare,* 48-53. Springer.

MagPi. (2018). Raspberry PI 3B+ specs and benchmarks. Retrieved May 2019 from https://www.raspberrypi.org/magpi/raspberry-pi-specs-benchmarks/.

MICROCHIP. (2015). RN42/RN42N, class 2 Bluetooth® module with EDR support. Retrieved May 2019 from http://ww1.microchip.com/downloads/en/DeviceDoc/50002328A.pdf.

Ministry of Health and Social Affairs. (2016). Vision for eHealth 2025. Retrieved May 2019 from https://www.government.se/information-material/2016/08/vision-for-ehealth-2025/.

Natarajan, R., Zand, P., & Nabi, M. (2016, October). Analysis of coexistence between IEEE 802.15. 4, BLE and IEEE 802.11 in the 2.4 GHz ISM band. *International Conference of the IEEE Industrial Electronics Society,* 6025-6032.

Premsankar, G., Di Francesco, M., & Taleb, T. (2018). Edge computing for the internet of things: A case study. *IEEE Internet of Things Journal, 5*(2), 1275-1284.

Ray, P. P. (2018). A survey on internet of things architectures. *Journal of King Saud University-Computer and Information Sciences, 30*(3), 291-319.

Reiter, G. (2014). Wireless connectivity for the internet of things. *Europe*, 433, 868MHz.

Santos, J., Rodrigues, J. J., Silva, B. M., Casal, J., Saleem, K., & Denisov, V. (2016). An IoT-based mobile gateway for intelligent personal assistants on mobile health environments. *Journal of Network & Computer Applications, 71*, 194-204.

Sarkar, S., & Misra, S. (2016). From micro to nano: The evolution of wireless sensor-based health care. *IEEE pulse, 7*(1), 21-25.

Shimmer, RG. (2018) shimmer3/LogAndStream/python_scripts/. Retrieved May 2019 from https://github.com/ShimmerResearch/shimmer3.

Shimmer. (2019). Consensys GSR Development Kits. Retrieved May 2019 from http://www.shimmersensing.com/products/gsr-optical-pulse-development-kit.

Tomasic, I., Petrović, N., Fotouhi, H., Lindén, M., & Björkman, M. (2017). Data flow and collection for remote patients monitoring: from wireless sensors through a relational database to a web interface in real time. *European Medical and Biological Engineering Conference (EMBEC) and the Nordic-Baltic,* 89-92. Springer.

United Nations. (2013). World population ageing. Department of Economic and Social Affairs, Population Division.

THE REVERSED MOSCOW METHOD. A GENERAL FRAMEWORK FOR DEVELOPING AGE-FRIENDLY TECHNOLOGIES

Luiza Spiru[1], Mircea Mărzan[1], Cosmina Paul[1], Magdalena Velciu[1] and Adrian Garleanu[2]
[1]Ana Aslan International Foundation
Spătarului Street, no 3, Bucharest, Romania
[2]Independent Researcher

ABSTRACT

In the last decade, the number of new technologies which target independent and active seniors at home tremendously increased, though many failed to address seniors' expectations and needs. The purpose of this study was to construct and test a new framework for developing age-friendly technologies for independent and active seniors at home. The proposed framework is the 'Reversed MoSCoW Method', where MoSCoW stands for must, should, could and have requirements to accomplish business needs. The tool is a prioritization for reaching a common ground among the technology developers and designers on not to do requirements, when conceptualizing new technologies for independent and active seniors at home. The proposed model was grounded on the research findings from Senior-TV. Hence, a questionnaire survey was conducted with a sample of 148 seniors aged 65 and over in Cyprus, Romania and Slovenia, between February and April 2019, and the results are transferred and tested in a current ongoing European AAL project, IOANNA. The results showed that understanding the specificities of the relationship between technology and the independent and active seniors is crucial for understanding their habits and dispositions towards adopting new technologies. The findings show the most frequent mistakes when developing new technologies for active seniors: 1) age stigmatization, 2) the tendency to improve already domesticated products or services, such as TV watching or communication by phone, as seniors do not engage on Internet at the expense of TV watching or on social media instead of communication by phone, 3) asking for a continue or frequent logging in, 4) health oriented services, 5) services' complexity and others. These were found as better predictors of failing technologies than the conventionally assumed attitudinal factors, such as ease of use, disinterest or lacking technological abilities.

KEYWORDS

Age-Friendly Technology, Independent Seniors, "Reversed Moscow Method', Gerontographics

1. INTRODUCTION

Senior population is ageing and its diversity is increasing (European Comission, 2018). Independent and active seniors at home are seen both as a demographic resource and as a societal concern for ageing well and so technologies are called to play a crucial role to ease the role of independent and active seniors in societies and communities, and to keep them ageing well. Even the paradigmatic understanding of aging as a natural process is challenged today, as traditionally ageing is associated with a reduced mobility, chronic diseases and mental health disorders. Gerontologists propose to look at the ageing process rather as a disease and the focus to be put on improving the health and wellbeing of the senior population (Bulterijs et al, 2015). Technologies of the near future for healthy aging are fast developing and they are rapidly moving into the home (National Research Council, 2011). New age-friendly technologies offer innovative solution for helping people to stay as healthy and independent as possible. The challenge is to develop the right technologies such as elderly would benefit greatly from products, and designing is crucial. Age-friendly technologies are used to promote healthy living, preserve physical and mental capacity and delaying normal brain aging (Spiru L., 2011), also enable seniors to continue to live well.

In order to develop an age-friendly technology to promote innovation and healthcare solutions targeted to the elderly, it is important to have a proper understanding of the elderly user requirements and their significance. An in-depth understanding of the elderly as a distinct group in all respects, such as the way of living (independent or not), status of health, physical and psychological support (physical, emotional, cognitive), socio-economical aspects, their relation with technology is key for a correct approach into the near future of technologies for ageing well. Willing to offer elders all they need, some projects are based on incorrect assumption and requirements and find later that the elderly requests are others that they believed.

We bring together our research findings from two projects part of Active Assisted Living (AAL) Programme, which aimed at promoting information technology for aging well and deliver healthy ageing solutions to market. By using these transferable results from various field-researchers conducted over the last decade, we want to illuminate the reality of interests, needs and habits of seniors related to technology use. The promoted AAL projects are: 1) IOANNA, which will develop six services: stores update tool, end-user interaction tool, route planner, optimum product or service choice, user-friendly interface and a mobile application [IOANNA, 2019], and 2) SENIOR-TV, which aimed at developing a platform for providing formal and informal caregiving services to older adults that live alone in their own homes, at low cost, and that focuses on the active prevention and the maintenance of relationships with friends, family, and the community [Senior-TV, 2019]

The paper's contribution resides on creating a framework based on the research findings and lessons learned from previous projects on the needs of independent and active seniors, when designing technologies which target them. There are necessary steps before putting down the elderly user requirements meaning what the elderly users expect the product or technology to do. Choosing the right requirements is essential for the success of the product, and avoid the failure. Especially in the case of elderly, the prioritization of the needs is as important as the prioritizations of refrains and anxieties, in order to reach technological adoption benefits. The process of prioritizing is an activity for defining what requirements to achieve or not to do, and in what sequence. It is a method to make the project more efficient by identifying the most wanted characteristics to offer.

2. BODY OF PAPER

We need to go behind the trite division of young and old seniors or other groupings based on age, as it has been shown that the heterogeneity of the senior population is much higher than that of other demographic groups (Nimrod, 2013; Sthienrapapayut et al, 2018). There are various types of segmentation of the elderly market which have been employed by economics and marketing when aiming at an understanding of this mature market. Though, due to the large heterogeneity of the people aged over 65, segmentations such as sociodemographic, behavioral or psychographic proved insufficiently explanatory (Nimrod, 2013, Moschis, 1996).

According to the French Silver Economy strategy, launched by the Ministère des Affaires Sociales et de la Santé in December 2013, the aging population is defined in 3 groups; active, fragile and dependent. Moschis developed the gerontographics approach in order to assess the high variability of the seniors' needs. The proposed gerontographics segmentation of the adult market is developed from the individual differences in aging, and the variances in the life experiences which influence psychographics factors. Four segments of the elderly are identified based on the type of aging experienced: healthy indulgers, ailing outgoers, healthy hermits and frail recluses (Moschis and Mathur 1993; Moschis 1996, 2003). One group is covering the segmentation of independent seniors living at home, and they are healthy indulgers. Healthy indulgers are seniors who experienced less negative life events. Comparative to others, they have a good health condition which supports their psychological and social aging. They continue to age well, staying independent, active, and socially integrated, enjoying life and sharing similar behavior with those younger. The other groups manifest either a psychological or a physical health decline, which either insulate them from society or develop various grades of dependency from their informal or formal caregivers.

The projects who fall under the scope of this article selected persons according to age criterion and included people soon-to-be seniors, aged 55 years and over, from three ranges: 55-65; 65-75 and 75+ years, with balanced distribution. According to health status, the target group is formed by healthy persons who are active or persons with some (minor) problems of health, independent or with physical disability.

2.1 The Relationship of Independent Seniors at Home with Technologies

The field research conducted during the project Senior-TV in Cyprus, Romania and Slovenia and which interviewed 57 persons, from a total sample of 148 which included also other categories of seniors based on their self-perceived health status. They were identified as healthy indulgers, namely independent and socially active seniors living at home, and two relevant findings concerning their relation with technology: 1) that their predisposition for using technologies decline with time, and 2) that is merely due to the perceived irrelevance of technology in their life.

Table 1. Independent seniors living at home and their relation with technology

	Enter houses	Average First encounter (years)	USAGE	
			Property	**Frequency of usage on average**
Radio	1940	-	Rather not	Few times a week
TV	1960	20	ALL HAVE	Daily
Landline Phone	1980	-	Rather Not	Few times a week
Mobile phone		59	WIDESPREAD	Few times a week
Computer	2000	60	Either have or don't have	Few times a week
Internet		62		Few times a week
E-mail			Rather not	Occasionally
Smart phones	2010		Either have or don't have	Either daily or never
Smart TV	2015		Rather not	Either daily or never
Smart Watch	2015		Inconclusive	-

Senior-TV project findings show that healthy indulgers are those who encounter various technologies earlier in life, comparative to the other three categories, though their anxiety towards technology increase in time and so does the perceived irrelevance of technology in their life. Healthy hermits keep a constant anxiety towards technology, and frail recluses' anxiety towards new technology decreases with time.

It is the 'perceived usefulness' of technology which matters for independents seniors at home and encourage them towards exploring and experimenting new technologies, comparative to other seniors who are embolden by the influences of their informal and formal caregivers and social entourage.

When developing new technologies, the ranking on seniors' future services preferences should be taken into account. According to the research conducted over the Senior-TV project, from February to Aprilie 2019 in Romania, Cyprus and Slovenia, the first concerns of the elderly refer to their safety, independency and socializing. Their top three technologies preferred are: 1) the sensors which warn about a tap or cooker left open; stove and oven safety control; 2) Sensors for preventing or detecting falls and 3) Organizing walking groups. Seniors from Cyprus and Slovenia add as one of their first interests also the 'online websites for seniors' services'. Therefore, the extent to which new technologies meet seniors' concerns should be of primary interest for scholars and practitioners from the conceptualization phase and not only over the developmental process.

2.1.1 The *Reversed MoSCoW Method* for Developing Technologies for Independent Seniors at Home

We have developed a 'Reversed MoSCoW Method' to serve as a general framework for guiding technology developers and designers for new technologies, which target independent seniors living at home. This framework is generally addressed for people over 65 years old but the self-perception of age, cultural and environmental settings also accounts for the adaptability of this framework.

MoSCoW approach refers to specific user requirements that depend on the development of a specific product or service targeting independent seniors living at home, specifically, to the must, should, could and would or won't of a product for market success. The prioritization of these requirements can be conducted according to MoSCow method. Because we discuss here universal requirements of the seniors, we have built requirements which are embedded in a "not to" framework, which we called "The Reversed-MoSCoW Method" and are equally relevant to technology development processes.

"Must NOT have" refers to critical requirements. "Should NOT have" are the important requirements though not critical, and for the case of elderly, are often as important as "Must NOT have" because social influences and societal environmental factors are much weaker than for the other categories of age, such as people on the job or youngsters. The difference consist in the fact that seniors would accept or adopt a technology which meet "the Must NOT" requirements but they will not continue to use it, or it will not became a natural habit for them if the "Should not have" requirements are not met. "Could NOT have" are desirable but not necessary, and could improve user experience or satisfaction. The least relevant requirements are listed at the end, in the "Won't have" category.

Must NOT have. *Age-stigma*: new technologies need to be promoted as universal technologies and not as gerontechnologies, meaning that when marketing new technologies to independent seniors it is important to address universality and not social stigma of ageing. These findings are in line with Moschis (2003) research findings on advertising universality. *Echo the domesticated technologies*: new technologies do not need to tend to either replace or duplicate already existent information, communication & entertainment services. For example, the following needs cannot be replaced by new technology devices because of the domestication process which they went through: the use of TV for information and phone for communication. Family and friends and television are mentioned as the first two most important sources of enjoyment and, moreover, studies indicate that the time spent on Internet and face to face interactions do not tend to replace TV watching (Östlund, 2010). *Continues or frequent logged in*: Participants refer to their usage of mobile technology as an activity that is performed 'only when it is necessary'. Participants explicitly state that they do not wish to be constantly connected; main reasons are privacy issues and the time needed to learn to use complex mobile devices. Additionally, they need to 'save' their sensory and cognitive resources for other things that require attention while on the way. Technology developers should also consider that independent seniors are not particularly interested in social networks (Senior-TV and T&T projects).

Should NOT have: *Health oriented services*: many times, services developed for independent elderly at home concern emergencies, supervision, monitoring health or help which are rather meant for assisted elderly and should be used by their carers. Independent and active seniors manifest interest in enjoyment and information and not in health preventive measures. Independent seniors living at home tend to appeal to professional medicals for their health management and not to new technologies, as they do not want to constantly be reminded of their age and possible frailty condition. This finding has its cultural limitation and should be tested against seniors with various educational and cultural backgrounds. *Development process without seniors' involvement*: Conceptualization, development and design conducted by middle-age people without the involvement of the elderly. Generally, the involvement of the seniors in the phase of conceptualization is ignored by the technology developers. *Complexity* (with a special emphasis on many choices): seniors dislike complexity and avoid the computer wherever possible. They keep a constant medium to high general anxiety towards technology, even though they used to integrate technology before in their social and working environments. Moreover, the participating seniors did not want many choices.

Could NOT have: *Ignoring physical impairments*. Research findings show that the 'Perceived ease of use' has an indirect impact on the behavioral usage of the independent seniors (Chen and Chan, 2013). Therefore, the ease of use considerations is integrated in 'could not have' category. Here, physical impairments should be factor in when developing and designing new technologies. Participants favor old remotes instead of the touch remote control, mention as desirable features big buttons, pleasant high-pitched voice, and lower complexity: few menu choices, low hierarchy of choices. The interface needs to be self-explicable, non-invasive and simple, perhaps at expense of other extra functionalities. *Presumed experience in general technology abilities*: the lack of experience of the independent seniors may be noticed in the small number of seniors who are familiar with GPS and many of them would not know how to connect to a Wi-Fi network. Moreover, a majority of seniors engage online only to do the necessary and do not check their accounts regularly. The previous AAL projects found that when participants plan a route, it is of equal relevance to them the user interface of the planning application and the output of this planning: the print, regular e-mail check, virtual contact instead of face to face contact. technological services without human contact as the trust decrease with age.

Would NOT be nice to have: *intrusive features, money management and gaming.* Most seniors are cautious of sharing personal data online, managing their bank accounts online and they refrain from playing games online, comparative to other groups of seniors who manifest more frailty and look for engaging online with games. Gaming online and on the phone is especially of interest to women who have a low psychological well-being, according to the Senior-TV research findings.

The below table present a systematic view of the above described research findings.

Must NOT Have	Should NOT Have	Could NOT Have	Would NOT Have
• Age-stigma • Echo the domesticated technologies • Requesting for continue or frequent logged in	• Health oriented services • Development process without seniors' involvement • Complexity	• Ignoring physical impairments • Presumed experience in general technology abilities	• Intrusive features • Money management • Gaming

Figure 1. The Reversed MoSCoW Method. A general framework for developing technologies for independent and active seniors at home

2.2 Using the Reversed MoSCoW Method to develop the integrated solution IOANNA

The IOANNA application is an integrated solution that helps seniors for searching for commercial offers, services and specific products near the home (new technological features). The application is designed to make elderly feel safer, stay active, find a more convenient route and stay in contact with persons and healthcare service, in case of emergency. The solution also helps social engagement through community service in paid or volunteer basis for experienced senior adults and assistive mobility with the navigation system aiding senior adults or other citizens to get to their destination in the most convenient way. Design technology was seen as crucial as well as the age-friendly interface, with special care for seniors' needs. Research on users' requirements was focused on the assessment of the characteristics of real needs as well as identifying their expectations and preferences and even the unconscious determinants of making decisions process.

The field research conducted during the IOANNA project involves people identified as independent seniors living at home, and it shows that seniors' self-perceived health status and self-determining movement are very important, also keeping them connected with family, if needed. The evaluation was based on quantitative methods for analyzing perceptions and ideas in a measurable approach. The data were collected using in-depth interviews with senior adults and the elderly (over 55 years) in order to understand their needs, requirements, and ideas for each service of the IOANNA solution. Information is analyzed by themes for qualitative insights or through numerical comparisons for quantitative results. In September-October 2018, we interviewed 25 persons from Romania and 22 persons from Cyprus. Regarding IOANNA application, seniors perceive that the idea of having all services in one product is very interesting and useful, and appreciate to have an easy to use application and save time. Independence and security are very important.

2.2.1 Applying the Reversed MoSCow Method for Elderly-User Requirements

After collecting the users' requirements, we employed the 'Reversed MoSCoW Method' to test the degree to which the product answer to the specified not requirements model. The MoSCoW method is a technique used in the development of new technology to identify the most important requirements to deliver and so, the involvement of targeted people is mandatory.

Must not have requirements have been met to a medium degree. The IOANNA application avoid age-stigma by developing an innovative advanced solution on mobile smart-phone in order to help people to stay active and by addressing elderly with respect. The findings show that the seniors' preference for using Internet on a personal computer and smartphone rather than a tablet may lead 1) to the perception that the already existing online information is duplicated and 2) to make seniors to prefer to connect on computer which is offering them the possibility of printing, which is shown as being highly appreciated by seniors, instead of using a permanent online connection, which refrains seniors from using a service.

Should not have requirements are met to a high degree**.** The IOANNA project followed the user needs analysis that was conducted in previous successfully Independent and seniors' involvement is a reliable mean to ensure that requirements and needs of elders are investigated and respected in every stage of the project in order to create a product which is useful, easy to use, with high acceptability. If addressing complexity, the project is meeting various expectations that decrease the acceptance of seniors who rather prefer simple and focused services.

Could not have requirements are met to a medium degree. IOANNA application is mobile which may affect the adoption of the application because of physical impairments. The fact that seniors prefer to have their app with them in order to be able to use the notification function if something happens to them implies the appealing to the service for emergency, which just a side feature. To permanently stay connected for checking the route planning is a presumption of experience in general technology abilities, which should not have been made.

Would not be nice to have requirements are met to a medium extent. Our results show that seniors order travel, transport ticket more frequently and less frequently for health issues, maybe because of the lack of trust. Tailoring IOANNA application that offers a verified network with a more secured link and limited access of unknown persons is just a presumed and not tested solution for countering seniors' lack of trust.

3. CONCLUSION

The present developed framework, 'the Reversed MoSCoW Method' refers to the prioritization of the 'Not to do requirements', when developing new technologies which target independent and socially active seniors at home. The here proposed method is meant to serve scholars and practitioners to deepen their understanding about developing technologies which potentate and maintain the well ageing of the independent and active seniors and their involvement in the society. Despite this positive contribution, there are two major limitations to the extant research. We discuss here just the universal requirements for the independent and active seniors at home, which are not transferrable to seniors who do not live independently and are insulated from society because of physical or psychological ill-being and who manifest a certain degree of dependency. Another limitation is that this framework does not account for culturally diverse factors and it might be relevant only to the Eastern and Southern Europe, as the Western Europeans seniors may exhibit a different relationship with new technologies. However, the wide application of the framework in new product development is of great relevance for supporting the decision process even from its very incipient phases.

ACKNOWLEDGEMENT

This work was performed in the frame of the EU project IOANNA (AAL-2017-077, with implementation period April 2018 - Sept 2020, funded by the AAL Programme, co-funded by the European Commission and the National Funding Authorities of Cyprus, Spain and Romania) and The Senior-TV project (AAL-2014-171, that is co funded by the EU AAL Joint programme and the related national Authorities in Spain, The Netherlands, Cyprus, Romania and Slovenia.

REFERENCES

Bulterijs, S. at al, 2015, It is time to classify biological aging as a disease. *Frontiers in genetics*, vol. 6, no. 205.

European Commission, 2017, *The 2018 Ageing Report*, Luxembourg: Publications Office of the European Union.

Ioanna AAL Projects. http://www.aal-europe.eu/projects/ioanna/ [Online]. [Accessed from: 25 May, 2019].

Ministère des Affaires Sociales et de la Santé, December 2013. Silver Economy Strategy. [Online] http://www.social-sante.gouv.fr/espaces,770/personnes-agees-autonomie,776/dossiers,758/silver-economie,2432/. [Accessed from: 29 May, 2019].

Moschis, G.P, 2003. Marketing to older adults: an updated overview of present knowledge and practice. *Journal of Consumer Marketing*, vol. 20 no 6, pp.516-525.

Moschis, G.P, 1996. *Gerontographics: Life-stage segmentation for marketing strategy development*, Greenwood Publishing Group.

Moschis, G.P. et al, 1997. Targeting the mature market: opportunities and challenges. *Journal of consumer marketing*, vol. 14 no 4, pp.282-293.

Moschis, G.P., Mathur A, 1993. How they're acting their age. *Marketing Management*, vol. 2 no 2, p.40.

National Research Council, 2011, *Health Care Comes Home: The Human Factors*. Washington, DC: The National Academies Press, pp https://doi.org/10.17226/13149.

Nimrod, G., 2013, Probing the Audience of Seniors' Online Communities, *The Journals of Gerontology*: Series B, Vol. 68, Issue 5, September 2013, pp 773–782, https://doi.org/10.1093/geronb/gbt059

Östlund, B., 2010. Watching television in later life: a deeper understanding of TV viewing in the homes of old people and in geriatric care contexts. *Scandinavian journal of caring sciences*, vol. 24 no 2, pp.233-243.

Senior-TV AAL Project. http://www.aal-europe.eu/projects/senior-tv/ [Online]. [Accessed from: 25 May, 2019]

Senior-TV Project. 2019. Deliverable 1.3 Part A "Analysis of existing services of formal and informal care".

Spiru L. and Turcu I, 2011, Nature and nurture: A golden concept for healthy brain aging, *Alzheimer's & Dementia: The Journal of the Alzheimer's Association*, Volume 7, Issue 4, pp.

Sthienrapapayut Th. at al, 2018, "Using gerontographics to explain consumer behaviour in later life: Evidence from a Thai study", *Journal of Consumer Marketing*, vol. 35, issue 3, pp https://doi.org/10.1108/JCM-02-2017-2083

ELECTRONIC HEALTH RECORD IN THE ERA OF INDUSTRY 4.0: THE FRENCH EXAMPLE

Sarah Manard, Nicolas Vergos, Simon Tamayo and Frédéric Fontane
Mines ParisTech – PSL Research University
60 boulevard Saint-Michel, 75272 Paris Cedex 06, France

ABSTRACT

The recent implementation of the Electronic Health Record (EHR) in France is part of a more general process of digitizing information flows, as the world enters the fourth industrial revolution in a phenomenon known as Industry 4.0. Behind this concept lies the concern to allow Man to remain permanently in control of his destiny, despite an increasingly interconnected world (Internet of Things, cooperative robots, augmented reality, etc.). Accordingly, the implementation of EHR must guarantee the respect for the private life of each citizen. From this perspective, healthcare professionals will therefore have to constantly ensure the protection of medical confidentiality during Electronic Data Interchange (EDI). This paper summarises the current state of the use of EHR in France. Based on a survey conducted by the European Commission to assess the deployment of digitalisation in the health sector in EU countries, this article aims to highlight the opportunities and perspectives that Industry 4.0 could bring to the health sector in France. However, this study also identifies a number of limits related to the application of such a system, the first of which is cyber threat or transhumanism. To this end, a SWOT matrix identifies the strengths and weaknesses related to the implementation of the French EHR.

KEYWORDS

Electronic Health Record, eHealth, European eHealth infrastructures, Industry 4.0

1. INTRODUCTION

The recent deployment of the Electronic Health Record represents a major challenge for France: among the 28 members countries and the 2 partner countries of the EU (EU (28+2)), i.e. the 28 members countries plus Iceland and Norway, it is lagging behind in terms of New Information and Communication Technologies (NICT) (Currie and Seddon, 2014). The digital gap has hit hard a country whose geographical variety and culture are ramparts to the optimal development of these NICTs. Thus, in a context of improving information flows through the digitalisation of processes, a phenomenon that is part of a more general evolution from industry and services to Industry 4.0, the health sector must adapt to the new requirements induced. The emergence of new medical applications on our smartphones follows this trend, with the possibility to book an online consultation, retrieve an electronic prescription or even simplify medical appointment bookings via direct access to doctors' schedules and consultation centres. The implementation of the Electronic Health Record (EHR) in France and Europe must therefore find the answers expected in order to guarantee access to reliable and confidential medical data for all. It will also make it possible, with the help of these new e-services, to respond effectively to the new time requirements induced by the increase in time spent at work or in public transport for workers. Similarly, this EHR should enable everyone, patients and healthcare professionals alike, to have an online connection in line with their needs (Allaert and Quantin, 2009).

The present study will first aim to establish a state of the art of France in terms of the progress of digitisation in healthcare since the launch of the "EHR for all". To do this, the methodology applied will consist in studying both the expectations of the French with regard to the EHR and the different possible uses and derivatives of the latter, such as epidemiological monitoring and/or follow-up (Haas *et al.*, 2017) as well as decision support to establish diagnoses. In a second step, a European benchmark of the hospital health chain will be made, in order to evaluate the relevance of the implementation of such a system in France.

Through the identification of patient and prescription drug information flows, medical research and the diagnosis and care of chronic and/or rare diseases (Finet *et al.*, 2018) can be improved and thus allow significant advances in medical progress. This will soon be possible with the various online medical records set up in European countries, which we will demonstrate in a summary photo at the end of this article. This French impetus towards eHealth, which is part of a European desire, aims to facilitate the exchange of medical data between the "28+2". The opening of borders would thus be generalized to information flows with an "e-systematization" of medical data directories. So far, no official computer system has been imposed by French political authorities, nor has a standardised model for entering computer data. Moreover, the risks associated with such an implementation should not be overlooked. Cyber threat is a reality that every actor in the healthcare chain will have to fight against. The reliability of computer servers, the security of data, or even the respect of medical confidentiality are concerns that will affect us all. Already today, questions are being raised about access rights to EHR with private and/or specific hospitals and institutions that have their own patient monitoring networks (Sabes-figuera, 2013), as well as the definition of need-to-know according to the specific specialities of each health professional.

2. EHR: EXPECTATIONS AND OPPORTUNITIES

These hopes and opportunities were identified based on a literature review of some 15 publications from 2010 to 2019.

2.1 Update on the Expectations of French People following the Implementation of the Electronic Health Record

From the beginning of the project, French authorities have ensured that the EHR aims to guarantee better continuity, better quality and better coordination of a patient's healthcare path. So far, the road to good health resembles a real quagmire, as each healthcare facility gets its own computerized medical data software, sometimes even developed internally. France wants to catch up with its European partners: by 2022, the FNHIF (French National Healthcare Insurance Fund) hopes to create 40 million EHRs (Burnel, 2018). Even if the opening of the EHR is optional, the French seem inclined to take the digital step. Between November 6th - the date of its very last launch - and December 13th, 2018, more than 3 million EHRs were opened (Assurance Maladie française, 2018). Indeed, against all popular beliefs, the French EHR has experienced implementation problems and has been relaunched several times (Séroussi and Bouaud, 2018), as shown in Figure 1.

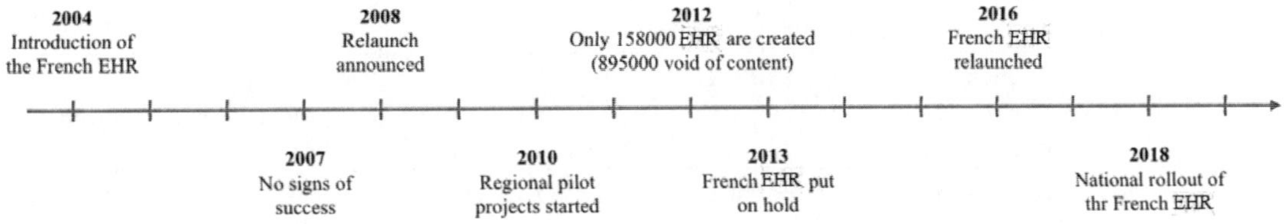

Figure 1. Timeline of the French EHR, data collected from (ASIPS, 2013)

Since its creation, the EHR has been regarded as a data warehouse by healthcare professionals; they can search for information about a patient or consult his healthcare background. All this should ultimately enable them to take better care of their patients. However, in order to guarantee the respect of a patient's private life, it will be possible to withdraw documents that the patient does not wish to divulge to others from this warehouse that a practitioner would have deposited there.

In the field of open data, France already has significant expertise. France has a considerable asset on which it can capitalize and benefit from good practices: since 1999, it has had the largest computer database in the world, the National System of Information between Healthcare Insurance Schemes (SNIIRAM) - managed by the FNHIF. In 2015, 8.9 billion healthcare sheets were managed and anonymized to supply the SNIIRAM warehouse. This database has a storage capacity of 600 terabytes, has 20 billion lines of services available, and manages 150 applications (Bezin *et al.*, 2017). In 2017, a law to modernize the French health system created the National Health Data System (SNDS). The SNDS collects and provides the databases that have existed independently until now:

- the SNIIRAM database containing health insurance data;
- the Information Systems Medicalization Program (IMSP) database containing data from the activities of health care institutions;
- the database Centre for Epidemiology on Medical Causes of Death (CépiDc), managed by NIHMR (National Institute of Health and Medical Research), containing data on causes of death;
- disability-related data from departmental homes for people with disabilities;
- data from "complementary health" (mutual health insurance, for example).

In addition, the FNHIF is now paying ever more attention to its healthcare expenses. This is why it introduced the Public Health Objectives Remuneration (PHOR) in 2011. These are incentive measures for physicians to change their practices to achieve the objectives. They are based on three components: the monitoring of chronic diseases, prevention and the efficiency of prescriptions. ROSP (French remuneration for public health purposes) could thus serve as a lever for healthcare professionals to encourage them to use the EHR. Pharmacists already benefit from payment each time they open an EHR. As local players and trusted third parties, they are called upon to play a major role in ensuring the successful deployment of the EHR. They are paid 1 euro each time they open an HER for a patient (Dormont, 2013).

Faced with the considerable challenge of securing the data entered on the EHR, the French Ministry of Health and Solidarity has chosen a highly secure server, whose name and location are kept confidential for security reasons. Physicians must enter their professional code to log in. If this is their first connection, an alert is immediately sent to the patient by SMS or e-mail. Until authorized, physicians cannot store any data in the EHR system. Currently, they are implementing medical data on patients' DMPs via software made available by the French government. These are linked to secure platforms that allow the official database to be incremented.

Thus, we have gathered the various previous information in Table 1, allowing us to compare the current state of the French patients' health journey with their expectations of the EHR.

Table 1. Expectations of the French regarding the implementation of the EHR

State of play	What the French expect
• Complicated and laborious healthcare journey • Each professional has his own patient follow-up file	• Better continuity, quality and coordination
• Medical data warehouse to which the patient does not have access or access rights	• Transparency of data to the patient • Respect for his privacy and his choices regarding the content of the EHR
• The world's largest warehouse of certified and computerized medical data	• Quick and concrete implementation of a reliable and efficient EHR in order to obtain better follow-up/management

2.2 Notable Opportunities

The implementation of the EHR represents considerable progress in the French people's healthcare journey. Indeed, the new system adapts to the lifestyle of a majority of the population, which is increasingly mobile. Between 2010 and 2011, 11.9% of 15-59 years olds changed homes, representing nearly 4.3 million mobile phones (FAR & MG, 2015). In addition, the family model is also changing, as more and more stepfamilies are formed: in 2011, 1.5 million children were living in stepfamilies, or 1 in 10 children (Breuil-Genier *et al.*, 2011).

In this context of profound change within our society, the EHR makes it possible to guarantee a follow-up of the patient's medical data anywhere in France or even abroad, thanks to the exchange of standardized medical data.

From birth till death, the patient will have a personal EHR. At each stage of his life, his EHR will offer him a number of advantages.

From birth, the results of tests for genetic diseases carried out on infants may be entered in this EHR. For example, the Guthrie test, performed at birth and used to detect cystic fibrosis, may be impacted in the EHR (Corne and Faure, 2018). A few years later, this will allow a doctor to confirm or invalidate a diagnosis if a patient is in doubt.

In addition, many children undergo minor operations during the first years of their lives: adenoids, appendicitis, etc. These surgical procedures included in the EHR will thus make it possible to avoid possible unnecessary, costly and even dangerous operations (Foucaud *et al.*, 1998).

When the patient is an adult, the EHR will act as a sentinel for healthcare professionals. Major seasonal epidemics such as the flu or gastroenteritis epidemic can be better contained, because the EHR will provide a clear and precise picture of the vaccinations the patient has received. Finally, this system may help to reduce vaccine hesitation, which is still very present in our society (Loubet and Launay, 2017). The EHR will also provide the most accurate experience feedback possible on previous epidemics (location, age, socio-professional category of patients, etc.). This will allow healthcare professionals to apply the right strategy to try to effectively combat these highly contagious diseases.

More generally, eHealth applications will facilitate cooperation between European doctors. The exchange of computer data between healthcare professionals will promote experience feedback and the definition of good practices, thus contributing to the advancement of medicine. In some specific cases, eHealth applications may contribute to the fight against bioterrorism (Hewson *et al.*, 2013).

To go further: based on the model of preventive maintenance in the industrial world, doctors could ultimately perform preventive medical procedures on patients, thanks to eHealth applications. Based on the experience of their British colleagues, French researchers could develop a computer program to detect women at risk of developing preeclampsia, a vascular-placental disease, during their pregnancy (de Moreuil *et al.*, 2018). This first step towards personalized medicine could be generalized to the fight against the two greatest causes of morbidity in France: cardiovascular diseases and cancer (Le Dantec, Chevailler and Renaudineau, 2015). Better still, the use of data obtained during clinical trials could enable health professionals to model therapeutic treatment, thus promoting better patient management (Le Fèvre, Poty and Noël, 2018). Finally, eHealth should contribute to increasing our life expectancy, thanks to the development of both preventive and predictive medicine, which will be increasingly personalized.

3. IMPLEMENTATION OF EHEALTH IN EUROPE

Since the launch of the project to create a European EHR in 2004, progress in this field has differed widely from one country to another. Within the EU (28+2), four groups have recently been identified based on their degree of maturity in establishing this project (Currie and Seddon, 2014). Leaders in eHealth in Europe are Finland, Denmark, the Netherlands, Sweden and the United Kingdom. France is currently at a neutral stage where the desire to create a national eHealth data network has been firmly imposed by the government (Brouard, 2015). However, the associated services and tools are only at the beginning of their definitive implementation and generalization in the professional medical environment.

At the twilight of the European Horizon 2020 project, the focus is on the results of the research funded during these six years, among others, to benefit eHealth innovation. Thus, group work (workshops) was able to highlight many disparities within the EU concerning the state of progress of eHealth policy, the heterogeneity of innovations in technology but also in legislation. The lack of a European legislative framework is at stake (Hägglund *et al.*, 2016).

3.1 Core Principles

In 2003, the ISHARE I project focused on the possibility of exchanging medical data between European countries and thus enabling healthcare professionals to share and consult the diagnoses and observations of their European colleagues (Ochoa *et al.*, 2003). For the implementation of this project, 13 countries were selected as well as national administrative regions identified as representatives of their country's healthcare system and benefiting from local policies conducive to the development of these exchanges. They were subjected to a questionnaire developed as part of the European Community Health Indicators Research (ECHI) to determine the most appropriate national level for exchanging medical data that is valid and usable by other regions.

The final result shows that the exchange of medical data between regions of the same country is possible, allowing the implementation of a compatible database. This project has made it possible to set the foundations for the launch of the European shared medical file by showing that national exchanges of medical data can be systematised and thus make it possible to envisage exchanges abroad. However, particular attention should be paid to the political and administrative context of the participating countries. At the time of the study, Greece had to be excluded because it had just set up a new system for processing medical data within its administration in 2001. Similarly, Finland could not be included in the project due to the specific administrative characteristics of each of its regions (Ochoa *et al.*, 2003).

Thus, the development of a shared European medical record requires first and foremost a national medical data processing unit before it can hope to share it across borders.

Since 2008, with the epSOS ("Smart Open Services for European Patients") project and the creation of the OpenNCP exchange platform, the problems of interoperability and security of medical data exchanges have been partially solved in compliance with national and European laws and policies (Staffa et al., 2018). The KONFIDO project, funded by the EU, was recently launched in order to successfully strengthen the trust and security of medical data exchanges between the different EU Member States. It uses advanced technologies tested in a realistic scenario by a few voluntary countries, such as Italy, and studies the results on the viability and security of the data exchanged (Nalin et al., 2019). Thus, the EU is moving forward to ensure the establishment by 2020 of reliable and secure exchanges of medical data between its member countries, but the road ahead is still full of obstacles.

3.2 French Progress Compared to European Countries

In 2012, an update on the progress of the development of eHealth and its associated services at European level EU (28+2) was carried out in order to obtain concrete figures. To this end, 1717 heads of acute care hospitals were interviewed and approximately 50 of them completed a targeted questionnaire on the availability and utilization of specific eHealth tools within their institution (Sabes-figuera, 2013).

Each hospital was evaluated individually on 13 criteria divided into 4 themes to evaluate the level of development of eHealth:

- Infrastructure: external and internal network connections.
- Applications: computer systems for recording patient medical data and examinations within the hospital (PACS) or prescriptions integrated into the patient's computer file.
- Integration: exchange of medical data between hospitals and other medical entities such as analytical laboratories or radiology centers.
- Security: structured frameworks and rules about data access and a system to recover data availability in the event of a technical incident.

We focused the 7 criteria that were most relevant to our benchmark (cf. Table 2). Thus, we only considered institutions with a network communicating with the outside world, ignoring those which did not even have an internal connection between their services. Similarly, only applications that allowed medical data to be communicated with entities outside these hospitals were selected, as well as the frequency of exchanges practiced by the institutions concerned and the reliability maturity of their server. In 2010, this survey had already been conducted, which allows us to assess the evolution of each country between 2010 and 2012.

In Table 2, we have therefore compared the evolution of the 7 criteria used for France and the 2 groups: "Leaders" (Finland, Netherlands, United Kingdom and Denmark) and "Laggards" (Cyprus, Romania, Poland, Greece, Malta, Lithuania, Latvian), in the European reference framework EU (28+2) (Currie and Seddon, 2014). Thus, we note that the changes within French hospitals are not remarkable, and are even declining for most criteria. Efforts are visible when it comes to the exchange between hospitals and external medical entities but lag behind in terms of network infrastructure.

Table 2. Indicators of the level of development of eHealth in France compared to that of the 4 leaders and 7 European "laggards" in the benchmark EU (28+2), data collected from (Sabes-figuera, 2013)

Domain	**eHealth indicators**	**7 laggards vs EU(28+2)**	**France vs. EU(28+2)**	**4 leaders vs EU(28+2)**
Infrastructure	Externally connected	-12,70%	-3%	3,25%
Applications	PACS usage	-12,14%	-25%	21,50%
	ePrescribing	-12,70%	-6%	36%
	Integrated system for eReferral	-4%	-8%	41%
Integration	Exchange of clinical care information with external providers	-14,86%	4%	34%
Security	Clear and structured rules on access to clinical data	-8,86%	-4%	11%
	EAS for disaster recovery in less than 24 hours	-8,57%	-5%	19,25%

Two results attract our attention: the important delta between the "laggards" and "leaders" for the criterion "integrated system for eReferral" and the strong French decline concerning the recording of medical data and examinations in computerized version within hospitals. The first delta can be explained by the fact that the countries considered as "laggards" in the field of eHealth development also correspond to the countries considered as being "less advanced" in terms of economic development in the EU reference framework. Thus, the acquisition of developed and expensive systems to integrate computerized medical data may not be a priority compared to investment in better quality healthcare equipment. As for the sharp decline in the recording of medical data in a computer system in France compared to the EU (28+2), this can be explained by the current political context, which is slow to define the limits of the implementation of an EHR in France. Once these limits are clearly set and the standardization of the EHR decided, the support of French healthcare professionals in the implementation of medical databases will probably be stronger.

In addition, France is just below European average, particularly when it comes to digital development in eHealth and data security and network access. When we compare France's progress with that of the 4 European leaders but also that of the 7 laggards, we see that it is far behind the precursors. In the same way, we note that it is closer to laggards than to leaders in this field.

In addition, these same indicators were used to assess the development of eHealth, its availability and use by hospital departments, allowing a comparison at national level but also with the EU (28+2) average. It shows that the French average is almost at the same level as the European EU (28+2) average with a score of 36% compared to 30% for Europe (Sabes-figuera, 2013). This study also reveals a great disparity between different French institutions, as some are at almost stage zero in the development of eHealth and its services while others are at the same level as hospitals in the leading countries. This can be explained by the strong regional disparities in France. Outside the major cities and their respective spheres of influence, small towns are medically disadvantaged with the desertification of healthcare professionals who do not wish to settle there (Alfano *et al.*, 2018).

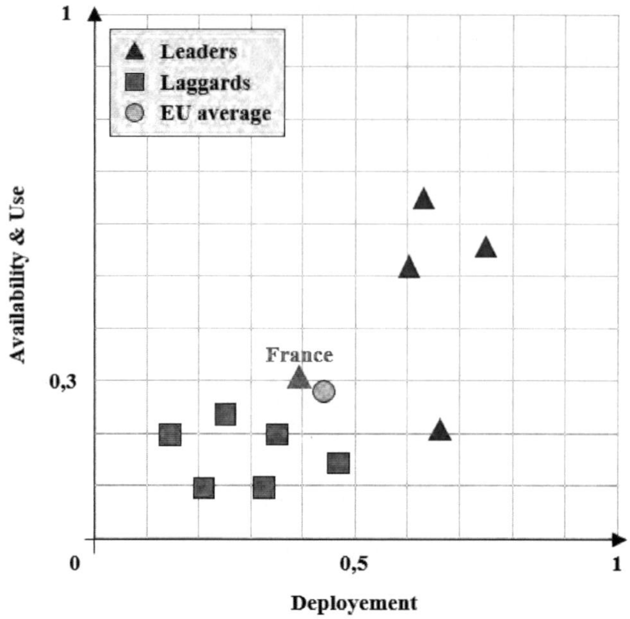

Figure 2. Average development and progress of eHealth in French and European hospitals within the EU (28+2), data collected from (Sabes-figuera, 2013)

In addition, we see in Figure 2 that the hospital averages (consistent with eHealth development criteria, availability and use within hospitals) of the "leaders" and "laggards" respect the ranking of these groups. The number of hospitals that could be evaluated for this comparison is the highest in France, compared to 10% for some of the "leading" countries (Sabes-figuera, 2013). This can be explained by the desire to create a French national unity before developing at European level. However, the "leading" hospitals remain well above the European average despite the few hospitals that could be selected for this graph. As for the "laggard" hospitals, although the number of hospitals selected for the latter is similar to that selected for the "leaders", their level of eHealth development remains low and well below the European average EU (28+2).

Thus, we can conclude that the level of development of eHealth within the EU (28+2) is directly related to the economic situation of the country (quality of life, welfare state, etc.) but also to the willingness of the government to enact a policy of open borders for medical data.

In addition, it is also important to note that many of the expected benefits of e-health in chronic disease prevention and care come from improved primary care, not necessarily hospital care. With the ageing of the population, these non-communicable diseases represent the financial burden of European health care and considerable efforts have been made in this area. However, our present study does not aim to reference medical and treatment advances and their impacts on the European health system (medical and financial) but rather to evaluate the development of eHealth in Europe that enables these new benefits.

4. DISCUSSION

Focusing on France, we have constructed a SWOT matrix of the French EHR (cf. Figure 3) which highlights the complexity of its implementation. We have identified only 3 strengths against 5 weaknesses which prevent the realization of this project in France. On the one hand, the French government's welfare state provides a sound and solid basis for the EHR with a robust and efficient health system. On the other hand, the French protectionist culture hinders the implementation of this new IT. Undeniably, the French people value privacy and consider EHR to be a significant intrusion that is not yet sufficiently standardized nor secured.

	POSITIVE	NEGATIVE
INTERNAL	**Strenghs** • Largest **certified medical database** in the world • Efficient and well meshed medical network in French cities • Good **dialogue with the European Union**	**Weaknesses** • **Late communication** of the government • **Cultural barrier** • **The French doubt** the reliability of the EHR (data security) • **Health professionals do not support** the concept • **No standardization** of medical I.S.
EXTERNAL	**Opportunities** • Better **follow-up/patient care** • Pooling of **research on diseases** • Mitigation of **medical desertification** • **Adaptability** of the health system to the contemporary lifestyle • **Exchanges** of best practices • Extended **Epidemiological monitoring** • Implementation of **connected objects** (consultations, prescriptions, patient history) • **Travel** peace of mind in Europe	**Threats** • **Cyber threat** • "Millefeuilles" of application with risk of loss of data standardization • **Transhumanism** • **Mercenarism** among health professionals • **Reduction of medical manpower**/loss of skills with exclusive teleconsultation • European **interference** in our health system

Figure 3. SWOT Matrix of the French EHR

However, the opportunities provided by the implementation of the EHR are attractive and address many of the problems faced by French citizens (cf. Table 1). Among other things, medical desertification, the ageing population and the high mobility of young workers are finding solutions with EHR-related applications. The arrival on the market of teleconsultations or even connected objects that implement vital data from the French meet the needs of the French but also those of healthcare professionals with easy access to data and better follow-up/management of patients.

In the same way, the data implemented in the databases will be accessible to all European healthcare professionals and will thus be able to prevent pandemics with epidemiological monitoring thanks to the contagion case directory. This will strengthen the links between practitioners and researchers from different countries and build trust in each other's skills. Undeniable progress can be envisaged with the exchange of research results on rare/orphan diseases and innovative treatments recently discovered.

However, as in any new project implemented, some drift is to be expected. Some of them have already been identified, such as cyber threats with the hacking and falsification of stored data, the mercenarism of health professionals with the pursuit of profit at the expense of the patient's wellbeing or telemedicine which destroys human relations between doctors and patients. The latter is to be seriously considered as a risk of gradually seeing a reduction in the medical workforce and patients treated based on the potential gain of the practitioner and not on the proven effectiveness of the treatment.

Before any European implementation, it is imperative to capitalize on French know-how in standardized, robust and certified databases in order to be credible to decision makers. This is accompanied by a strong commitment on behalf of political leaders (cf. Figure 4). As such, the NSDS is a proven prerequisite that can serve as a solid starting point for a rapid and relevant dissemination of the EHR in France.

In order to summarize the study presented in this article, Figure 4 proposes a synthesis of the different fields of application of EHR: it is at the heart of a system of exchange and consultation of computerized medical data of which the patient remains the owner.

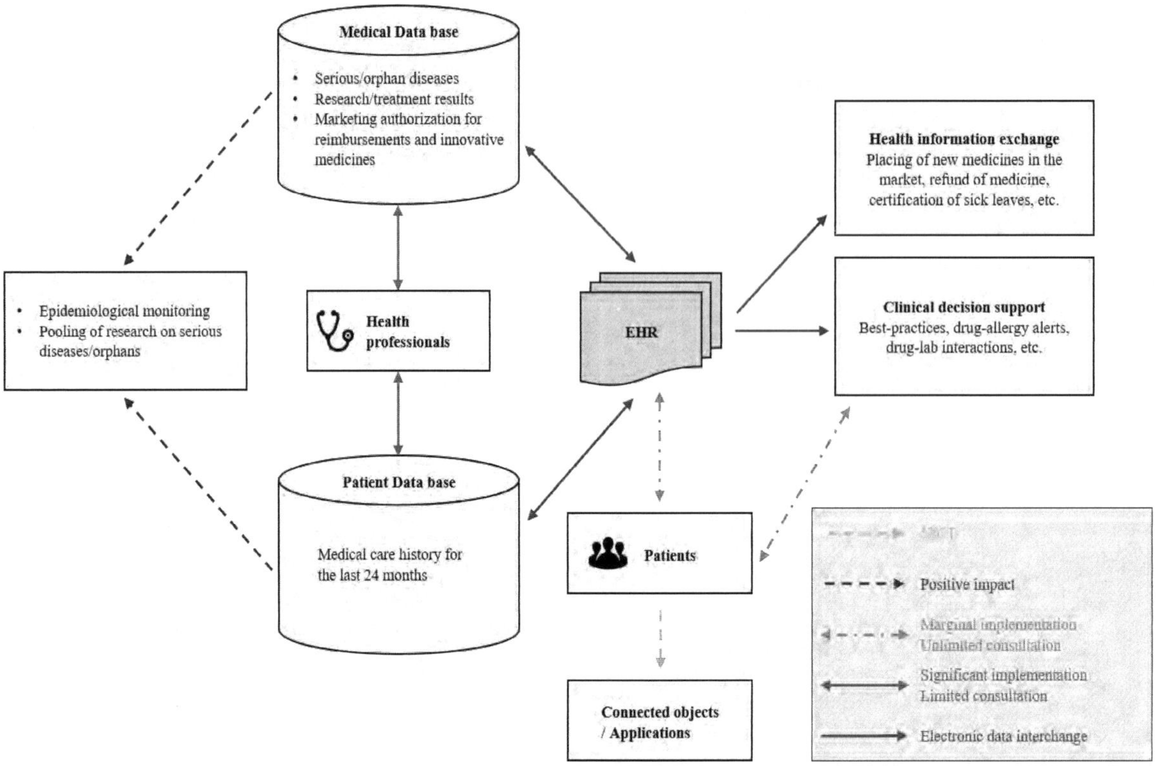

Figure 4. Summary of the implementation of the EHR, the prerequisites for its implementation and its applications

Two types of databases feed the EHR: one concerns only the patient with his medical history; the other contains more general information such as decisions to market and/or reimburse innovative medicines/treatments, the results of research on serious/orphan diseases or the reporting of new cases of contagion (Ebola, Plague, Cholera, Flu, etc.), which allows for better epidemiological monitoring. As a result, healthcare professionals can consult these databases and have a decision-making tool at their disposal to better care for their patients. Healthcare professionals feed data into both databases, which are then consolidated in the EHR. The links between healthcare professionals and the databases are constrained to limited consultation (but they can perform queries about any patient) in order to comply with privacy regulations. As a matter of fact, patients decide which healthcare professionals can consult their full data. On the other hand, patients have unlimited consultation access to their own information, but cannot consult data from other patients.

Finally, the Internet of Things completes the field of possibilities in the eHealth domain by collecting vital patient data in real time (connected watches, mobile applications, etc.), or telemedicine (online consultations, ePrescriptions, etc.), which should make it possible to solve the problem of medical deserts and respond to the new lifestyles of the workers (increased mobility, daily commutes, etc.).

It is essential to highlight here that an implementation like the one seen in Figure 4 can only take place if 4 main prerequisites are met:

1. Standardization of national data.
2. Certified and consistent databases.
3. Stable economic environment.
4. Strong willingness of governments to implement EHR.

As was demonstrated in this article, the EHR project is no longer a simple utopia (Cacot, 2016). However, it will be necessary to ensure that the implementation of the EHR does not drag on too long and get lost in a heap of complex variations, otherwise it risks becoming a chimera.

5. CONCLUSION

This article proposed an analysis of the fields of application of eHealth in France and Europe, in particular through the implementation of the EHR. The SWOT matrix highlighted a number of opportunities, such as the mitigation of medical deserts, better epidemiological monitoring and the adaptability of the healthcare system to the new contemporary way of life.

France has considerable assets enabling it to implement the EHR under good conditions: a reliable and relatively exhaustive medical database, a high-performance healthcare system envied throughout the world, good engineering skills, innovative start-ups, etc. However, France still seems to lag behind its European partners in terms of eHealth applications and use.

By placing greater trust in its engineers and healthcare professionals, France will be able to deploy its EHR, and more generally eHealth applications, in a consistent and efficient manner.

With the opening of borders and cooperation between European countries, an eHealth exploratory laboratory could be created. This would promote research and development (R&D) of new common and interoperable eHealth applications for all European patients.

In addition, the establishment of a European EHR will also require stable economic contexts in all participating countries and a standardization of the data implemented in this dossier.

The benefits of this new medical tool are numerous and revolutionary. It will be able to provide epidemiological monitoring over a wide area by identifying cases of contagion, pool the results of innovative research on serious and orphan diseases, and help doctors in complex cases to decide on the most appropriate treatment. From the patient's point of view, the first expectations are total transparency with access to all their medical data but also the primary right to possess their data by having the power to decide whether or not to disclose them. In the same vein, the EHR will make it possible to instantly and securely disseminate information on the marketing of innovative medicines but also on their reimbursement by public bodies. In this way, patients will be able to benefit from the latest medical advances for their treatments without being constrained by the restriction of their right to know.

In the age of Industry 4.0, new telemedicine-related applications such as e-consultations via smartphones, online medical prescriptions and connected objects allowing real-time patient tracking are beginning to appear on the market. It is a revolution in a context of high geographical mobility of populations. With these applications, the follow-up of patients at home is also facilitated for healthcare professionals and makes it possible to push back the limits of medical desertification and the ageing of populations constrained in their movements. Thus, patient care would be facilitated by implementing medical data from the EHR to these connected objects and applications.

However, these applications, designed to save time for both healthcare professionals and patients, should not become a simple "cash machine" whose yield would be the only concern of physicians using such tools. Used to the extreme, such applications would involve risks of transhumanist drifts (tracking by chips implanted under the skin, vital organs duplicated by 3D printing, etc.) and could eventually lead to a commodification of the human body. Thus, the patient would become a simple object to be maintained in an optimal state of functioning, with preventive operations/treatments, without any ethical consideration. He would be tracked throughout his life. In such a system, one could then ask oneself: what are a patient's rights to privacy and possible treatment choices? This would raise the question of access to appropriate care based on reliable and well-considered diagnoses.

REFERENCES

Alfano, N. *et al.* (2018) 'Accès aux soins en contexte de desertification medicale : un rôle tampon joué par la « chaîne de secours » ? Le cas des sapeurs-pompiers français Mots-clés Key-words'.

Allaert, F. A. and Quantin, C. (2009) 'Le dossier médical personnel du patient : réflexions sur le portail d'accès unique et le masquage du dossier', *IRBM*, 30(3), pp. 114–118. doi: 10.1016/j.irbm.2008.12.001.

ASIPS (2013) *IHE Implementation Case Study : French Electronic Health Record Program*.

Assurance Maladie française (2018) 'Le DMP, un succès qui ne se dément pas'.

Bezin, J. *et al.* (2017) 'The national healthcare system claims databases in France, SNIIRAM and EGB: Powerful tools for pharmacoepidemiology', *Pharmacoepidemiology and Drug Safety*, 26(8), pp. 954–962. doi: 10.1002/pds.4233.

Breuil-Genier, P. *et al.* (2011) 'Enquête famille et logements adossée au recensement de 2011'. INSEE.

Brouard, B. (2015) 'Les nouvelles technologies de la communication au service de la santé', *Actualités Pharmaceutiques*. Elsevier Masson SAS, 54(544), pp. 18–22. doi: 10.1016/j.actpha.2014.12.012.

Burnel, P. (2018) 'The introduction of electronic medical records in France: More progress during the second attempt', *Health Policy*. Elsevier Ireland Ltd, 122(9), pp. 937–940. doi: 10.1016/j.healthpol.2018.07.022.

Cacot, P. (2016) 'Les enseignements d'une histoire : l'utopie du DMP', *Les tribunes de la santé*.

Corne, C. and Faure, P. (2018) 'Le dépistage néonatal en France : approche biologique'. Revue Francophone des Laboratoires.

Currie, W. L. and Seddon, J. J. M. (2014) 'A cross-national analysis of eHealth in the European Union: Some policy and research directions', *Information and Management*. Elsevier B.V., 51(6), pp. 783–797. doi: 10.1016/j.im.2014.04.004.

Le Dantec, C., Chevailler, A. and Renaudineau, Y. (2015) 'Épigénétique Et Médecine Prédictive', *Revue Francophone des Laboratoires*. Elsevier, 2015(473), pp. 71–75. doi: 10.1016/S1773-035X(15)30162-3.

Dormont, B. (2013) 'Le paiement à la performance : contraire à l'éthique ou au service de la santé publique ?', *Les Tribunes de la santé*, 40(3), p. 53. doi: 10.3917/seve.040.0053.

FAR & MG (2015) 'Livre blanc de la mobilité géographique des salariés'.

Le Fèvre, C., Poty, L. and Noël, G. (2018) 'Les big data, généralités et intégration en radiothérapie', *Cancer/Radiotherapie*. Elsevier Masson SAS, 22(1), pp. 73–84. doi: 10.1016/j.canrad.2017.04.013.

Finet, P. *et al.* (2018) 'Interoperable Infrastructure and Implementation of a Health Data Model for Remote Monitoring of Chronic Diseases with Comorbidities To cite this version : HAL Id : hal-01815338 Interoperable Infrastructure and Implementation of a Health Data Model for Rem'.

Foucaud, P. *et al.* (1998) 'Reflux gastro-oesophagien du nourrisson: une médicalisation excessive?'

Haas, J. *et al.* (2017) 'PCN320 - Relevance of real world data in german AMNOG submissions in oncology', 20, p. 2017. doi: 10.1016/j.jval.2017.08.403.

Hägglund, M. *et al.* (2016) 'Report on the HEC 2016 Workshop : Opportunities and Challenges of Consumer - centric eHealth Services – An Interdisciplinary Workshop', (December). doi: 10.13140/RG.2.2.16767.82081.

Hewson, R. *et al.* (2013) 'Biosafety Level-4 Laboratories in Europe : Opportunities for Public Health , Diagnostics , and Research', 9(1), pp. 1–4. doi: 10.1371/journal.ppat.1003105.

Loubet, P. and Launay, O. (2017) 'Adult immunisation: General points, hot topics and perspectives', *Revue de Medecine Interne*. Société Nationale Française de Médecine Interne (SNFMI), 38(11), pp. 749–759. doi: 10.1016/j.revmed.2017.01.007.

de Moreuil, C. *et al.* (2018) 'Pre-eclampsia prevention in 2018 in general population and in lupic women: At the dawn of a personalized medicine?', *Revue de Medecine Interne*. Société Nationale Française de Médecine Interne (SNFMI), 39(12), pp. 935–941. doi: 10.1016/j.revmed.2018.06.001.

Nalin, M. *et al.* (2019) 'The European cross-border health data exchange roadmap: Case study in the Italian setting', *Journal of Biomedical Informatics*. Academic Press, 94, p. 103183. doi: 10.1016/J.JBI.2019.103183.

Ochoa, A. *et al.* (2003) 'Health indicators in the European Regions.', *Eur J Public Health*, 13(3 Suppl), pp. 118–119. doi: 10.1093/eurpub/13.suppl.

Sabes-figuera, R. (2013) *European Hospital Survey: Benchmarking Deployment of e-Health Services*. doi: 10.2791/55973.

Séroussi, B. and Bouaud, J. (2018) 'The (Re)-Relaunching of the DMP, the French Shared Medical Record: New Features to Improve Uptake and Use..."Building Continents of Knowledge in Oceans of Data: The Future of Co-Created eHealth," EFMI, Medical Informatics Europe (MIE), April 24-26th, 2018', *Studies in Health Technology & Informatics*, 247, pp. 256–260. doi: 10.3233/978-1-61499-852-5-256.

Staffa, M. *et al.* (2018) 'An OpenNCP-based Solution for Secure eHealth Data Exchange', *Journal of Network and Computer Applications*. Academic Press, 116, pp. 65–85. doi: 10.1016/J.JNCA.2018.05.012.

A GENERAL MODEL FOR ASSESSMENT OF CHANGE IN PROFESSIONAL PRACTICES WHILE SHIFTING TO TELE-MEDICINE

Nathalie Bricon-Souf[1,2] and Hervé Pingaud[2]
[1]*IRIT, Université de Toulouse, CNRS, INPT, UPS, UT1, UT2J, France, 118 Route de Narbonne, Toulouse, France*
[2] *CNRS-LGC, INU Champollion Université de Toulouse, Castres, France*

ABSTRACT

Implementing telemedicine in a broad scale requires good practices recommendations. In this paper we are describing an assessment model proposed to assess changes in professional practices while shifting to telemedicine. Two dimensions are explored: the care process and the feelings experienced by the involved actors. We proposed a method based on knowledge framework and providing a qualification of the differences about practices and their impacts. Care process is represented through process modeling and helps analyzing practices *before* and *after* the introduction of telemedicine. A two dimensions matrix (effects/participants) identifies the impacts of such an operation. This work is performed in the context of a secondary care pilot study for elderly people that implements teleexpertise or teleconsultation. Process analyses highlight important variations according to the institutions and their actual organization of care. The introduction of telemedicine modified the involvement of the care team, lightened the travels for patient and needed reentering of data for health professionals. A matrix evaluation was built in order to ask professionals their experiences. The professional assertions analysis lead to propose 12 criteria highlighting the nature and the challenges emerging with this innovation.

KEYWORDS

Telemedicine, Professional Practice, Evaluation

1. INTRODUCTION

Telemedicine (TM) practice increases nowadays due to several important reasons. The technology is mature enough to spread TM in a wide dimension. The legal context has evolved and most of the barriers seem to be removed now. Among many motivations, in particular in some country areas, the lack of caregivers induces remote medical practices that triggers the need of such tools in order to provide effective solutions. Primary-secondary care interfaces are important (Winpenny et al. 2017). Moreover the organization of care in conformity with public savings is becoming sensitive in terms of refunding. Heavy transportation costs should for example be reduced by the use of TM. Before implementing TM in every day medical practice in a major scale it is crucial to detect the positive and negative aspects of such relatively new usage and to give some good practices information/recommendation within existing organizations.

We worked on a public project driven by the Regional Agency of Care in south of France (ARS Occitanie) that aims at developing a pilot study to understand better how the use of TM for secondary care could improve the access to care considering elderly's follow-up and could be considered as a smart innovation for all stakeholders. Picking up on the last point, a special attention was paid to the impacts of telemedicine on the activities and practices of the healthcare practitioners. In a nutshell, our contribution tends to capture real changes and to assess accurately the impacts of TM on the care service. This *a posteriori* evaluation has been performed with a TM field study at the scale of one department (Tarn, 81). According to the latter results TM could be progressing towards new territories in the best conditions.

If TM has often been studied in order to evaluate the benefit for the patient for some precise pathologies, very few papers are dealing with the evaluation of the impact of such tools in the organization of care and in particular referring to healthcare professional practices (PP) (Esterle & Mathieu-Fritz 2013)

We proposed a new method in order to make such an evaluation. At first, we built a specific knowledge framework to objectively catch all features of change in such PP. Secondly, we qualified the differences within the practices and their impacts before and after the use of TM through direct observations during TM acts and thanks to interviews of practitioners. Hence a data analysis led to choose evaluation criteria reflecting the many facets of changes. The interpretation of their values drew conclusions on whether or not progress has been achieved and to what extent. Each step of this approach implies developing tools or artefacts for the acquisition or interpretation of evaluation criteria. Actually, the aim of this paper is to decribe this new approach based on our knowledge framework to assess PP changes induced by healthcare innovation.

The next part of this paper describes the objectives of the performed work, third part describes the materials and methods used, it deals with the medical and technological context of the study, the rationale used to describe and evaluate PP with a specific focus on the use of TM, and the proposed models, fourth part comments the obtained results and a discussion-conclusion ends the paper.

2. OBJECTIVES

The introduction of TM will change the preexisting organization of care (learning and training; new practices acquisition; less partition between specialties; optimization of working time; respect of quality procedures; reduction of transports; planned activity for all the professionals and so on) (HAS report, 2017). The progress expected by this innovation has to be assessed on the field as an integral part of public health policies. However the assessment of the impact of TM can be a bit tricky since on one hand it links different aspects of health services delivery and on the other hand it raises clinical and organizational issues that are often overlapping.

In a regional context of TM innovation in the south of France, the Regional Health Care Agency (ARS Occitanie) has a global strategy to reinforce health care trajectories and to propose an optimization of such trajectories for all the pathologies and populations. One of the main objectives for such a program concerns elderly people and in particular the development of the caring for elderly dependent people living in nursing home. These patients are suffering from geriatrics and psychiatrics pathologies. According to the ARS, introducing new teleconsultation or teleexpertise modes of TM should limit the fragmentation of care trajectory, help the care coordination, limit the professional isolation of the physicians or coordinators and limit trips for elderly people. In this context, a project named PÂSTEL (for Aged People Services TeLehealth) aims at developing the use of TM, complementary to conventional care, in a coordinated care trajectory. A better understanding and an assessment of the impacts of the introduction of telemedicine in professional practices was then needed and therefore grounded the work presented in this paper.

It is important to spot on the aim of our study: it focused only on the *professional* practices which are linked to how the professionals manage the information and the communication in order to build an efficient care trajectory. It was absolutely not an evaluation of the *clinical* practices which are linked to the know-how of the professionals implementing a care strategy. We made the assumption that the purely medical part has the same content with or without TM, which is quite allowable in the case of secondary care. We have been building and using a framework to capture 2 dimensions. From our point of view the two main dimensions that needed to be considered were:

- **the care process** which can highlight factual change *in the organization*,
- **the experience about the execution of the care process** which can be done by capturing *actor's feelings* about aspects of such a new organization (such as positive or negative, easy or difficult).

3. MATERIAL AND METHODS

3.1 Medical and Technological Context

A first experimentation phase of the PÂSTEL project has been performed on the department of Tarn (a French and rural department in Occitanie). It involves 4 requested institutions offering 8 types of specialized advices and 16 requesting units of which 15 are from primary care. In this pilot, 2 TM acts were concerned: teleconsultation and teleexpertise and they are both synchronous. In each case, getting advices on some patient health status was often needed, hence primary care staff (the requesting physician is the general practitioner) and secondary care staff (the requested specialist is a geriatric or a psychiatric specialist) were brought to cooperate. Two distant sites are connected: the *requesting site* where the patient lives and the *requested site* where the medical secondary care team works. Only health care professionals are having a meeting using TM technologies when teleexpertise whereas during teleconsultation the patient is attending the meeting.

TM is based on a digital technology with a set of functionalities provided by a software provider that will support the TM process execution. Generally, the system performs a connection between remote actors using the available communication network. It operates the display of sounds, images, data, files and medical content that has meaning for any medical act practice. Some information has to be shared for interactivity purpose and stored for traceability. We named "Infra" the level that alludes to all the previous aspects. This "Infra" level cannot be tuned since the TM tool is already installed and is linked to IT performances. However, even if it is an important factor, its evaluation is not in the scope of our study. In the frame of our project, a preliminary work package has selected and validated the adequacy of the applications for the targeted operations. A software (named synapse) is installed to take in charge the communication in synchronous (during the acts of TM) or asynchronous (to prepare or conclude the visit) way. The process manages a workflow which allows a functionality to be used (as soon as the preconditions are fulfilled during process execution), prepares information and publishes it when needed by each health care professional.

The "Infra" level has to be complemented by a "Supra" level. Considering this level, the objectives are to represent the rationale of the different caring steps (for the patient, for the health care professionals or for the administrative professionals) with an explicit vision about what coordination is (when should I work and with who?). The relevance of coordination is facilitated by the workflow implemented with the TM software.

3.2 Rationale

According to the MeSH (Medical Thesaurus provided by the National Library of Medicine) Professional Practice is defined as "*The use of one's knowledge in a particular profession. It includes, in the case of the field of biomedicine, professional activities related to health care and the actual performance of the duties related to the provision of health care.*" PP are concerned with actual activities performed in a given applicative context and by different actors. Knowledge about these performed actions as well as information about the patient management according to the care context are consequently important. Moreover knowledge about PP is often implicit and therefore difficult to capture. The four following aspects illustrate this issue as PP representation should take into account:

1. what is made which is not expressed (because it is not compliant to the prescribed task)
2. what professionals would like to do which is not done (the professional ideal)
3. what professionals do but don't like to
4. what professionals are performing without being conscious of it (subjective unconsciousness or forgetting due to the habit).

Different ways of analyzing PP are then proposed (Boutin 2002) as the supervision of case studies or as the analysis of simulation of actual situations or routine. Cases studies have been used for our work.

The area of teaching habits is one of the well investigated domain. A 7 dimensions model named ECORSET (*E*fficacy, *C*omfort, *O*rganization, *R*esponsibility, *S*ecurity, *E*conomy and *T*ransmissions) is commonly used for nursing education as it sums up main features about care professional individual practices that should be currently improved by reflexive attitudes. It refers to quality of care and satisfaction of patient (E and C); communication as part of the team coordination (O and T); regulation and ethics (R); risk evaluation (S) and efficiency for the nurse aid decision process (E). Nevertheless, the medical area has been slightly involved until recently in professional practices studies (Lagadec 2009)

Clinical guidelines for good PP are pillars of the organization of care deliveries. The French HAS (National Authority for Health) for example provides lots of best-practice guidelines to manage optimal care processes. Unfortunately, TM is not yet concerned by such guidelines as being relatively new in care service deliveries. We found some literature reviews on this subject as the one proposed by (Flodgren et al. 2015) onto "*Interactive telemedicine: effects on professional practice and health care outcomes*". But they mainly concern the forthcoming outcome of TM and they are therefore more dedicated to an *a priori* assessment of the expected effectiveness than to the evaluation of the actual practice. Ethics and deontology are moreover fixing rules for actual practices and related information is widely disseminated among many sources. PP are more and more subject to consideration within the growing interest for health innovation, the last evolutions of public health system and the new constraints law, on legal practices about caregiving. It is then still prime challenging to propose conceptualization of healthcare PP covering a large range of situations. In order to help care organizations to develop TM, the ANAP French agency (National Agency for the development of the Performance of medical and socio medical facilities) published a guide providing a general approach of the telemedicine domain (ANAP report, 2016) which defined a generic representation of an operating process for a TM medical act as shown in Figure 1.

Original version	*Text Translation*
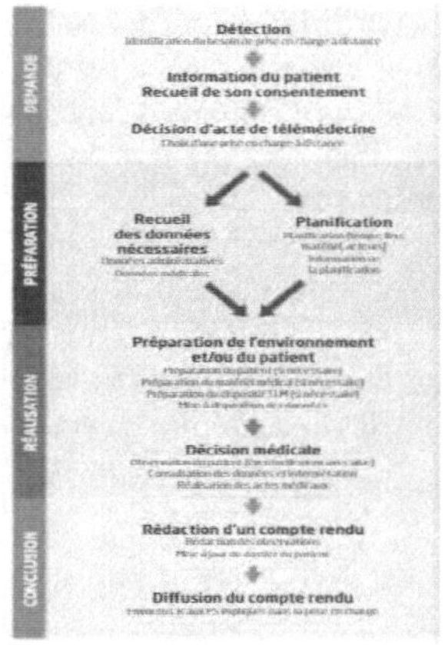	**Demand :** Detection Patient information and consent Decision of TM act **Preparation** Data entry <> planification **Realization** Setting and or patient preparation Medical decision **Conclusion** Report redaction Report diffusion

Figure 1. The generic process of TLM act: the original figure is shown on the left and texts are translated on the right

In the very specific range of medical imaging area, the HAS (HAS Report, 2013) defined criteria which should be observed to assess the effects of TM innovation and organized them within a framework defined by a two dimensions matrix. The first dimension splits TM impacts in four categories: (A) Accessibility to care; (B) PPs/organization of care; (C) Quality of care/ security of care; (D) Costs. The second dimension informs on who are the people concerned: (1) Patient / Aids and family; (2) Physicians and allied health professionals; (3) Health facilities; (4) State /Medical insurance/ Collectivities. This precise description helps to focus on the different aspects which are concerned by TM innovation and therefore helps to perform a better estimation of the associated impacts. For example, regarding the (B, 2) cell of the matrix which focused on (PPs/organization of care impacts from the Physician and allied health professionals point of view) some important aspects can be observed: timings used to launch the system (protocols, teaching, meeting and so on), coordination between health care professionals (protocol, quality and security of the transmitted data), time required by TM for the professionals, instructions about the system use, cooperation between health care professionals (delegation, transfer, competence sharing) and optimization of the medical flow management.

Our objective was then to reuse the principle of such a matrix for an evaluation purpose in the area of telemedicine PP changes. To keep in mind the dynamics of change, the reference to a TM care process evolution will be used as a link between past and future. In the particular context of our study we had to derive some precise representations to represent better the two acts of TM complementing the general representation of TM proposed by the ANAP. We used these results to build a new framework facing with this complexity and perform our PP assessment, it could ease any studies on the use of TM tools.

3.3 Proposed Models

We focused on the two dimensions presented above for the evaluation that is to say the care process and the experience about the execution of the care process. A first phase of work was planned, it led to a model of representation for the care process. The second phase of work concerned the experience about the execution of the care process and permitted an evaluation of the impacts on professional practices through the use of an evaluation matrix.

3.3.1 The Framework Model for the Representation of the Care Process

In modeling processes it is very important to understand how, when, and by whom, actions are performed and which resources are used in order to reach a global goal. We are dealing with digital transformation of healthcare and it is important to notice that processes and existing software applications are intricated. The digital transformation of the practices is our matter of concern in the case of TM.

It is therefore of prime interest to provide a representation of this transformation onto the care process before ("as is") and after ("to be") the introduction of such a tool. The way TM will be performed is varying only at the *supra* level as the software workflow has been defined and proposed previously at the *infra* level. The supra level is therefore the right one for the evaluation of PPs. Our objective was to model the supra level of our study while detailing the different activities linked to TM. Such a model is useful to point out what is changing in practice between "as is" and "to be" situations. We paid attention to be compliant with the representation of the TM process proposed by the ANAP as it provides a general framework for definition and assessment of such processes. The "demand"; "preparation"; "realization"; "conclusion" steps of the ANAP representation can be refined further and instantiated to any specific medical acts. We performed indeed such refinements in the case of our study.

Many languages for process representation are known and used effectively in the management science. However the culture of process modelling in healthcare system is still in its infancy. We have chosen to use the well-known Business Process Model (BPM) for care process representation. BPM is fulfilling the requirements of many quality standarts and is convenient in our context. We adopted a three level model architecture that allowed a top down representation of an organization with an increasing amount details:

- The upper domain level presents the model area
- The second level is a process mapping level that describes macro processes involved in the domain
- The third level is the activity level. It details how the different activities are tackled and sequenced, while reporting on objects which are transformed by activities (mainly information on patient and the patient him/herself) as well as on resources which are needed and on people who are acting.

Care is modelled as information, domain of the process; patient state; requirement; requested decision; activity or process; role of an actor; risk and event. Actors are modelled as care professionals, requesting doctor, requested specialist; authorized person; assistant. Four levels of granularity are used for the description of activity. For example level0 contains "carry out a TM act"; level1: "practice the act"; level2: "take the decision"; level3: "deliver a prescription". The nature of the act (teleconsultation or teleexpertise) is mentioned and its implication with the infra level is indicated. Exception could also be represented. We were then able to model the different care processes for the different studied institutions, both before and after the introduction of TM.

Using graphical representations has been nowadays recognized as an efficient mean for sharing knowledge about activities in an organization inside a pool of actors non-specialists of BPM. A graphical representation of the processes was accordingly implemented using the Aris™ software as illustrated in Figure 2.

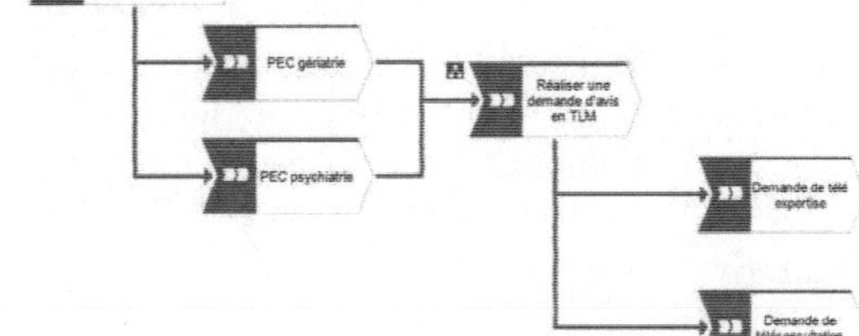

Figure 2. Example of graphical representations of care process

3.3.2 The Evaluation Matrix for the Impacts of Telemedicine according the Experience about the Execution of the Care Process

Three dimensions are proposed for the structuration of the evaluation of the impact of TM on the PP: organization; performance and steering process (see Figure 3a).

The first two dimensions are inspired by the work proposed by the HAS (cf above). (i) The *organization* dimension allows understanding how TM will be practiced. Appropriation of the processes by the different care participants is a main issue for this dimension. It is linked to the representation of the process, explaining the nature and the chaining of different actions. (ii)The *performance* dimension helps to qualify the value of the TM act with two subdimensions reflecting on quality of care and risk control. These two dimensions should at least capture whether or not there is a regression between the practices with or without TM and to determine which developing new services opportunities could be generated while introducing TM.

The *steering process* dimension focuses on the coordinated nature of the care trajectories using TM. Since all the implied people including the patient him/herself have to contribute to build the decision of care, the competency of each actor is insufficient to propose a global efficiency of a care system. Information sharing is needed to help coordination, this is then the fourth point on which we decided to focus.

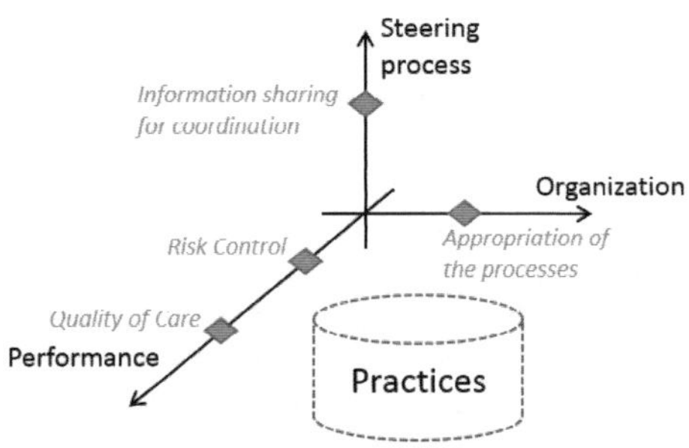

Dimension	Institution viewpoint	Remarks and suggestions onto the impacts
Process		
Quality		
Risk		
Information Sharing		

a) Criteria used : four points and three axes

b) Matrix of evaluation built on the four points: each one is collecting either viewpoints and remarks.

Figure 3. a: Organization of the criteria used for PP measure of impact: the 3 practices representation axes and the 4 studied aspects (in red/italic); b: The associated Matrix

These four criteria were therefore proposed to a panel of care professionals through an evaluation matrix: appropriation of the processes; risks; quality; Information sharing for coordination. Two kinds of advices were requested on each of these four criteria: the first one concerned the specificities which can be highlighted by the professionals regarding this criterion; the second one requested remarks or suggestions on the associated impacts (see Figure 3b).

3.4 Methods

Observations and interviews were used to model the different processes. An observation of the experimental field was conducted on two sites during six months by the authors. A proposition of model was done then and refined during different working brainstorming sessions. Interviews were reported on the management of care with or without TM. Seven sessions of TM were observed concerning both processes of teleconsultation and teleexpertise from two different requesting sites. The observation did not concern the clinical area but was focused on how to build the care process model, reporting for example on from where the information is obtained or on which action is performed by whom. All these materials were then used to model the different processes in a unified view. The resulting model contains four macro processes which are the *consultation demand*, the *preparation of the visit* to the specialist, the *carry out of this visit* and the *conclusion*. Each macro process was then detailed according to if it was performed in a conventional way or it was using TM.

All the institutions involved in the project were invited to share their experience and feelings about the execution of the care process. A collective session was organized gathering participants from eleven institutions involved in the PÂSTEL project: two of them were *requested units* and the others were *requesting sites*. Each institution was represented by one or several participants who were asked to write their comments on the prepared matrix of evaluation. One of the first open questions was targeting the specificities felt for the appropriation of the processes. All this handwritten information was gathered and retyped on a computer. More than 230 assertions were so collected and organized according to the proposed evaluation matrix dimensions.

4. RESULTS

4.1 Processes Analyses Using the Framework Model

In consequence of using the graphical process representations, it was easy to ask practitioners if they validated the representation and to ask them about differences they were feeling between “as is” and “to be” situations. They could comment the care process using this model which was also useful to investigate the other focus points: *quality*, *risk* and *information sharing*.

The processes analysis helped us to comfort our understanding of the impact of TM innovation:

- The observed processes of care were compliant with a derivation of the global process proposed by the ANAP.
- At a high level of details some important variations are found according to the institutions and their actual organization of care: the role distribution could be different; the practice of the TM act could be performed in a more or less multidisciplinary dimension of the caring team, the synergy between the general practitioner and the coordinator could differ in intensity.
- Other variations are linked to the change of practice while introducing TM. When a patient stops travelling, it eases the preparation phase for the patient and eludes some stress previously due to the travel. Creation and transfer of medical record for TM are meanwhile inducing some important data (re)entering for the professionals. Yet the preparation phase for TM advice helps to better involve the whole team during the process of requesting advices.

4.2 The Experience of Care through the Evaluation Matrix of Telemedicine Impacts

All the assertions were analyzed and 3 topics have been derived from the annotations concerning each of the 4 dimensions. This led to a set of 12 criteria to overall assess the changes in practices. The evaluation matrix has helped to focus on topics which are important when analyzing the PP. While investigating answers collected from the matrix focusing on organizational dimension for example, three types of opinions could be synthesized: kind of interaction, induction of workload and factors for practice anchoring. Moreover we obtained information about their impacts:

- Kind of interaction between requested/requesting participants: telexpertise organization differs from teleconsultation one.
- Induction of workload: the evaluation of the impact is highly concerned by the difficulties encountered by care professionals in terms of workload.
- Factors for practice anchoring: some specificities could come from the experimental aspect of our pilot study.

We have been completing this work by an assertion analysis. The previous descriptions were enriched by details on the positive or negative evaluation contained in each assertion on each criteria. If the health care professional mentioned an improvement for example then the assertion was marked as a positive one, while if the assertion mentioned a fear, it was marked as a negative one. Cross analyses of each assertions (Topic * Evaluation) have been performed for each matrix dimension, as illustrated in Table 1. By performing such analysis, it was possible to retrieve and explore the opinion of the different participants on the impact of the introduction of TM in their practices.

We can sum up the main ideas provided by this study as follow: TM has a positive impact onto coordinated care trajectory, TM is massively improving the performance of the care in the perimeter of this study, TM is new and there is a high level of risk of a degradation of care unless experience feedbacks and pilot phases have been well conducted and are finished, TM has a good impact on coordination for care trajectory.

Table 1. Qualitative analyses: Example of the "appropriation of the processes" (organizational dimension): the topics for annotations are shown in column1, the counts of negative and positive assertions are shown columns 2 and 3

Assertions Analyses	-	+	**Total**
Interaction Requested/Requesting	1	16	**17** *(26.6%)*
Workload	13	8	**21** *(32.8%)*
Anchoring	10	16	**26** *(40.6%)*
TOTAL	**24** *(37.5%)*	**40** *(62.5%)*	**64**

5. CONCLUSION AND DISCUSSION

Nowadays, it is crucial to understand why and how TM could be introduced to efficiently help PP. In this paper we covered a two stages work conducted to analyze TM in the context of secondary care. At first, we have provided a reliable tool to characterize the care process and thereby comparing and expliciting the differences before and after the TM innovation act. Later on, we have proposed an evaluation matrix built as an organized artifact which give the professional actors the opportunity to express themselves in natural language and without constraints. The structured environment proposed by this framework helped to organize the answers. We are already detecting very interesting changes due to the introduction of TM as the one offered by the proximity of all the member of a care team during the TM session. Aid nurses, nurses, doctors can easily integrate the consultation or the expertise and give their specific points of view. It seems to be one of the potential huge improvement of coordination for care trajectory: the so called pluriprofessional way of practices.

The proposed assessment method gave us a rigorous setting to analyze the different remarks and helped us to detect the main topics which are important for health care professionals. Several perspectives of improvement of our evaluation method should be proposed as for example enlarging the number of studied sites, completing the panel of participants by involving patients or improving the interpretation of the natural language assertions by adding some NLP tools.

ACKNOWLEDGEMENT

The authors want to acknowledge the Agence Régionale de Santé Occitanie for its support during this research in the frame of the Pastel project, as well as all project partners for their open minds and real involment in our experiments.

REFERENCES

ANAP Report, 2016, « la télémédecine en action : construire un projet de télémédecine ».

Boutin, G., 2002. Analyse des pratiques professionnelles : de l'intention au changement. Recherche & Formation, 39(1), pp.27–39. Available at: https://www.persee.fr/doc/refor_0988-1824_2002_num_39_1_1739.

Esterle, L. & Mathieu-Fritz, A., 2013. Teleconsultation in geriatrics: Impact on professional practice. International Journal of Medical Informatics, 82(8), pp.684–695. Available at: http://dx.doi.org/10.1016/j.ijmedinf.2013.04.006.

Flodgren, G. et al., 2015. Interactive telemedicine: effects on professional practice and health care outcomes. Cochrane Database of Systematic Reviews, (9), p.CD002098. Available at: http://www.ncbi.nlm.nih.gov/pubmed/26343551

HAS Report, 2007, Élaboration de critères de qualité pour l'évaluation et l'amélioration des pratiques professionnelles.

HAS report, 2013, « Elaboration de matrices d'impacts des effets attendus de la télémédecine ».

Lagadec, A.M., 2009. L'analyse des pratiques professionnelles comme moyen de développement des compétences: ancrage théorique, processus à l'?uvre et limites de ces dispositifs. Recherche en soins infirmiers, 97(2), p. 4. Available at: http://www.cairn.info/revue-recherche-en-soins-infirmiers-2009-2-page-4.htm

Winpenny, E.M. et al., 2017. Improving the effectiveness and efficiency of outpatient services: A scoping review of interventions at the primary-secondary care interface.Journal of Health Services Research and Policy, 22(1), pp. 53–64.

QUERY PERFORMANCE EVALUATION OVER HEALTH DATA

Sultan Turhan and Ozgun Pinarer
Galatasaray University Ciragan Cad. No:36 34349 Istanbul, Turkey

ABSTRACT

In recent years, there has been a significant increase in the number and variety of application scenarios studied under the e-health. Each application generates an immense data that is growing constantly. In this context, it becomes an important challenge to store and analyze the data efficiently and economically via conventional database management tools. The traditional relational database systems may sometimes not answer the requirements of the increased type, volume, velocity and dynamic structure of the new datasets. Effective healthcare data management and its transformation into information/knowledge are therefore challenging issues. So, organizations especially hospitals and medical centers that deal with immense data, either have to purchase new systems or re-tool what they already have. The new data models so-called NOSQL, its management tool Hadoop Distributed File Systems is replacing RDBMs especially in real-time healthcare data analytics processes. It becomes a real challenge to perform complex reporting in these applications as the size of the data grows exponentially. Along with that, there is customers demand complex analysis and reporting on those data. Compared to the traditional DBs, Hadoop Framework is designed to process a large volume of data. In this study, we examine the query performance of a traditional DBs and Big Data platforms on healthcare data. In this paper, we try to explore whether it is really necessary to invest on big data environment to run queries on the high volume data or this can also be done with the current relational database management systems and their supporting hardware infrastructure. We present our experience and a comprehensive performance evaluation of data management systems in the context of application performance.

KEYWORDS

Healthcare Data Analysis, e-Health, SQL, HiveQL, Hadoop, Relational Databases

1. INTRODUCTION

Aging world population and increasing hospital care costs have lead to allocation of large resources for e-health research studies. In today's world, healthcare data becomes the most valuable asset for the health organizations. To fulfill the customer requirements, researchers focus on analyzing the immense data: Big Data. Big Data has changed the way we manage, analyze and leverage data in any industry. One of the most promising areas where it can be applied to make a change is healthcare. Healthcare analytics have the potential to reduce costs of treatment, predict outbreaks of epidemics, avoid preventable diseases and improve the quality of life in general. Not only in health sector but, according to the new business world's dynamics, any organization may become one of the leading organizations in their sector, if they transform this infinite volume of data to fit their interests and transform it into added value for their management (Loebbecke, 2015).

As the volume of data is becoming increasingly large, it poses a challenge to executives in the analysis area. All the decision makers want to process and analyze this voluminous data in order to reveal the valuable knowledge for the healthcare data management. Capturing, storing, managing, retrieving and processing large scale data in an acceptable time frame (often real time) is one of the crucial challenges in health domain. Besides, there are several different ways of analyzing data; from querying on a traditional relational database system to machine learning algorithms. Of course each of them requires its own infrastructure investment of different sizes. As the complexity of analysis that the executives want to get increase, needs for specific data storage and processing infrastructure increase also (Wixom, 2014).

The digitization of the healthcare industry is happening fast. A major result of this transformation from paper to electronic records is the proliferation of healthcare data. It is really hard to choose the right database and business analytics systems to satisfy the organization's needs and working scenario. Various database management systems are in use today. Majority of the systems relies on the high-end hardware and/or special-purpose architecture to deliver the desired performance. Health organizations are faced to finance higher capital expenditure for the acquisition of infrastructure and higher licensing cost because of the competition in the market, even they don't really need them.

Proper database system choice has a great impact on user friendliness and scalability of data analytics applications (Chen, 2014). The traditional relational database model (RDBM) is the most common and proven approach to storing and querying data in various forms. However, the major disadvantage is the need to pre-design the exact field structures of the data, which is necessary in the normalization process of the database to ensure data consistency. A relational database storing this type of data will contain many empty fields, which will result in inefficient storage and poor query performance On the other hand, querying high level normalized data on real time basis is a highly costly operation. For this reason, database programmers denormalize the data before running the queries but this may result sometimes with data inconsistency. Besides, tackling the large scale data require a distinct approach that sometimes runs counter to traditional models of storage which provides good scalability and desired level of performance with insignificant or little cost (Zhao, 2017).

Limitations of the traditional relational database system gave birth to a new concept called "NOSQL database". There are a substantial number of projects as an alternative to traditional database system. Google's BigTable (Chang, 2008), Amazon's Dynamo (Giuseppe, 2007) Apache's Cassandra (Cassandra, 2014), Hypertable (Judd, 2008), Apache's CouchDB (Anderson, 2009), LinkedIn's Project Voldermort (Sumbaly, 2012), MongoDB (Kristina, 2010) are just to name a few. The Apache Hadoop-based project - HBase (Team, 2016) is one such approach. These applications are mainly served on big data analytics projects.

In this study, we try to explore whether it is really necessary to invest on a NOSQL database environment to run the queries on the high volume data or this can also be done with the traditional relational database management systems and their supporting hardware infrastructure? In order to answer this question, a bunch of queries are run on two well-known databases, one from NOSQL domain; Hadoop and one from the domain of relational databases SQL Server 2014 and their performances are evaluated. For NOSQL domain, the queries are written with HiveQL and for the relational database domain with SQL. The databases are populated with actual depersonalized prescriptions' record from a public hospital, written between 2015 – 2016.

Rest of the paper is organized as follows: Section 2 describes the methodology and the database platforms and working environment compared in this study. Section 3 describes the dataset, queries used for performance evaluation and presents the experiment results. Finally, Section 4 discusses the results and presents the future works.

2. METHODOLOGY

2.1 Experimental Setup

In order to evaluate the performance of the queries, two environments are set up. To simulate the traditional relational database environment, SQL Server 2014 is used. The hardware infrastructure on which SQL Server 2014 is running, consists of a virtual server equipped with a 2.6 GHz Intel Xenon E5-2690 v4 processor, 16.00 GB of RAM. The queries are written with SQL []. For NOSQL database environment, Apache Hadoop-based project and its database HBase are chosen. HBase is a distributed, fault-tolerant, highly scalable, NOSQL database, built on top of Hadoop Distributed File System (HDFS). The hardware infrastructure supporting the system with Hadoop architecture, includes 9 physical servers and a total of 1 TB of RAM. The queries are written with HiveQL (Lydia, 2015).

As all these NOSQL environment's products mentioned here, have recently entered the market and become prevalent, following subsections give brief technical details about them.

2.2 NOSQL DATABASES

2.2.1 Hadoop

Hadoop is an open-source distributed processing framework for large-scale distributed data storage and high-performance computation on a clustered system network of inexpensive pieces of commodity hardware (Khetrapal, 2006, Dorin, 2010, Taylor, 2010). It is at the center of a growing ecosystem of leading data technologies that are primarily used to support advanced analytics initiatives, including predictive analytics, data mining, and machine learning applications. It can handle various forms of structured and unstructured data, giving users more flexibility in collecting, processing, and analyzing data than relational databases and data warehouses. Hadoop framework developed in java programming language rely on two major components: Hadoop Distributed File System (HDFS) and Map/Reduce.

2.2.2 Hadoop Distributed File System (HDFS)

The Hadoop Distributed File System is the primary data storage system used by Hadoop applications. It uses a NameNode and DataNode architecture to implement a distributed file system that provides high-performance data access through highly scalable Hadoop clusters (Narayan, 2012). In HDFS, data is organized into files and directories. Files are divided into uniform sized blocks and distributed across cluster nodes and thereby removing the file size restriction. Also blocks are significantly larger than block sizes in standard file systems to minimize the cost of seeks and thereby enhancing the application-performance. HDFS adopts a master-slave architecture. NameNode has the master role and it maintains the file namespace including metadata, directory structure, files' list, blocks' list for each file, location for each block, attributes, authorization and authentication information. DataNodes play the slaves' role. They are responsible of creating, deleting or replicating the actual data blocks based on the instructions received from the NameNode and they report back periodically to NameNode (Narayan, 2012).

The Hadoop Distributed File System is specially designed to be highly fault-tolerant. The NameNode is a single point of failure for the HDFS cluster and a DataNode stores data in the Hadoop file management system. The file system replicates, or copies, each piece of data multiple times and distributes the copies to individual nodes, placing at least one copy on a different server rack than the others. As a result, the data on nodes that crash can be found elsewhere within a cluster. This ensures that processing can continue while data is recovered.

2.2.3 Map/Reduce

Map/Reduce a linearly scalable programming model to process large scale data stored in Hadoop File Distribution System (Pol, 2016). As it can be easily understood from the title, the model consists of two phase; The map phase corresponds to the map operation, whereas reduce phase corresponds to the fold operation. Data may be structured or unstructured. Map/Reduce performs two essential functions: it filters and distributes the work between the different nodes of the cluster or the map. The operation logic is based on "divide & conquer" principle and it partitions the large problem into smaller subproblems to the extent that the sub-problems are independent and they can be tackled in parallel by different slaves. To process the data, the programmer writes two functions; a function sometimes called mapper, and it organizes and reduces the results of each node in a coherent response to a request, called reducer. Each of these functions defines a mapping from one set of key-value pairs to another. The same functions can be used for a small as well as for a large-scale database without any modification since these functions are written in a fashion that is independent from data size as well from the cluster-size.

2.2.4 HBase

HBase is an Apache open-source project which presents a new data model similar to Google's big table. It is a column-oriented, distributed fault-tolerant and highly scalable database management systems, running on HDFS. To realize real-time read/write operations on a large scale database, HBase is a powerful tool [19]. The data in HBase is organized in labeled tables with rows and columns. Each row has absolutely one sorting key but the number of columns that it owns may differentiate.

Each cell in the table are versioned by a timestamp auto-assigned at the time of insertion. Its content is uniquely identified with a special set consisting of Table Name, Row-Key, Column Family, Column Name and the Timestamp. Each table uses the Row Key as primary key and they are easily accessible via this primary key. Thanks to this table structure, parallel scan in terms of Map/Reduce operations results into faster query response time and better overall throughput. Similar to HDFS and Map/Reduce, HBase also adopts master/slave architecture (Team, 2016).

2.2.5 Apache Hive – HiveQL

Apache Hive is a data warehouse solution for Hadoop environment (Huai, 2014). Hive provides data analysis operations, data summarization operations and querying while managing large datasets residing in distributed storage. It is one of the easiest to use of the high-level MapReduce (MR) frameworks. The Hive queries is written with a SQL-like language called HiveQL which runs over Hadoop Map/Reduce framework itself but hides complexity from the developer. HiveQL is composed of a subset of SQL features. It has also special extensions which are useful for batch processing systems. Hive supports analysis of large datasets stored in Hadoop's HDFS as well as easily compatible file systems and perfectly fits low level interface requirement of Hadoop.

Apache Hive supports a SQL-like query language known as the Hive query language over one or multiple data files located either in a local file system or in HDFS. Hive query language runs over Hadoop map-reduce framework itself, but hides complexity from the developer, Hive query language (HiveQL) supports SQL features of data definition language and data manipulation language. It supports also all types of joint operations as well as aggregate functions and operations and provides always good results on primitive as well as complex data types. HiveQL adds a dialect of SQL and JDBC bindings for HBase.

HiveQL does have some limitations compared with traditional RDBMS SQL. In Hive, HBase automatically partitions tables horizontally into regions. Each region comprises a subset of a table's rows called partitions. Each table can have one or many partitions in Hive which allows insertion of data in single or multiple tables but does not allow deletion of updating of data. For read operations (SELECT) the complex filtering operations such as HAVING are not supported by HiveQL. SQL offers hundreds of built-in functions whereas HiveQL has dozens which obliges the programmer to have more efforts while querying the data. Besides, SQL supports the transactions and their management as well as indexes but HiveSQL does not support neither transactions nor the indexes. Finally, latency can be measured by subseconds in SQL but by minutes in HiveQL (Ahmed, 2017).

3. PERFORMANCE EVALUATION

3.1 Description of the Data

Actually this study is the preliminary work of the development of a web service designed to reveal the interactions between the drugs written in a single prescription on a real time basis. During the development of this web service, in order to measure the performance of the data on different environment, the queries of this study are written and executed. The data subject to this study contains the prescription information written for inpatients and outpatients in a public hospital between the years 2015 – 2016. For the patients who undergo an outpatient treatment, minimum 1, maximum 5 medicine can be described per prescription. However, for inpatients, dozens of medicine can be prescribed per prescription, or even hundreds of them can be prescribed per patient. The datasets consist of text-based structured data. For the purpose of this performance evaluation study, the datasets are stored in both the tables on SQL Server environment and the files on Hadoop environment. Prescriptions contain 35 million records (approximately 3 GB of text). Figure 1 shows the relational table format and a brief example of records containing in them.

	BASVURU_ID	RECETE_ID	RECETE_TARIHI	RECETE_RENK	RECETE_TURU
1	1_1_199699818	1_1_199700168	2015-01-01	Normal	Ayaktan ReÃ§etesi
2	1_1_199699829	1_1_199699960	2015-01-01	Normal	Ayaktan ReÃ§etesi
3	1_1_199700219	1_1_199700347	2015-01-01	Normal	Ayaktan ReÃ§etesi
4	1_1_199700484	1_1_199700733	2015-01-01	Normal	Ayaktan ReÃ§etesi
5	1_1_199702117	1_1_199702194	2015-01-01	Normal	Ayaktan ReÃ§etesi
6	1_1_199702797	1_1_199703334	2015-01-01	Normal	Ayaktan ReÃ§etesi
7	1_1_199702922	1_1_199702988	2015-01-01	Normal	Ayaktan ReÃ§etesi
8	1_1_199703041	1_1_199703164	2015-01-01	Normal	Ayaktan ReÃ§etesi
9	1_1_199715634	1_1_199715714	2015-01-01	Normal	Ayaktan ReÃ§etesi
10	1_1_199716868	1_1_199717953	2015-01-01	Normal	Ayaktan ReÃ§etesi

RECETE.csv

	RECETE_ID	ILAC_BARKODU	ILAC_DOZ	ILAC_FIYAT	ILAC_PERIYODU
1	1_1_199700168	8699541270922	"2	00"	"2,89",1
2	1_1_199700168	8699809897038	"4	00"	"2,56",1
3	1_1_199699960	8699541270922	"2	00"	"2,89",1
4	1_1_199699960	8699717570061	"3	00"	"4,95",1
5	1_1_199700347	8699717280168	"2	00"	"6,5",1
6	1_1_199700347	8699717570061	"3	00"	"4,95",1
7	1_1_199700733	8697330094258	"2	00"	"16,12",1
8	1_1_199700733	8699506012055	"2	00"	"1,42",1
9	1_1_199700733	8699508090020	"2	00"	5,1
10	1_1_199702194	8699514159179	"3	00"	"5,62",1

RECETE_ILAC.csv

Figure 1. Data format on RDBS. (PatentID, PrescriptionID, PrescriptionDate, PrescriptionColor,PrescriptionType) (PrescriptionID, DrugBarcode, DrugDosage, DrugPrice, DrugPeriodicity)

3.2 Hardware Features

Two database systems are developed respectively to implement the database approaches involved in the study. The traditional database system is built using a virtual server equipped with an Intel Xenon E5-2690 v4 2.6 GHz processor, 16.00 GB of RAM. The system with Hadoop architecture includes 9 physical servers and a total of 1 TB of RAM.

3.3 Queries

The query time of the two databases is evaluated by making three different queries with varied complexity. The queries are designed to perform read operations. The first query is responsible to execute a single SELECT operation on several tables with "inner join". The second and third queries are also using aggregate functions and group by clause which put a serious burden on the query runtime. Figure 2 shows the queries executed in the study.

```
select re.BASVURU_ID, re.RECETE_ID, re.RECETE_RENK,
re.RECETE_TARIHI, re.RECETE_TURU, il.ILAC_BARKODU
il.ILAC_DOZ, il.ILAC_FIYAT, il.ILAC_PERIYODU
from drug.dbo.RECETE as re
inner join drug.dbo.RECETE_ILAC as il
on re.RECETE_ID=il.RECETE_ID
```

```
select il.ILAC_BARKODU, count(*) as num
from drug.dbo.RECETE as re
inner join drug.dbo.RECETE_ILAC as il
on re.RECETE_ID = il.RECETE_ID
group by il.ILAC_BARKODU
order by num desc
```

```
select RECETE_TARIHI, COUNT(*) as num
from drug.dbo.RECETE
group by RECETE_TARIHI
order by num desc
```

Figure 2. Queries

3.4 Results

The variation of the query time with the size of the database is also studied. For each of the two database approaches, the time required to perform the queries with a variable complexity specified above is measured with databases containing respectively 1.8, 10, 20, 30 and 37 million records. At each parameter, the query is performed 10 times to calculate the average query time and the standard deviation. The query synchronization performance is given in Figure 3 respectively.

Query I - (seconds)				
	1,8 millions	10 millions	20 millions	30 millions
SQL	27	96	186	260
Hadoop	8	13	25	43

Query II - (seconds)					
	1,8 millions	10 millions	20 millions	30 millions	37 millions
SQL	13	21	31	37	43
Hadoop	12	15	19	22	23

Query III - (seconds)					
	1,8 millions	10 millions	20 millions	30 millions	37 millions
SQL	2	3	4	5	6
Hadoop	12	14	15	16	15

Figure 3. Query Synchronization Performance

The query time on the first query (shown in Figure 4) in the SQL database of more than 30 million records is quite disappointing. The reason is that the result of the query contains about 30 million records. The database with the Hadoop architecture has a very good performance on the same query and the same number of records. With Hadoop, parallel processing capability, query results can be returned faster than traditional databases.

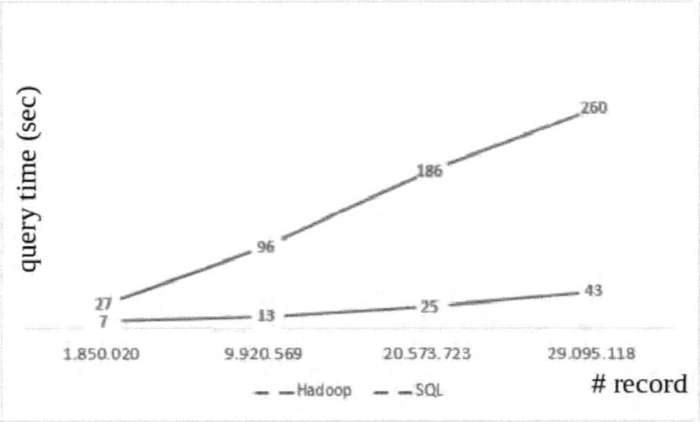

Figure 4. Query time according to row number for 1st Query

Although the amount of data in the second query was the same, the fact that the results contain about 15,000 records reduced the performance gap between the databases (see the Figure 5).

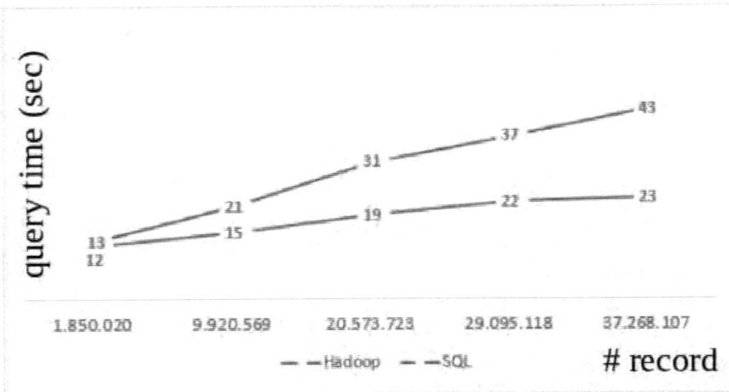

Figure 5. Query time according to row number for 2nd Query

In both queries, the Hadoop architecture has responded better to the increase in the amount of data. With the reduced amount of data, the Hadoop architecture performance on the third query is worse than the SQL database. The reason is that the result of the query contains about 800 records. The Map and Reduce operations of the Hadoop architecture have become useless with the reduction of the amount of data generated in the result of the query. So the performance of the Hadoop architecture in the third query is worse than the SQL database as shown in Figure 6.

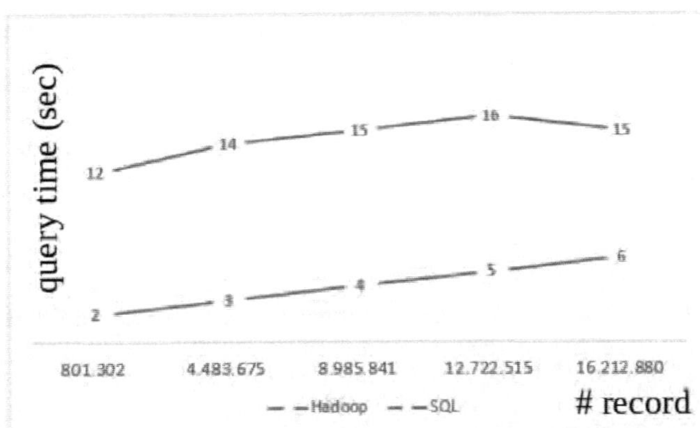

Figure 6. Query time according to row number for 3rd Query

4. CONCLUSION AND DISCUSSION

An examination of current approaches to prescription data storage indicates that the Hadoop approach is a viable alternative to relational database design because it provides better query performance.

As seen in the results of the first and second queries, Hive is designed specifically for large dataset analysis and works well for a range of complex queries. Hive is the most accessible way to quickly query and inspect datasets already stored in Hadoop. But in the third query, we see that the performance of Hadoop decreases when the amount of data decreases. Hive works ideally in large datasets, but MySQL works much better with smaller datasets and can be optimized in a variety of ways.

This study attempts to explore the vast opportunities of Hadoop and Sql query technologies in the management of prescription data. The prototype system developed is initially tested with a maximum of 30 million records only. Further evaluation using larger datasets, or even multiple databases and a data warehouse, will provide a more complete results on the performance of the Sql and Hadoop databases.

ACKNOWLEDGEMENT

This work is supported by Galatasaray University Research Foundation under the Grant No. 19.401.002 and Consortium of Galatasaray University.

REFERENCES

Ahmed, Nadeem, et al. 2017, "Data processing in Hive vs. SQL server: A comparative analysis in the query performance." 2017 IEEE 3rd International Conference on Engineering Technologies and Social Sciences (ICETSS).

Anderson, J. et al., 2009, “CouchDB: The Definitive Guide” , 1st edition, O'Reilly Media.

Cassandra, A., 2014. Apache cassandra. Website. Available online at http://planetcassandra. org/what-is-apache-cassandra, p.13.

Chang, F. et al., 2008. Bigtable: A distributed storage system for structured data. ACM Transactions on Computer Systems (TOCS), 26(2), p.4.

Chen, C.P. and Zhang, C.Y., 2014. Data-intensive applications, challenges, techniques and technologies: A survey on Big Data. Information sciences, 275, pp.314-347.

Dorin Carstoiu et al, 2010, "Hbase - non SQL Database, Performances Evaluation", International Journal of Advancements in Computing Technology, 2(5).

Giuseppe De Candia et al., 2007 “Dynamo: Amazon's Highly Available Key-Value Store”, in the Proceedings of the 21st ACM Symposium on Operating Systems Principles, Stevenson.

Huai, Y., et al., 2014, June. Major technical advancements in apache hive. In Proceedings of the 2014 ACM SIGMOD international conference on Management of data (pp. 1235-1246). ACM.

Judd, D., 2008. Scale out with HyperTable. Linux magazine, August 7th, 1.

Khetrapal, A. and Ganesh, V., 2006. HBase and Hypertable for large scale distributed storage systems. Dept. of Computer Science, Purdue University, 10(1376616.1376726).

Kristina Chodorow, Michael Dirolf, 2010, "MongoDB: The Definitive Guide", 1st edition, O'Reilly Media.

Loebbecke, C. and Picot, A., 2015. Reflections on societal and business model transformation arising from digitization and big data analytics: A research agenda. The Journal of Strategic Information Systems, 24(3), pp.149-157.

Lydia, E.L. and Swarup, M.B., 2015. Big data analysis using hadoop components like flume, mapreduce, pig and hive. International Journal of Science, Engineering and Computer Technology, 5(11), p.390.

Narayan, S. et al, 2012, Hadoop acceleration in an openflow-based cluster. 2012 SC Companion In High Performance Computing, Networking, Storage and Analysis (SCC),(pp. 535-538). IEEE.

Pol, U.R., 2016. Big data analysis: comparison of hadoop mapreduce, pig and hive. Int. J. Innov. Res. Sci. Eng. Technol, 5(6).

Sumbaly, R. etal., 2012, February. Serving large-scale batch computed data with project voldemort. In Proceedings of the 10th USENIX conference on File and Storage Technologies (pp. 18-18). USENIX Association.

Taylor, R.C., 2010, December. An overview of the Hadoop/MapReduce/HBase framework and its current applications in bioinformatics. In BMC bioinformatics (Vol. 11, No. 12, p. S1). BioMed Central.

Team, A.H., 2016. Apache hbase reference guide. Apache, version, 2(0).

Wixom, B. et al., 2014. The current state of business intelligence in academia: The arrival of big data. CAIS, 34(1), pp.1-13.

Zhao, W. et al., 2017, November. Querying big data from a database perspective. In 2017 4th International Conference on Systems and Informatics (ICSAI) (pp. 1433-1437). IEEE.

RoC-BAC: ROLE AND CONTEXT BASED ACCESS CONTROL MODEL FOR THE HEALTHCARE ENVIRONMENT

Zanifa Omary
Department of Computer Science and Mathematics
The Institute of Finance Management, Tanzania

ABSTRACT

This paper presents a new context-based access control model that allows healthcare professionals to bypass access rules in an accountable manner in case of unexpected (emergency) situation so as to provide continuity of care. The proposed RoC-BAC model is an extension of the conventional, policy-neutral, Role-Based Access Control (RBAC) model with two variables: contexts and obligations. In addition to being fine-grained, RoC-BAC model is dynamic, flexible and scalable and incorporates procedures for emergency access requests.

KEYWORDS

Access Control, Model, Role, Context, and Roc-Bac, Healthcare

1. INTRODUCTION

This paper aims at designing a new context-based access control model that allows healthcare workers to bypass access rules in an accountable manner in case of unexpected (emergency) situation. The proposed RoC-BAC model is an extension of the conventional, policy-neutral, Role-Based Access Control (RBAC) model with two variables: contexts and obligations. In addition to being fine-grained, RoC-BAC model is dynamic, flexible and scalable and incorporates procedures for emergency access requests. The reason behind a decision to incorporate health-related contexts into RBAC lies on accumulating evidence available on the existing literature. Many researchers in Ubiquitous Computing (Ahmed and Zhang, 2008; Covington et al., 2002; Tolone et al., 2005; Trnka and Cerny, 2016), where context has been thoroughly studied, point out that, when context is appropriately incorporated into an existing access control solution, the resulting model may improve security in critical applications. With this in mind, consider the significance of contexts in a military base as an example of improvement in security. Normally in a military base, a user is allowed to fire a missile if that user is assigned a certain role (for example, a top secret commander) and a user is in a specific high security location at a certain time only. Consequently, a missile can be fired only when it is in a certain *location* at a certain *time*. By adding location and temporal constraints in access decision, additional verification checks are henceforth included that must be satisfied before an individual user is granted access to fire a missile.

To ensure the same cross checking, Break-The-Glass (BTG) procedure has been introduced for the healthcare domain (Deshmukh and Croasdell, 2008; Petritsch, 2014; Preuveneers and Joosen, 2014; Souza et al., 2017) BTG is a significant and quick approach used in the domain to extend access privileges for a user who does not have access rights to a certain information to gain access when necessary (Marinovic et al., 2011; Petritsch, 2014; Petritsch and Brucker, 2009). The procedure draws its name from breaking-the-glass to activate a fire alarm. The healthcare systems that contains primary source data (information) for treatment must therefore develop, document, implement and test BTG procedures that would be used in the event when healthcare professionals require unexpected (emergency) access to those records. These systems must have a clearly stated and widely understood procedure for allowing access using an alternate or manual method. It should be noted that, the usage of emergency access rights facilitated by BTG approach needs to be documented for later audits and reviews. As discussed by NEMA et al., (2014), examples of infrequent situations that may require BTG and logging could be

1. When user experiences account problems such as forgotten username or password or both (after an extended absence or vacation), locked password (as a result of mistyping many times) and even no user account (e.g. a clinician from another organisation or a new clinician is assisting in facility during an emergency)
2. When there is an authentication problem including failure of the Central Authentication System (CAS) server, smart card or biometrics reader failure (that is, when a smart card reader or biometric is damaged), and
3. When there are authorization problems, including an emergency situation which thrusts an individual user into a role where he or she lacks sufficient access rights (for example, an administrative assistant is entering orders during emergency).

As acclaimed by Ferreira et al., (2009), another reason for an implementation of BTG procedure in healthcare is that access control policies are defined with "maximum freedom of access" and "maximum user responsibility" in order to ensure that nothing interferes with the delivery of care. To guarantee maximum freedom of access, information systems in healthcare must provide mechanisms for users to access requested information at any time, whenever the need arises. With maximum responsibility, systems in healthcare must provide a mechanism that offers additional information such as an alert making users requesting access to be aware that, the user is trying to access unauthorized information under normal healthcare settings or even through a call for help from other qualified people. Abstractly, this makes an individual user requesting such access to information responsible for what user is doing and may subsequently be held accountable. The system must also provide a mechanism that automatically notifies all responsible parties in relation to that access so that user's actions can be justified afterwards. To address this and other significant access control requirements for the healthcare domain, this paper discusses a new context-based access control model called Role and Context-Based Access Control (RoC-BAC), that allows healthcare professionals to bypass access rules in an accountable manner in case of unexpected (emergency) requests. With the proposed model, a new concept called health-related contexts is introduced. Section 2 discusses a new Role and Context-Based Access Control (RoC-BAC) model.

2. ROLE AND CONTEXT-BASED ACCESS CONTROL (ROC-BAC) MODEL

The Role and Context-Based Access Control (RoC-BAC) model is an extension of the traditional Role-Based Access Control model with the notion of health-related contexts and obligations. The model has been designed to allow healthcare professionals to bypass access rules in an accountable manner in case of unexpected (emergency) requests. This section presents the model in three parts: Section 2.1 presents a formal definition of components and relations of RoC-BAC model while Section 2.2 presents a formal RoC-BAC model. Finally, Section 2.3 presents a model augmented with obligations.

2.1 RoC-BAC: Definition of Components and Relations

This section defines components as well as relations in RoC-BAC model. To allow context-based access in healthcare, the RoC-BAC model incorporates health-related contexts in its access decisions so as to allow healthcare professionals to bypass access rules, and hence to support the continuity of care.

RoC-BAC model comprises of the following components:

i. Users (U) denote entities requesting access to objects. They may include human beings, processes, machines, networks or autonomous agents.
   Mathematically, $U= \{u_1, u_2 ... u_i\}$ denotes a set of users.

ii. Roles (R) represent organizational job functions with clear definition of inherent responsibility and authority. The concept of roles, as is used in this paper, is adopted from the RBAC model where users are assigned to roles (user roles) and roles are associated with permissions (permission roles). Furthermore, in RoC-BAC, roles represent a combination of user roles (UR) and context roles (CR), as such R= 2(UR × CR)

iii. An object (OBJ) is a passive entity that contains or receives information. A set of objects (OBJ) is represented mathematically as, OBJ= $\{obj_1, obj_2...obj_j\}$

iv. An operation (OPS) is an execution of a program-specific function that is invoked by the user. OPS= $\{ops_1, ops_2...ops_k\}$, denotes a set of operations recognized by the system.

v. Context (C) is defined as a set of variables that define a situation (Federenko et al., 2004). Contrary to traditional access control models that do not take into account contexts in determining whether access should be allowed or not, RoC-BAC offers a context-centric access control solution, that allow system administrators to define permissions based on the relevant contexts. Mathematically, C= $\{c_1, c_2...c_x\}$. For access control purposes, contexts in this paper are categorized in four groups:

- Subject Contexts (SC): define contexts that must be exercised by the subject in order to obtain access rights to an object or resource. In RoC-BAC, subject contexts are used to determine access rights for an entity requesting access privileges. They may include the subject's role, identity, credentials, name, organisation that the subject is affiliated with, activity, location, and task (Omary et al., 2011).

- Object Contexts (OC): An object is a passive entity that is acted upon by a subject. Similar to subjects, objects contain contexts that can be used to control their accesses. Object Contexts (OC) refer to any object-related information that can be used to characterize the situation in which the protected object was created and its current status, which is relevant for making an access control decision.

- Environment Contexts (EC): These contexts describe operational, technical, and even situational contexts at the time a transaction takes place. They may include current date and time, temperature, network's security level, air quality, light level, noise level. These factors are neither associated with a subject nor a resource, but may nevertheless be relevant in controlling access (Omary et al., 2011).

- Health-related Contexts (HC): these variables define the healthcare domain specific contexts that would be evaluated by an access control system during access decision. Some of the specific health-related contexts (based on the region when the service is offered) may include Gender-Based Violence (GBV), Female Genital Mutilation (FGM), non-surgical amputation of limbs for people with albinism, clashes between: pastoralist groups, pastoralist and farmers, among others. Even though the main aim of this paper is to extend the traditional RBAC model with health-related contexts; these four groups can further be used to define context roles in RoC-BAC, and thus extending the traditional RBAC model with contexts.

vi. User Roles (UR) represent a set of user roles. User role is similar to "role" or "subject role" as used in a traditional Role-Based Access Control (RBAC) model.

vii. Context Roles (CR) describe a set of context roles. It is used to capture relevant contexts that can be used in RoC-BAC policies. In addition to health-related context roles, RoC-BAC also contains time-based context roles, location-based context role and location and time-based context roles, which are part of environment contexts. Typically, CR⊆UR

viii. Permission (P) is an approval to perform an operation on one or more objects. Permission (P) = $\{p_1, p_2... p_w\}$, is a set of permissions in a system such that $P = 2^{OPS \times OBJ}$. PermOperation: ops (p: P)→{ops⊆ OPS}, is a permission-to-operation mapping which gives a set of operations associated with a permission p. PermObject: obj (p : P) → {obj ⊆ OBJ}, a permission-to-object mapping which gives a set of objects associated with permission p.

ix. Session (S) represents a set of sessions. Usually, a role is activated for a user during each session. Activated Role (AR) is a mapping between user roles (UR) and Context Roles (CR).

The RoC-BAC model consists of several relations, including User Assignment (UA), Permission Assignment (PA) and Context Role Assignment (CRA). These relations define associations between users, user roles, permission roles and context roles.

- User Assignment (UA): $UA \subseteq U \times R$, many-to-many mapping that assigns a user role to a user
- Context Role Assignment (CRA) is a mapping that assigns contexts to a user role, such that $CRA \subseteq UR \times C$
- Permission Assignment (PA): $PA \subseteq P \times R$, many-to-many mapping that assigns permissions to a role
- assigned_users (ur: UR) $\rightarrow 2^U$, a mapping of a user "ur" onto a set of users. Formally assigned_users (ur) = {u ∈ U | (u, ur) ∈ UA= {u ∈ U | (u, ur) ∈ UA }
- assigned_permissions(r: R) $\rightarrow 2^P$, mapping of role "r" onto a set of permissions. Formally, assigned permissions = { p ∈ PRMS | (p, r) ∈ PA }
- User Sessions: user_sessions (u: U) $\rightarrow 2^S$, a mapping of user u onto a set of sessions
- Session Roles: session_roles (s: S) $\rightarrow 2^R$, mapping of session "s" onto a set of roles. Formally, session_roles (s) = {r ∈ R→| (user_sessions (s_i), r) ∈ UA}
- avail_session_perms (s: S) $\rightarrow 2^P$, denotes permissions available to a user in a session $\bigcup_{r \in session\_roles(s)}$ assigned_permissions(r)

2.2 Formal Definition

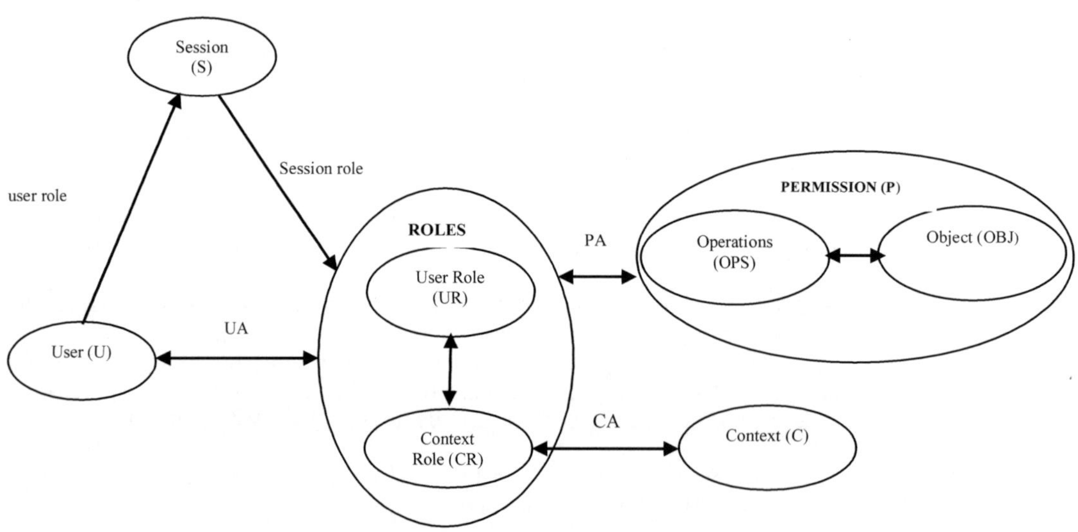

Figure 1. Role and Context Based Access Control (RoC-BAC) Model

The RoC-BAC model, presented in Figure 1, has the following components:

- U, R, C, UR, CR, OPS, OBJ, P and S (represents users, roles, context, user roles, context roles, operations, objects, permissions and sessions respectively)
- $UA \subseteq U \times R$, a many-to-many mapping that assigns a user role to a user

- assigned_users (ur: UR) $\rightarrow 2^U$, a mapping of a user "ur" onto a set of users. Formally assigned_users (ur) = { u ∈ U | (u, ur) ∈ UA}
- R⊆ 2 (UR × CR), set of roles
- PA ⊆ P × R, many-to-many permission to role assignment relation.
- assigned_permissions (r: R) $\rightarrow 2^P$, mapping of role "r" onto a set of permissions. Formally, assigned_permissions = { p ∈ PRMS | (p, r) ∈ PA}
- user_sessions (u : U) $\rightarrow 2^S$, a mapping of user u onto a set of sessions
- Session Roles: session_roles (s: S) $\rightarrow 2^R$, mapping of session "s" onto a set of roles. Formally, session_roles (s_i) = {r ∈ R→| (user_sessions (s_i), r) ∈ UA }

Definition 1: Context Role (CR) = {L_CR, T_CR, Hr_CR}

CR represents a set of Context Roles. They are used to capture context-relevant information for RoC-BAC access policies. In addition to health-related context roles, introduced in this paper through a notion of Health-related contexts, RoC-BAC model also supports the specification of Location-based Context Roles (L_CR) and Time-based Context Roles (T_CR). The context roles share similar characteristics with user roles, which are defined in a traditional RBAC model, such as role activation and deactivation.

Definition 1.1. L_CR (Location-based Context Roles)
Location-based Context Roles L_CR) is a pair (r, l) where r denotes a user role name and l is the current position of a user role.

Definition 1.2. T_CR (Time-based Context Roles)
Similar to location, time-based Context Roles is a pair (r, time) where r is a user role name and t is the current time of a user role.

Definition 1.3. Hr_CR (Health-related Context Roles)
A Health-related Context Role (Hr_CR) is a pair (r, hr) where r is a user role name and hr represents a patient's current health condition that requires healthcare professionals' attention, thus referred as a health-related reason for access.

Definition 2: Authorizations

Location-based Authorizations are policies that define accesses that users may have over certain locations. While temporal authorizations limit periods of time during which an authorization is valid, health-related authorizations define accesses that users may have in case of an unexpected emergency. Formally, they are defined as follows:

Definition 2.1: L-Auth (Location Authorization)
A location authorization is a pair ((s, r), l) where

- s denotes a subject who requests authorization
- l is a location
- r is a role that subject s can be assigned when in location l

A location authorization ((s, r), l) means that a subject s is authorized to gain role r when in location l. ((Anne, Nurse), Ward) denotes that Anne has a role Nurse when she is in a hospital ward.

Definition 2.2: Ti_Auth (Temporal Authorization)
A temporal authorization is a quadruple ((s, r), t ((t^i_s, t^i_e], (t^o_s, t^o_e]), n) where

- Time interval [tis, tie] is an entry duration during which a subject can gain a role r to access object o
- Time interval [t^o_s, t^o_e] denotes an exit duration during which a subject s will revoke a role such that $t^o_s \geq t^i_s$ and $t^o_e \geq t^i_e$
- n is the number of accesses that subject s can exercise n ∈ [1, ∞)

A temporal authorization ((Anne, Nurse), (8, 17], (17, 8], 50) means Anne is allowed to activate role Nurse between (8:00, 17:00) and a role will be revoked fifty times during (17:00, 8:00) interval.

Definition 2.3: LoT_Auth (Spatial Temporal Authorization)
RoC-BAC also specifies spatial temporal authorization where policies contain both location and time-

related information. A spatial temporal authorization is defined as ((s, r), ops, obj, t ((t^i_s, t^i_e], (t^o_s, t^o_e], l) where

- s, r, ops and obj represent subject, role, operation and object respectively
- l is a location
- A spatial temporal authorization ((s, r), ops, obj, ftime, etime, l) semantically means a subject s with role r can perform operation ops on object obj under temporal condition ftime and etime and location condition l.

Definition 2.4: Hr_Auth (Health-related Authorizations)
A health-related authorization is a pair ((s, r), hr) where

- s represents a subject who requests an authorization and
- r represents a role which s can activate
- hr is a health-related context that subject s can specify to access records of the patient

A health-related authorization ((Anne, Nurse), FGM) means Anne has a role Nurse but can also activate contextual role FGM when there is a new case to be attended in a respective health facility.

Definition 3: RoC-BACA (RoC-BAC Authorization)
RoC-BAC authorization is defined as RoC-BACA (Hr-Auth (L_Auth (T_Auth))) where

- L_Auth denotes location authorization
- T_Auth denotes temporal authorization
- Hr_Auth denotes health-related authorization.

Definition 4: RoC-BAC Authorization rule
The RoC-BAC authorization rule is defined by the tetrad (s, R_Auth, obj, m)

- S denotes a subject (user) requesting access
- R_Auth is a RoC-BAC Authorization
- Obj is an authorized object
- M represents an access mode which can either be read or write

2.3 Augmenting Obligations

One limitation of the traditional access control mechanisms is their inability to provide users with obligations as a result of executing permitted action. This major weakness may result in the compromise of integrity or security of the system as system integrity, for instance, requires that certain action(s) performed in a system should always be followed by other action(s) within a limited time frame (Minsky & Lockman, 1985). The RoC-BAC model is, therefore, augmented with obligations that are often imposed by operational rules, modern corporation regulations or privacy laws. Obligations specify actions that must be fulfilled in conjunction with an authorization decision (permit or deny) when an access rule is triggered to maintain accountability. This section presents how RoC-BAC model could be augmented with obligations, as shown in Figure 2 in order to maintain accountability in the system. Section 3 presents related work.

The following four enhancements were applied to the formal RoC-BAC model:

1. A new basic element called OBGS representing a set of valid obligations is introduced. In this paper, obligation is defined as task(s) that must be fulfilled by the system, and they are associated with permissions allocated to roles. They are categorized as pre-obligations, post-obligations,

conditional obligations or repeating obligations. In this regard, the RoC-BAC model, which has been designed for the healthcare domain, is augmented with post-obligations to allow for healthcare professionals to access medical records in time, and thus ensure the continuity of care.

2. A new relation called Obligation Permission (OPRMS) is introduced. It represents a relation between permissions (P) and obligations (OBGS), such that OPRMS $\subseteq$ P $\times$ 2OBG. For oprms = (prms, obgs) $\in$ OPRMS, means oprm is an obligation augmented permission which specifies if permission prm is exercised, a set of obligations obg should be fulfilled.

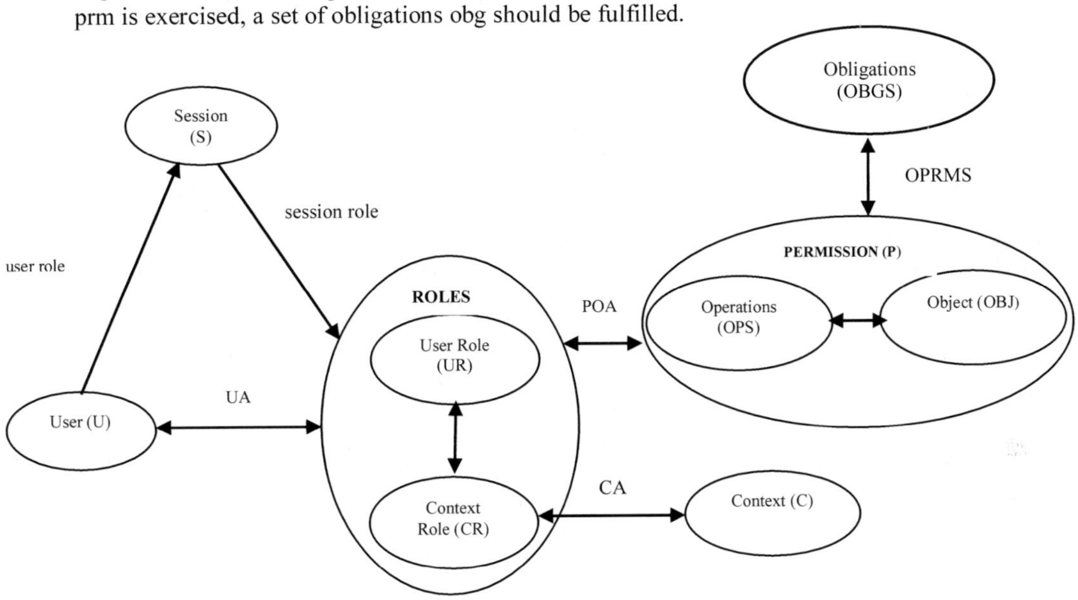

Figure 2. An augmentation of RoC-BAC model with obligations

3. Permission Assignment (PA) relation is modified and a new relation called Permission-Obligation Assignment (POA) is defined. Formally, POA $\subseteq$ ROLES $\times$ OPRMS. For p = (r, oprms) $\in$ PA, it means role r is allocated with obligation augmented permission oprm.
4. To retrieve obligations with authorization decisions, Check Access function is modified into (Zhao et al., 2007)
CheckAccess: SESSIONS, OPS, OBJ $\rightarrow$ BOOL $\times$ 2^{OBG}

The possible combination between authorization elements and results from CheckAccess function are as follows:

- (FALSE, ø) → DENY access to a resource
- (FALSE, 2OBG) → DENY access to a resource AND perform obligations on DENY
- (TRUE, ø) → GRANT access to a resource
- (TRUE, 2OBG) → GRANT access to a resource and perform obligations on GRANT

3. RELATED WORK

Role-Based Access Control (RBAC) is a well-established access control model with widely-recognized advantages; and it is supported by a large number of software products and platforms (Sandhu, 1996), (Ferraiolo et al., 1995, 2001; Lodderstedt et al., 2002). The model addresses access control requirements of commercial organizations through an introduction of the role concept (Ferraiolo et al., 2003; Hu et al., 2006; Sandhu, 1996). In RBAC, permissions are attached to roles and users must be assigned to roles to get the permissions. A role represents a specific task competency, such as that of a medical doctor or lecturer, or it can embody the authority and responsibility (Sandhu, 1996). Roles in RBAC define both specific individuals allowed to access resources as well as the extent to which those resources can be accessed. Furthermore, permissions determine what operations can be carried out on resources under access control. A user must establish a session to activate a subset of roles to which the user is assigned. Each user can activate multiple sessions; however, each session is associated with only one user. The operations that a user can perform in a session depends on the roles activated in that session and permissions associated with those roles. The RBAC model supports roles hierarchy that define an inheritance relationship between roles. To prevent conflict of interest that may arise in an organisation, RBAC allows the specification of static and dynamic separation of duty constraints. Several works exist in the literature that improves RBAC functionality for different application domains. This section discusses the research works that are closely related to the access control model proposed in this paper, and particularly how RBAC has been extended with different concepts or notions to make the overall approach "context-based" or "context-aware".

As part of RBAC's extensions, Sampemane et al., (2002) introduced a new access control model for active spaces. The authors define an active space as "a computing environment that integrates physical spaces and embedded computing software and hardware entities. The active space allows interactive exchange of information between the user and the space". Environmental aspects are adopted into the access control model for active spaces, and the space roles are introduced into the implementation of the access control model based on RBAC. The proposed model supports specification of both Discretionary Access Control (DAC) and Mandatory Access Control (MAC) policies. While DAC policies allow users to create and update security policies for their devices, the MAC policies allow system administrators to maintain the access matrix. In addition to Sampemane et al., (2002), a development infrastructure for access spaces has been proposed by other researchers including Cerqueira et al., (2001).

(Covington et al., 2001) introduced a Generalized RBAC (GRBAC) model to help control access to private information and resources in ubiquitous computing applications. In fact, the GRBAC model extends the traditional RBAC model by incorporating the notion of object and environment roles. Even though environment roles differ from the subject roles in RBAC but they do have similar properties, including: role activation, role hierarchy and separation of duty. In an access control framework enabled by environment roles, each element of permission assignment is associated with a set of environment roles, and environment roles are activated according to changing conditions specified in environmental conditions. In this way, environmental properties such as location and time are introduced to the access control framework. Temporal Role-Based Access Control (TRBAC) is another extension of RBAC, proposed by (Bertino et al., 2001). This model adds a time dimension to the RBAC model whereas its authors introduce the concept of role enabling and disabling. The temporal constraints in the model determine when the roles can be enabled or disabled. A role can be activated only if it has been enabled. (Joshi et al., 2005) extend the TRBAC model by proposing the Generalized Temporal Role-Based Access Control (GTRBAC) model. In this work the authors introduce the concept of time-based role hierarchy and time-based separation of duty.

Several other researchers have also extended RBAC to incorporate spatial information. To cope with spatial requirements, Hansen and Oleshchuk, (2003) extend the conventional RBAC model and propose Spatial Role-Based Access Control (SRBAC) model. The SRBAC model utilizes location in its security policy definitions. Another important work in this regard is GEO-RBAC model proposed by Damiani et al., (2007), which is probably the most expressive location-based access control model. The role activation is based on the location of the user. Consider the following as a usage example of GEO-RBAC in a real world where a user can acquire a role of a patient when in a hospital grounds and a role of a teacher when in a school compound. The GEO-RBAC model supports the notion of role hierarchies but does not deal with separation of duties. Another work incorporating spatial information is by Ray et al., (2006). In this work, authors analyze how each component of a traditional RBAC is influenced by location. The authors define their model using Z specification language. The Location-Based Access Control has been addressed in some other works not pertaining to RBAC including (Leonhardt and Magee, 1998), (Hengartner and Steenkiste, 2004), and (Ray et al., 2006)

There are also other works that incorporate both location and time in to the traditional RBAC. The research work by Chandran and Joshi, (2005) combines the main features of GTRBAC and GEO-RBAC. In their LoT-RBAC model, role is enabled by time constraints. A user can activate a role if such a role is enabled and an individual user satisfies location constraints associated with role activation. Another feature worth mentioning from their work is that, when a role is activated all permissions associated with such a role can be invoked. Kumar and Newman, (2006) proposed the Spatial Temporal Role-Based Access Control model that is suitable for pervasive computing applications. With their access control model, the authors show the association of each component of the conventional RBAC model with spatial-temporal information.

In addition to location and time, the traditional Role-Based Access Control (RBAC) model has also been extended using purpose information. Ni et al., (2010) proposed a purpose-based access control model for privacy protection. Their proposed model is based on the notion of purpose roles through which purpose is categorized into intended purposes and access purposes. In addition to role attributes and system attributes, Ni et al., (2010) also incorporate purpose hierarchy into their model. RBAC can also be extended by the notion of trust. Bhatti et al., (2005) proposed a trust-enhanced version of their XML-based access control (X-RBAC) framework for web services. The proposed model, which is an extension of RBAC with trust and contexts, incorporates context-based access control. Chakraborty and Ray, (2006) propose a trust-based access control model called TrustBAC. The model extends the conventional role-based access control model with the notion of trust levels. Users are assigned to trust levels instead of roles based on a number of factors, including user credentials, user behavior history, and user recommendation, and trust levels are assigned to roles which are assigned to permissions. Baracaldo et al., (2015) propose a Geo-Social RBAC, which is a new role based access control model that allows the inclusion of geo-social constraints as part of the access control policy. Besides capturing the locations of a user requesting access and her social connections, geo-social includes cardinality constraints that dictate how many people related by a particular social relation need to be present in the required locations at the time of an access

4. CONCLUSION

Despite a large number of contributions from various access control models in the existing literature, majority of these models do not deal with vital privacy-aware elements, such as obligations and complex conditions while providing access for unexpected situations, which form an essential part of privacy protection. With this limitation, this paper argues that, to gain full advantage of contexts in an access control solution, an access control solution should incorporate context conditions that are specific to the application domain. In this case, RoC-BAC model has been designed by extending the traditional Role-Based Access Control (RBAC) model with the notion of health-related contexts, from the healthcare domain.

REFERENCES

Ahmed, A., Zhang, N., 2008. A context-risk-aware access control model for ubiquitous environments. Presented at the 2008 International Multiconference on Computer Science and Information Technology, IEEE, pp. 775–782.

Baracaldo, N., Palanisamy, B., Joshi, J., 2015. Geo-social-rbac: A location-based socially aware access control framework. Presented at the International Conference on Network and System Security, Springer, pp. 501–509.

Bertino, E., Bonatti, P.A., Ferrari, E., 2001. TRBAC: A Temporal Role-based Access Control Model. ACM Trans Inf Syst Secur 4.

Bhatti, R., Bertino, E., Ghafoor, A., 2005. A trust-based context-aware access control model for web-services. Distrib. Parallel Databases 18, 83–105.

Cerqueira, R., Hess, C.K., Román, M., Campbell, R.H., 2001. Gaia: A development infrastructure for active spaces, in: Workshop on Application Models and Programming Tools for Ubiquitous Computing (Held in Conjunction with the UBICOMP 2001). p. 11.

Chakraborty, S., Ray, I., 2006. TrustBAC: integrating trust relationships into the RBAC model for access control in open systems, in: Proceedings of the Eleventh ACM Symposium on Access Control Models and Technologies. ACM, pp. 49–58.

Chandran, S.M., Joshi, J.B., 2005. LoT-RBAC: a location and time-based RBAC model, in: International Conference on Web Information Systems Engineering. Springer, pp. 361–375.

Covington, M.J., Fogla, P., Zhan, Z., Ahamad, M., 2002. A context-aware security architecture for emerging applications. Presented at the 18th Annual Computer Security Applications Conference, 2002. Proceedings., IEEE, pp. 249–258.

Covington, M.J., Long, W., Srinivasan, S., Dev, A.K., Ahmad, M., Abowd, G.D., 2001. Securing context-aware applications using environment roles, in: Proceedings of the Sixth ACM Symposium on Access Control Models and Technologies. ACM, pp. 10–20.

Damiani, M.L., Bertino, E., Catania, B., Perlasca, P., 2007. GEO-RBAC: a spatially aware RBAC. ACM Trans. Inf. Syst. Secur. TISSEC 10.

Deshmukh, P., Croasdell, D., 2008. HIPAA: Privacy and security in health care networks, in: Information Security and Ethics: Concepts, Methodologies, Tools, and Applications. IGI Global, pp. 2770–2781.

Ferraiolo, D., Cugini, J., Kuhn, D.R., 1995. Role-based access control (RBAC): Features and motivations, in: Proceedings of 11th Annual Computer Security Application Conference. pp. 241–48.

Ferraiolo, D., Kuhn, D., Chandramouli, R., 2003. Role-Based Access Control. Artech House Publishers.

Ferraiolo, D.F., Sandhu, R., Gavrila, S., Kuhn, D.R., Chandramouli, R., 2001. Proposed NIST standard for role-based access control. ACM Trans. Inf. Syst. Secur. TISSEC 4, 224–274.

Ferreira, A., Chadwick, D., Farinha, P., Correia, R., Zao, G., Chilro, R., Antunes, L., 2009. How to securely break into RBAC: the BTG-RBAC model, in: Computer Security Applications Conference, 2009. ACSAC'09. Annual. IEEE, pp. 23–31.

Hansen, F., Oleshchuk, V., 2003. Spatial Role-Based Access Control Model for Wireless Networks, in: VTC 2003-Fall IEEE 58th Vehicular Technology Conference. pp. 2093–2097.

Hengartner, U., Steenkiste, P., 2004. Implementing access control to people location information, in: Proceedings of the Ninth ACM Symposium on Access Control Models and Technologies. ACM, pp. 11–20.

Hu, V., Ferraiolo, D., Kuhn, D., 2006. Assessment of Access Control Systems (No. NIST 7316), US Department of Commerce. National Institute of Standards and Technology.

Joshi, J.B., Bertino, E., Latif, U., Ghafoor, A., 2005. A Generalized Temporal Role-Based Access Control Model. IEEEE Trans. Knowl. Data Eng. 17, 2–23.

Kumar, M., Newman, R.E., 2006. STRBAC-An approach towards spatio-temporal role-based access control. Commun. Netw. Inf. Secur. 155.

Leonhardt, U., Magee, J., 1998. Security considerations for a distributed location service. J. Netw. Syst. Manag. 6, 51–70.

Lodderstedt, T., Basin, D., Doser, J., 2002. SecureUML: A UML-based modeling language for model-driven security. Presented at the International Conference on the Unified Modeling Language, Springer, pp. 426–441.

Marinovic, S., Craven, R., Ma, J., Dulay, N., 2011. Rumpole: a flexible break-glass access control model, in: Proceedings of the 16th ACM Symposium on Access Control Models and Technologies. ACM, pp. 73–82.

National Electrical Manufacturers Association, European Coordination Committee of the Radiological and Electromedical Industry, Japan Industries Association of Radiological Systems, 2014. Break-Glass – An Approach to Granting Emergency Access to Healthcare Systems. NEMA/COCIR/JIRA Security and Privacy Committee (SPC).

Ni, Q., Bertino, E., Lobo, J., Brodie, C., Karat, C.-M., Karat, J., Trombeta, A., 2010. Privacy-aware role-based access control. ACM Trans. Inf. Syst. Secur. TISSEC 13, 24.

Petritsch, H., 2014. A generic break-glass model, in: Break-Glass. Springer, pp. 37–50.

Petritsch, H., Brucker, A.D., 2009. Extending access control models with break-glass.

Preuveneers, D., Joosen, W., 2014. Federated Privileged Identity Management for Break-the-Glass: A Case Study with OpenAM. Presented at the PAHI, pp. 37–52.

Ray, I., Kumar, M., Yu, L., 2006. LRBAC: a location-aware role-based access control model, in: International Conference on Information Systems Security. Springer, pp. 147–161.

Sampemane, G., Naldurg, P., Campbell, R.H., 2002. Access Control for Active Spaces. IEEE, p. 343.

Sandhu, R., 1996. Access Control: The Neglected Frontier. Inf. Secur. Priv. Springer 219–227.

Souza, S., Gonçalves, R., Leonova, E., Puttini, R., Nascimento, A., 2017. Privacy-ensuring electronic health records in the cloud. Concurr. Comput. Pract. Exp. 29, e4045.

Tolone, W., Ahn, G.-J., Pai, T., Hong, S.-P., 2005. Access control in collaborative systems. ACM Comput. Surv. CSUR 37, 29–41.

Trnka, M., Cerny, T., 2016. On security level usage in context-aware role-based access control. Presented at the Proceedings of the 31st Annual ACM Symposium on Applied Computing, ACM, pp. 1192–1195.

Zhao, G., Chadwick, D., Otenko, S., 2007. Obligations for role based access control, in: Advanced Information Networking and Applications Workshops, 2007, AINAW'07. 21st International Conference On. IEEE, pp. 424–431.

ANALYSIS OF OPEN HEALTH DATA QUALITY USING DATA OBJECT-DRIVEN APPROACH TO DATA QUALITY EVALUATION: INSIGHTS FROM A LATVIAN CONTEXT

Anastasija Nikiforova
Faculty of Computing, University of Latvia
Rainis Boulevard 19. Riga, LV-1586, Latvia

ABSTRACT

This research focuses on the analysis of the quality of open health data that are freely available and can be used by everyone for their own purposes. The quality of open data is crucial as it can lead to unreliable decision-making and financial losses, however, the quality of open health data has even more critical role. Despite its importance, this topic is rarely discussed. Therefore, the previously proposed data object-driven approach to data quality evaluation is applied to open health data in Latvia in order to (a) evaluate their quality, highlighting common quality issues that should be considered by both, users and data publishers, (b) demonstrate that the used approach is suitable for given purpose as it is simple enough, and ensures the involvement of users even without IT and data quality knowledge (domain experts) in the data quality analysis examining data for their own purposes. The proposed solution seems to be useful in establishing communication between data users and publishers, improving the overall quality of data.

KEYWORDS

Open Data, Health Data, Data Quality, Data Object-Driven Approach

1. INTRODUCTION

Data quality issue was the topic of interest for many years, and, nowadays, since data are everywhere, and their amount continuously increases, this issue is still popular and topical. One of the reasons of its topicality is the popularity of e-Government projects. However, despite the high number of open data sets all over the world, there is no confidence that published data are of high quality. More and more researches indicate the existence of data quality problems in open data. This issue is especially important in the case of health data, as it can lead to unreliable decisions, financial losses or even tragedies. As health and healthcare data are very broad concepts (Cabitza & Batini, 2016), this research focuses on one subdomain - open health data. Despite open health data don't have sensitive data on people, i.e. patients, they are needed for health(care) planning and administrative purposes (Kerr et al., 2007). As the level of details of "open" data might be lower in comparison with "closed" data stored in databases, it is assumed quality checks could be simpler.

In addition, as open data are usually used by wide audience that may not have deep knowledge in IT or data quality areas, a solution that should be applied to open data sets should be simple enough ensuring particular users with possibility to take part in the analysis of open data. To achieve this aim previously proposed user-oriented data object-driven approach (Bicevskis, 2018a; Nikiforova, 2019) should be examined in order to ensure it is applicable on such data sets providing all necessary functionality making its quality analysis.

The paper is structured as follows: a short overview of the state of open [health(care)] data quality (Section 2), a description of data object-driven approach to data quality evaluation (Section 3), an analysis of open health(care) data (Section 4), conclusions (Section 5).

2. OPEN DATA QUALITY

Nowadays, open data become more and more popular all around the world. As an example, European Data Portal *https://www.europeandataportal.eu* collects more than 800 thousand data sets. Moreover, open data significantly impact economy, as in accordance with (Tinholt, 2013), the aggregate economic impact from applications based on open data across the EU27 economy is estimated to be €140 billion annually. At the same time, the 2013 McKinsey Global Institute report estimated that open data could add over $3 trillion annually in total value to the global economy (Castro et al., 2015). In accordance with G8 Open Data Charter, Open Government Data (OGD) impact economic growth, improving government services, reducing fraud and waste. It is obvious, that data quality has the crucial role in achieving these aims.

According to 8 principles of open data (SunlightFoundation, 2007), in order to admit data as open data they must be: complete; primary; timely; accessible; machine-processable; non-discriminatory; non-proprietary; licence-free. The main idea of every principle won't be discussed here, however, the main point is that these principles don't focus on data quality (Nikiforova, 2018a). The same tendency is observed in the case of evaluation of OGDs, as in accordance with (Klein et al., 2018), the quality aspect takes only the 4th place by popularity after policy, benefit and risk in spite of the fact that quality can impact these aspects. At the same time, in accordance with European Data Portal, data quality appears as one of most problematical dimensions for open data portals.

Importance of data quality was discussed for multiple times and, unfortunately, the results are disappointing. The list of researches indicating the existence of data quality problems in open data is quite impressive (Ferney et al., 2017; Kerr et al., 2007; Kuk and Davies, 2011; Martin, 2014; Nikiforova, 2018a, 2018b; Nikiforova and Bicevskis, 2019; Vetrò et al., 2016 etc.).

2.1 The State of Open Data in Latvia

In 2017 the Latvian Ministry of Environmental Protection and Regional Development has launched the new Latvian Open Data Portal (European Data Portal, 2018) - *data.gov.lv*. At the moment of its launch it had 33 data sets from 13 data publishers, in July of 2018 it had 139 data sets from 41 publishers (Nikiforova, 2018b), but in June of 2019 – 217 data sets from 62 publishers. It demonstrates how quickly the number of data sets increases. Moreover, it isn't the only one prove of Latvian OGD development and progress.

Latvia is one of 70 countries participating in the Open Government Partnership (2017) - an international platform for domestic reformers that committed to making their governments more open, accountable, and responsive to citizens. At the same time, in accordance with Open Data Maturity report (European Data Portal, 2018), that measures open data maturity across Europe, Latvia belongs to the fast-trackers (among beginners, followers, fast-trackers, trend-setters). Moreover, Latvia has the highest rate of open data maturity in comparison with neighbourhoods from Baltic States and Scandinavian countries. Furthermore, in recent years open data maturity of Latvian open data portal continuously improves, as in 2016 Latvia was ranked 31st, in 2017 – 20th, however in 2018 - the 12th. As for the quality aspect, that was added to the series of indicators only in 2018 data, Latvia takes the 11th place with 370 of 520 points. An important point is that in comparison with average rate for all analysed countries, the state of the quality of open data maturity for Latvia is the worst aspect among impact, policy, portal, and quality (only 62% while the average is 71%).

2.2 Open Health(care) Data

In last decades, the volume of health(care) data continuously increasing, and it is expected to grow dramatically in the years ahead (Raghupathi, 2014). Nowadays, open health data is one of the most popular categories of open data, and, according to European Open Data portal, in Europe it takes the 6th place among 13 categories. As for Latvia, health data take the 6th place among 14 categories, however, in 2014 they took the 7th place among 9 categories (Bojārs and Liepiņš, 2014). It means that their popularity in Latvia increases.

Aims and possible uses of open health(care) data can be very different, as health data and information are characterized by multiple number of possible applications, uses and users (Cabitza and Batini, 2016). Some of them can be compared with (Schmidt et al., 2015) as open health data can (1) form the basis for health and medicines authority's hospital statistics, or health economic calculations, (2) provide authorities with data to support hospital planning, (3) provide data to support the authorities responsible for hospital inspection,

(4) monitor the frequency of various diseases and treatments, (5) provide a sampling frame for medical research, (6) facilitate quality assurance of the health(care) services etc.. Moreover, according to Andreassen et al., (2007), the US studies have found that between 56% and 79% of Internet users seek health information online. For Latvia this proportion is 35%, for Poland - 42% with the lowest proportion of Internet health users in the Southern countries - 30% in Portugal and 23% in Greece. It means that health(care) data must be of high quality, as in spite of the fact that some open health(care) data sets contain mainly numerical data that, possibly, doesn't make sense to common user, there are such data sets that can be very useful searching data on medications, their dose, contraindications and other information available for the wide audience.

Despite data quality has crucial role, the current state of the [open health(care)] data quality is not reassuring. As an example, the existence of data quality issues was detected in the Danish National Patient Registry (Schmidt et al., 2015), even in spite the authors analysed data set using only two measures (dimensions), namely validity and completeness. Moreover, despite analysed data were "closed", identified data quality problems are common for "open" data, too. The list of researches discussing quality of health(care) data in many countries such as New Zealand - (Kerr et al., 2007), (Raghupathi, 2014), Denmark - (Schmidt et al., 2015), Brazil - (Oliveira, 2016), USA, UK and Japan - (Yi, 2019), Switzerland - (Wanner et al., 2018), Sweden - (Tomic, 2015), Colombia – (Prieto Rodríguez, 2018), comes to the one conclusion – health(care) data have data quality problems (with the sole exception of Norway – (Larsen, 2009)).

3. DATA OBJECT-DRIVEN APPROACH TO DATA QUALITY EVALUATION

The use of open data brings significant benefits data providers as because of multiple number of possible use-cases, data users address various challenges that can rarely be solved by data providers alone. As a result, a simple and constructive approach to data quality evaluation is needed in order to provide users with possibility to take part in open data analysis even if they don't have deep knowledge in IT and data quality areas. The data object-driven approach to data quality evaluation previously proposed in (Bicevskis, 2018a; Nikiforova, 2019) is considered as highly appropriate for given purpose.

It doesn't use "quality dimension" concept, that became traditional for data quality solutions, as there is no universal classification that should be followed by data quality solutions' developers and users, despite the high number of attempts. In accordance with Scannapieco et al. (2005), even if there is agreement on dimensions that are universally important, there is no agreement on their meanings. In different proposals, the same name is often used to indicate semantically different things and vice versa. In addition, each dimension can be supplied with one or more metrics that varies from one solution to another. As a result, the number of different data quality dimensions, their definitions and grouping are often useful for only particular solution. As an example, Kerr et al. (2007) analyses the quality of New Zealand's healthcare data, and despite they admit the problem of different meanings assignation to the same dimensions, they choose 6 data quality dimensions, supplying them with 24 characteristics and 69 data quality criteria. Such high number of criteria defined by authors minimise a possibility that this solution will be used, especially by non-IT and non-DQ experts. Moreover, while some researches propose too high number of dimensions, certain researches limit them to accuracy and completeness ignoring other dimensions (Dahbi et al., 2018; Weiskopf and Weng 2013). The proposed approach doesn't respect such limitations and provides possibility to define data quality requirements that depends on the use-case. Taking into account the common data quality division in syntactic and semantic accuracy (Scannapieco, et al., 2005) as well as multiple issues appearing from "quality dimension" concept, the proposed approach avoids it, respecting wider concept "data quality requirement" where "data quality dimension" is its subset. Syntactic accuracy of the proposed solution can be found in (Bicevskis et al., 2018a; Nikiforova, 2018b), however the semantic accuracy was presented in (Nikiforova and Bicevskis, 2019). As the comprehensive discussion on the existing solutions, rationale and detail of the given approach were discussed previously, they won't be discussed in detail here, explaining only the basics of the solution that should be understood by the users.

The data object-driven approach consists of 3 main components: (1) data object, (2) quality requirements and (3) quality measuring process. All components are defined using graphical flowchart-based diagrams, that are based on the developed Domain Specific Languages (DSL) that are easy to read, interpret and modify even for non-IT experts. In scope of particular research, these components are demonstrated by the example of one of analysed data sets - Medicinal Product Register of Latvia.

Data object defines data which quality should be analysed. Two types of data objects are possible: (a) primary – data object which quality is analysed, (b) secondary – data object against which primary data object can be analysed ensuring contextual/ semantic analysis of primary data object (Nikiforova and Bicevskis, 2019). Data object "Medicinal_Product" is depicted with 9 of 41 parameters (Figure 1). It is primary data object which quality is analysed against three secondary data objects, namely, "Country", "Country_LV" and "ATC", that are provided by other open data publishers.

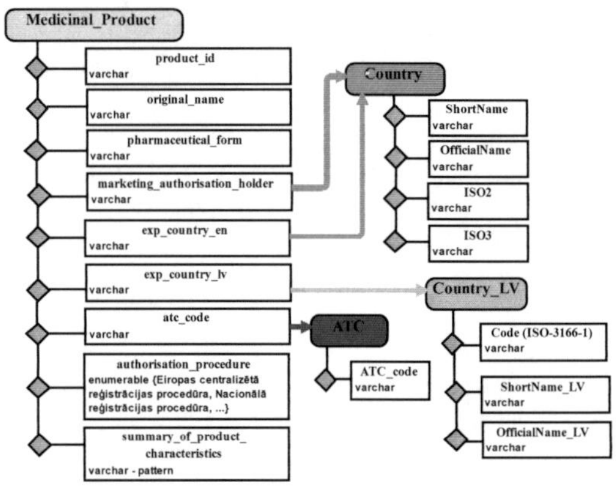

Figure 1. Data object (DO) "Medicinal Product"

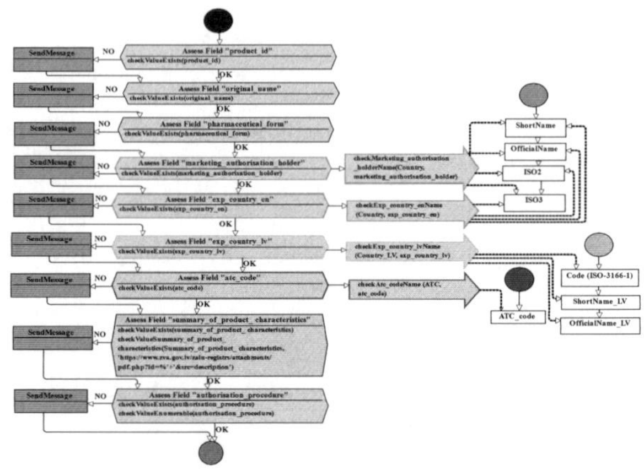

Figure 2. Quality specification for DO "Medicinal Product"

As for data quality specification, conditions that must be met in order to admit data as qualitative, are formulated using logical expressions where operand of every condition is a name of the data object's attribute (Figure 2). Data quality specification as well as data object can significantly vary depending on the use-case. Data quality conditions are formulated for both, attributes in context of one data object, and attributes in context of multiple data objects, depicting relations of appropriate attributes by arrow with logical condition on how these attributes are interconnected. Every parameter of the primary data object can be related to an arbitrary number of parameters of the secondary data object.

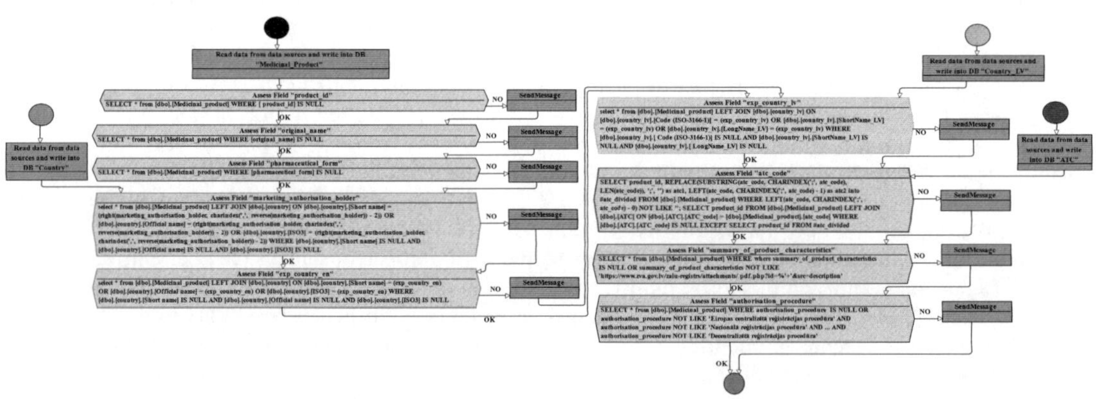

Figure 3. Data quality evaluation process of data object "Medicinal Product"

The last step is description of quality measuring process (Figure 3). It defines a procedure that must be performed to evaluate data quality, from data reading from a data source and their loading to database for further processing to replacing of quality conditions defined at the previous step by executable SQL queries (or another executable artefact, as its nature depends on the skills, experience and preferences of a user involved at this stage). This step is supposed to be done by an IT-expert. The result of the execution of this diagram is a protocol containing all records that don't conform defined quality requirements.

4. DATA QUALITY ANALYSIS OF OPEN HEALTH(CARE) DATA

All open health(care) data sets in Latvia were firstly published in 2018 and now they are represented by 15 data sets from 7 different data publishers (the 1st and 2nd columns in Table 1).

Despite the part of data sets is supplied with metadata, specified by CKAN, the majority of data publishers do not fill all metadata or do not use standard values for them as only 8 out of 15 data sets are supplied with explanation of parameters. This issue is typical not only for Latvia but also for other countries, for instance, Brazil (Oliveira, 2016). Moreover, only 6 out of 15 data sets are updated as frequently as it is promised. Possibly, identified problem occurs because by the term "frequency of updates" data publishers understand the frequency of data updates in their systems instead of data publishing as open data sets. This issue is also widely spread among other countries, as, according to Tinholt (2013), 22 out of 23 analysed countries and even the UK (Kuk and Davies, 2011) publish data that are not regularly updated.The most popular open data format in Latvia is *.xlsx* as 53.3% of data sets are provided in this format, while 26.7% in *.zip* archive including data sets in *.xlsx* and *.csv* format, and 1 data set in *.html* that cannot be considered open data in the traditional meaning as it is link on the statistics ("Database of official statistics"). The positive point is that almost all available data sets are provided in machine-readable format.

As for data quality, that is the focus of the current research, every parameter of every data set was analysed in order to make comprehensive data quality analysis. Apart 15 primary data objects, 11 secondary data objects were involved in data quality analysis and applied on 35 parameters of primary data objects.

The most popular and frequently occurred data quality issues in analysed data sets are: (a) contextual data quality issues, (b) empty values (completeness); (c) multiple notation for the same object in scope of one data object and even parameter; (d) issues in interrelated parameters. Table 1 provides an overview of these data quality issues by number of parameters in which they occur for every data set.

Table 1. The most common data quality problems by data set

Publisher	***Dataset***	***Context issues/ context total***	***Empty / Total***	***Multiple notation/ Total***	***Clean/ Total***
Centre for Disease Prevention and Control	Incidence of 2nd type diabetes in Latvia	0/0	0/6	0/6 (0)	6/6
Ministry of Welfare	Distribution of persons receiving tech aid by AT	2/2 (100%)	3/7 (43%)	0/7 (0)	2/7
	Number of social service providers	2/2 (100%)	22/27 (82%)	10/27 (37%)	4/27
	Persons with disabilities by the severity of the disability and AT	2/2 (100%)	0/23 (0)	0/23 (0)	20/23
	Number of children with disabilities by AT	2/2 (100%)	0/10 (0)	0/10 (0)	8/10
State labour inspectorate	Accidents at work	(0-1/1) (0-100%)	1/10 (10%)	0/10 (0)	8/10
	Occupational diseases confirmed	4/5 (80%)	2/11 (18%)	1/11 (0.09%)	9/11
National Blood Donor Centre Statistics	National Blood Donor Center Statistics	0/0	0/4 (0)	0/4 (0)	4/4
State Agency of medicines	Register of licensed pharmaceutical companies	1/2 (50%)	17/38 (45%)	0/38 (0)	19/38
	Medicines consumption statistics	3/3 (100%)	5/8 (63%)	2/8 (25%)	0/8
	Medicinal Product Register of Latvia	4/9 (44%)	21/41 (51%)	1/41 (2%)	14/41
Food and veterinary service	Food supplements register	2/2 (100%)	30/35 (86%)	4/35 (11%)	5/35
	Dietary foodstuffs register	2/2 (100%)	19/22 (87%)	4/22 (18%)	3/22
	Veterinary medicinal product register	1/3 (33%)	16/26 (62%)	0/26 (0)	8/26

One of the most commonly appeared data quality issue was empty values (the 4th column of Table 1). More precisely, 136 out of 167 (81.4%) analysed parameters at least one value was empty. The number of empty values per data set and parameter varies from 1 to all values of a certain parameter. However, the total number of empty values in analysed data sets is 15%. In some cases, problem of empty values appears for the primary data of the data sets, for instance, in the case of "Dietary foodstuffs register", 4 records don't have *[Name]* and *[ProducerName]* despite these parameters are primary for a particular data object. However, sometimes, an empty value may be allowed, for instance, in the case of the "Food supplements register", the

[Reason for cancellation] parameter must have value only if the value of parameter *[Status]* is "*cancelled*". In other words, if a parameter should have value only when some condition is true, or it is not mandatory, it may not be considered a data quality problem (depending on the use case). 28 of 136 detected empty values may not be considered as data quality issues, however, while there are no any notes from the data publisher regarding the nullability of these parameters, there is no certainty, that there are no any problems there, as empty values may have different interpretations. The issue of completeness is almost "traditional" and it occurs not only in the case of open data (Oliveira, 2016; Yi, 2019) but also in "closed" data (Schmidt et al., 2015; Tomic, 2015) in many sectors and countries (Wanner et al., 2018). However, some researches demonstrate a high level of data completeness can be achieved – Cancer Registry of Iceland (Sigurdardottir, 2012) and Norway (Larsen, 2009) (however, both are examples of "closed" data).

One more widely occurred data quality problem is multiple notations for a single object within a single data set and even a parameter (the 5th column of Table 1). This problem occurs in 6 out of 15 data sets (40%) – in 22 out of 167 parameters (13.2%). It may appear in different ways such as a different name: (a) for one country, (b) for the type of preparation, ingredient or unit size, (c) for different patterns for one value, for instance, phone or registration number (with or without (1) code or (2) delimiter; type of delimiter etc.). For 2 data sets it is represented by two different notations indicating the absence of a value: *NULL* and '*0*' (often called "heterogeneity"). On the one hand, this data quality problem could be considered as insignificant as the value stored in the appropriate field may point on an absence of value. However, there is no confidence that both values *NULL* and '*0*' have the same meaning, as '*0*' can point out to the value that is equal to zero, while *NULL* can mean that the value isn't known. It means, that even if the user has noticed the presence of both of these values, it is difficult to make a decision on how to process these data. As for "type of preparation, ingredient, or unit size", one of the most expressive examples are values of *[Component]* and *[Measure]* parameters, where the same value can appear in multiple ways: (a) singular, (b) plural, (c) shortened form, (d) some of the previous forms but with a spelling mistake. Moreover, while this problem occurs 8 times (2 of them relates to multiple values denoting value doesn't exist), in 5 cases it could be solved, involving the mechanisms, controlling the list of permissible values. At this point, users, inserting data, choose value according to their viewpoint. This problem is also widely spread for many sectors and even countries, and one of the most expressive examples is OGD of the UK (Kuk and Davies, 2011).

As for the contextual data quality analysis within multiple data objects (3rd column in Table 1), only 1 data set among 12 analysed (8.3%) didn't had any data quality issues ("Accidents at work"), however, some manipulations were needed in order to achieve this result. The data set "Accidents at work" was analysed against the secondary data object ("Work codes") in order to check the correctness of provided codes by comparing them with the codes conforming the standard. When this check was made for the first time, comparing the codes in both data sets, it appeared, that the values of the *[Occupation of the injured person (code)]* parameter don't correspond to the work codes, until the moment when the patterns of both values from both data sets were analysed and compared. The pattern of the values of the analysed data set was modified: the code value "*88.3332-03*" corresponds to the values "*8332*" – "*03*" (two parameters) in the secondary data object. In result of this comparison, any quality problems weren't detected, however, it was a relatively complex check as the comparison of these data sets can be performed only after multiple modifications of the parameters' values. As for data quality, such manipulations should be avoided, ensuring homogeneity between data sets. Another example relating to this category, is the case of 4 data sets published by the Ministry of Welfare, where 3 records have data quality issues in the *[ATTU code]* and *[City, county]* parameters. These parameters are supposed to store the code of the administrative territory and city that must correspond to the secondary data object "Classification of Administrative Territories and Territorial Units". However, 3 values of the primary data objects are invalid – aren't available in the secondary data set. However, possibly, the data publisher is aware of this data quality problem, as the appropriate values of the *[City, county]* parameter - "*Total*", "*Abroad*" and "*Address isn't specified*" make sense. Unfortunately, the fact that data publisher is aware of this issue doesn't appear, as, even if he knows about it, he should inform the data consumers about it, at least by making notes about it together with a description of the data sets' parameters. In other cases, this data quality problem can be easily unnoticed and can lead to inaccurate data analysis results. One more observation is that in spite that the codes in the primary and secondary data objects can be the same, the name corresponding to these codes can vary depending on the data set. However, the difference in names may occur just in notation rather than in the meaning. As an example, "Occupational diseases confirmed" data set was analysed against three secondary data objects. The number of quality defects in *[Classification of Occupational Disease (code)]* parameter is 5, however, in *[Classification of*

Occupational Disease] – 871. While 5 of them are invalid as the code is invalid, 866 out of 871 values can be admitted as of low quality only in the cases, when values should be equal one to one. However, the other 10 out of 15 data sets have more significant data quality problems, storing values that don't correspond to secondary data objects. In total 25 out of 35 parameters (71.4%) had at least few data quality issues.

Another problem (or possibly just anomaly) was detected in the "Number of social service providers" data set in the case of three parameters - *[Service with accommodation]* and *[Service without accommodation]* and *[Service with and without accommodation]*. From the users' viewpoint, it is almost obvious, that the total number of *[Service with accommodation]* and *[Service without accommodation]* for every particular record must be equal with *[Service with and without accommodation]* value, however, for 95 records this assumption is not in force. There are at least two possible explanations of such result: (1) there are 95 data quality problems; (2) these field aren't interconnected, and the sum of values of the first two parameters not necessarily should be equal with the value of the third parameter. However, as information on how these values were calculated isn't provided, it is not possible to understand, which of these options is correct. A similar anomaly was observed in the "Number of children with disabilities by administrative territory" data set. In accordance with the principles of open data, data publishers must provide a brief explanation of the parameters and how numerical data was gotten, in order to avoid such misunderstandings. Another problem appearing for 4 out of 15 data sets (26.7%) is the different number of interrelated values. It may appear in different ways: (a) values in different languages, (b) ID number and name, (c) name and supplementary data such as type, country, phone number of representatives.

Few inaccuracies appeared in the result of analysis of the 3 data objects relate to an inaccurate choice of the values delimiter in the *.csv* files listing more than one value per parameter. The delimiter that separates multiple values of one parameter is the same with delimiter of parameters. As a result, the processing of these data sets became more complicated as it was necessary to understand in which cases the delimiter separates multiple values of one parameter, and in which – multiple parameters, since both types of delimiters are the same. It is highly recommended to use different delimiters for parameters and multiple values of one parameter. This issue is also common for open data sets of many countries (Kuk et al., 2011).

5. CONCLUSION

Despite the importance of data quality, the quality of open data is not always one of the main areas of analysis and evaluation of open data. The paper demonstrates and summarizes a quality analysis of Latvian open health(care) data, highlighting commonly appeared data quality problems and some specific problems that should be avoided by data publishers and taken into account by data users. Unfortunately, results of the research demonstrate open health(care) data have a high number of different data quality problem, however, data publishers (who provides data used in their IS), probably, don't even aware of them. The most frequently occurred of them are: (a) contextual data quality issues, (b) empty values even for primary data; (c) multiple denotation for the same object within one data object and even a parameter; (d) issues on interrelated parameters. These quality issues become a tendency for the most part of the analysed items.

Obviously, such an analysis and use of a data object-driven approach to data quality evaluation can be applied not only to open health(care) data but also to other structured and semi-structured data - this solution is effective in many domains. The advantages of the used approach are that (a) it can be applied to "foreign" data sets without any information on how data were accrued and processed – it is an external mechanism with a higher level of abstraction, (b) it can be used even by users without IT and DQ knowledge that is highly important in order to improve data quality by increasing participation and inherent feedback on the quality of data sets (corresponds with (Attard et al., 2015; Ruijer et al., 2019)). Moreover, the quality of data in this way can be improved not only at the national level, but also at the international level as it was already demonstrated in (Bicevskis et al., 2018b, Nikiforova and Bicevskis, 2019).

ACKNOWLEDGMENT

This work has been supported by University of Latvia Faculty of Computing project AAP2016/B032 "Innovative information technologies".

REFERENCES

Agrawal, S. et al., 2013. A survey of Indian open data. In *Proceedings of the 5th IBM Collaborative Academia Research Exchange Workshop* (p. 2). ACM.

Andreassen, H. et al., 2007. European citizens' use of E-health services: a study of seven countries. *BMC public health*.

Attard, J. et al., 2015. A systematic review of open government data initiatives. *Government Information Quarterly*.

Bicevskis, J. et al., 2018a. An Approach to Data Quality Evaluation. In *2018 Fifth International Conference on Social Networks Analysis, Management and Security (SNAMS)* (pp. 196-201). IEEE.

Bicevskis, J. et al., 2018b. Data quality evaluation: a comparative analysis of company registers' open data in four European countries. In *Communication Papers of the FedCSIS* (pp. 197-204).

Bojārs, U., Liepiņš, R., 2014. The State of Open Data in Latvia: 2014. *arXiv preprint arXiv:1406.5052*.

Cabitza, F., Batini, C., 2016. Information quality in health(care). In *Data and Information Quality*, Springer, Cham.

Castro, D. Korte, T., 2015. Open Data in the G8: A Review of Progress on the Open Data Charter.

Dahbi, K. Y. et al., 2018. Toward an Evaluation Model for Open Government Data Portals. In *International Conference Europe Middle East & North Africa Information Systems and Technologies to Support Learning*. Springer, Cham.

Ferney, M. M. J. et al., 2017. Assessing data quality in open data: A case study. In *2017 Congreso Internacional de Innovacion y Tendencias en Ingenieria (CONIITI)* (pp. 1-5). IEEE.

Kerr, K. et al., 2007. Data quality information and decision making: a health(care) case study. *ACIS 2007 Proceedings*.

Klein, R. H., et al., 2018. Open government data: concepts, approaches and dimensions over time. *Revista Economia & Gestão*, 18(49), 4-24.

Kuk, G., Davies, T., 2011. The roles of agency and artifacts in assembling open data complementarities.

Larsen, I. K. et al., 2009. Data quality at the Cancer Registry of Norway: an overview of comparability, completeness, validity and timeliness. *European journal of cancer*, 45(7), 1218-1231.

Martin, C., 2014. Barriers to the open government data agenda: Taking a multi-level perspective. *Policy & Internet*, 6(3).

Nikiforova, A., Bicevskis, J., 2019. An Extended Data Object-driven Approach to Data Quality Evaluation: Contextual Data Quality Analysis. In *21th International Conference on Enterprise Information Systems, ICEIS 2019*.

Nikiforova, A., 2018a. Open Data Quality. In *Baltic DB&IS 2018 Joint Proceedings of the Conference Forum and Doctoral Consortium, Trakai, Lithuania* (Vol. 2158, pp. 00742158-1).

Nikiforova, A., 2018b. Open Data Quality Evaluation: A Comparative Analysis of Open Data in Latvia. *BJMC*, 6(4).

Oliveira, M. I. S. et al., 2016. Open government data portals analysis: The brazilian case. In *Proceedings of the 17th International Digital Government Research Conference on Digital Government Research* (pp. 415-424). ACM.

Open Government Partnership, 2017. Latvia National Action Plan 2015-2017. *https://www.opengovpartnership.org/documents/latvia-national-action-plan-2015-2017-updated-changes*

Prieto Rodríguez, J. D., Suárez Hurtado, V., 2018. Medical Records Digital Perspective Colombian: Safety, Quality, and Management of Data. *Journal of Alternative Perspectives in the Social Sciences*, 9(3).

Raghupathi, W., Raghupathi, V., 2014. Big data analytics in health(care): promise and potential. *Health information science and systems*, 2(1), 3.

Ruijer, E., Meijer, A., 2019. Open Government Data as an Innovation Process: Lessons from a Living Lab Experiment. *Public Performance & Management Review*, 1-23.

Scannapieco, M. et al., 2005. Data quality at a glance. *Datenbank-Spektrum*, 14(January), 6-14.

Schmidt, M. et al., 2015. The Danish National Patient Registry: a review of content, data quality, and research potential. *Clinical epidemiology*, 7, 449.

Sigurdardottir, L. G. et al., 2012. Data quality at the Icelandic Cancer Registry: comparability, validity, timeliness and completeness. *Acta oncologica*, 51(7), 880-889.

SunlightFoundation, 2007. Ten Principles For Opening Up Government Information.

Tinholt, D., 2013. The Open Data Economy: Unlocking Economic Value by Opening Government and Public Data. *Capgemini Consulting*.

Tomic, K. et al., 2015. Evaluation of data quality in the National Prostate Cancer Register of Sweden. *European journal of cancer*, 51(1), 101-111.

Wanner, M. et al., 2018. Indicators of data quality at the cancer registry Zurich and Zug in Switzerland. *BioMed research international*.

Weiskopf, N. G., Weng, C., 2013. Methods and dimensions of electronic health record data quality assessment: enabling reuse for clinical research. *Journal of the American Medical Informatics Association*, 20(1), 144-151.

Yi, M., 2019. Exploring the quality of government open data: Comparison study of the UK, the USA and Korea.

NEW SYSTEM ECONOMIC EVALUATION OF EHEALTH INNOVATIONS

Angela Martin[1], Delphine Cosandier-Rimele[1], Corentin Canu[2] and Gabriële Breda[1]
[1]*ALTRAN Research Department, 2 Rue Paul Dautier, Vélizy, France*
[2]*ENSAM, 151, bd de l'Hôpital, Paris, France*

ABSTRACT

The aim of this study is to provide a new research framework for the economic evaluation of eHealth innovations by using system approach and value analysis. It offers stakeholders an explorative method to find the optimal organization following economic, social and quality of care value creation. It helps them design new process by exploring the link between resources, activities, indicators and expected values. Method application and main findings are discussed for EclairAge tele-expertise case study.

KEYWORDS

ICT, Evaluation, Innovation, Value Analysis, System Complex Modelling, Tele-Expertise Case Study

1. INTRODUCTION

Nowadays, rethinking eHealth business models is crucial. The value proposition of solutions should rely on the expected benefits of the innovation. Modelling the shared value of innovation is important for the decision-making process of both internal and external stakeholders. It also allows them to compare their strategies. Therefore, a systemic vision of the ecosystem is necessary to foster the emergence and the spread of original value propositions.

Facing technical and market uncertainty, one of the key issues of entrepreneurs is to anticipate value creation models at a very early stage. Consequently, there is a need to design business models that may enable the potential of a new technology or service to be fully exploited (Chesbrough *et al.*, 2009).

Among the frameworks provided, the business model canvas (BMC) introduced by Osterwalder and Pigneur (2010) is a tool that specifically helps entrepreneurs to design business models. Based on creativity methods such as design thinking, BMC are effective especially in the early stage of projects or in start-up environment. However, the BMC designed does not provide the support for field experimentation of innovation. Yet, this is the only way to identify and create the necessary inputs to justify new business models (Chesbrough *et al.*, 2009). Therefore, we need a method to successfully combine thinking and modelling with field innovation initiatives (Chanal, 2011).

Following a systemic approach, methods and tools of industrial engineering are suitable for the exploration of case studies from an economic point of view, while taking into account the creation of values pursued by all stakeholders (internal or external to the project). The value emerges at the intersection of performance, costs, planning and risk management. Process value means maximum quality at a minimum cost (Chase, 2001).

For the assessment of stakeholders' satisfaction, we add to the economic the social value and the quality of service/care to the global performance of the system. By expanding the range of indicators available, we design business models by taking into account the specific characteristics of the ecosystem and its stakeholders. The value creation is intrinsically subjective and contextual (Hahn and Kuhn, 2011). Yet, it drives the design and management of projects based on multiple values and no longer on cost minimization.

The goal of this paper is to provide a research framework for designing new telehealth systems based on economic shared value creation. Using complex system modelling, we tackle the complexity of evaluating an eHealth innovation according to the objectives set out by stakeholders. In the next section, we describe the methodology followed by its application to the case study and the discussion of results. Finally, we conclude.

2. METHODOLOGY

Our research framework is based on SCOS'M (Systemic for Complex Organizational Systems' Modelling) method aiming at integrating ex nihilo value creation for a maximum of stakeholders in the modelling of multi-criteria performance (Schindler, 2007). It allows the identification of shared value while taking into account stakeholder's expectations and their impact on system constraints at every phase: feasibility, conception, deployment, exploitation, end of life.

To transform shared value, milestones and expected deliverables, process must be designed to capture the set of redefined values. Resources (human, material, financial) are then mobilized to execute processes. This method also allows the steering of the system.

To tackle the complexity of eHealth ecosystems, we use Multiple Domain Matrix (MDM) applied to SCOS'M method to identify the link between resources, process, indicators, expected values and stakeholders (Jean, 2013). We build four matrix of OCSM (Organization Complex System Modelling) to answer to the following four questions:

1) What are the values expected by which stakeholder? => Stakeholders' Expected Value Analysis (OCSM1)
2) What indicator quantifies which expected value? => Indicators for Expected Value Measurement (OCSM2)
3) What activity is measured by which indicator? => Activity Analysis for Value Creation (OCSM3)
4) What resource finances which activity? => Identification of Resources financing activities (OCSM4).

The 4 matrix help design new process aiming to satisfy all stakeholders.

3. OCSM APPLIED TO A CASE STUDY

We use OCSM to model a telemedicine process for EclairAge project. It is an innovative project conducted by the Research Department of Altran Group. Selected and financed by the French Government, the goal is to prove its efficiency as an eHealth demonstrator in telemedicine on the French territory. The project aimed to create and test innovative services to improve elderly permanence of care in long-term dependency care centres (called EHPAD). Nowadays, French Emergency Services System (SAMU) assures the continuity of care. Two major challenges arise. Firstly, SAMU affiliated physicians are not specialized in elderly care. Secondly, SAMU may become quickly congested. To face these challenges, the tele-expertise service EclairAge is a geriatric hotline for EHPAD's residents during weekdays. We conducted an in situ and in vivo non-randomized trial in five French EHPADs situated in the South of the Essonne department (Ile de France region) for 18 months (August 2014- February 2016).

In order to design EclairAge new process, we started by collecting data. Firstly, we conducted a literature survey to map all the stakeholders of the system and to identify a first list of expected values. Every stakeholder expects different values from the deployment of the innovation. We have consulted over 100 publications: international telemedicine survey (Davalos *et al.*, 2009, Bergmo (2009), public reports (FIEEC and ASIP, 2011, Lasbordes, 2009) and private reports (CSC, 2001, OPIIEC, 2011). Secondly, in order to complete the list iteratively (Miles and Huberman, 2003), we conducted two workshops and individual interviews on expected values with different stakeholders of the local experimentation. These stakeholders are the EclairAge Project Manager, the CEO of the software company that developed the EclairAge solution, a geriatrician, a physician from SOS Doctors, a SAMU representative, an economic expert, an eHealth specialist and 4 elderly patients. Consequently, we mapped sixteen stakeholders into several categories (with subcategories): patients, caregiver, healthcare professionals, industrial, financer, EHPAD (long-term dependency care centres), health institutions (see the columns of the matrix OCSM1, in table 1). The categories of values (see the rows of the matrix OCSM1, in table 1, from *A01* to *A15*) are quality of care, social and economic to facilitate the creation of future process. If the stakeholder expects this value, then the intersection between values and stakeholders is "1".

In the matrix OCSM1 (table 1), we observe that telemedicine expected values are:

- For the patient (EHPAD's residents in the case of EclairAge), we notice an improved access to healthcare and to specialists (e.g. geriatrician) (A01), telemedicine development for improving continuous care (A06), risk reduction related to unnecessary transportation and hospitalizations (A14), and also a new way to interact with healthcare professionals and help shared medical decision (A03).
- For the healthcare institutions and personnel requesting telemedicine (EHPAD in our case study), the stakes are on the collaboration among HP (*A02*) and shared medical decision (*A03*).
- For national or regional health institutions for instance, the quality of access to healthcare (*A01*), the healthcare professionals' collaboration (*A02*), shared medical decision (*A03*), improving medical knowledge (*A09*), better control of health spending (*A11*), and the number of unnecessary hospitalizations or avoidable transportation (*A14*).
- Almost all stakeholders expect the creation and evolution of jobs and medical practices (*A05*), and emphasize issues concerning data security and safety (*A08*).

Table 1. Stakeholders' Expected Value Analysis

OCSM N°1 : Stakeholders' Expected Value Analysis

				STAKEHOLDERS															
				PATIENT		CARE-GIVER	HEALTHCARE PROFESSIONALS (HP)				INDUSTRIAL				EHPAD*	FINANCER		HEALTH INSTITUTION	
				Elderly person	Legal guardien		Geriatricien	Nurse and care assistant	Emergency physician	Other	Manufacturer	Digital	Medical Transportation	Other		Public	Private	National	Regional
EXPECTED VALUES	QUALITY OF CARE	A01	Improving access to healthcare	1	1		1	1	1	1	1	1		1	1			1	1
		A02	Improving colaboration among HP	1		1	1	1	1	1	1	1		1	1			1	1
		A03	Shared medical decision	1		1	1	1	1	1		1		1	1	1	1	1	1
		A04	Work delegation				1	1		1					1				
		A05	Creation/ Evolution of jobs and medical practices	1			1	1	1	1	1	1		1	1	1		1	1
		A06	Improving continous healthcare	1	1		1	1	1	1	1	1		1	1	1		1	1
		A07	Transport risk reduction	1		1									1				
		A08	Data security and safety	1	1		1	1	1	1	1	1	1	1			1	1	1
	SOCIAL	A09	Improving medical knowledge			1	1	1	1	1		1			1			1	
		A10	Improving the quality of life	1	1							1						1	
	ECONOMIC	A11	Better control of health spendings	1		1	1	1		1	1	1		1	1	1	1	1	1
		A12	Activity development								1	1	1	1				1	1
		A13	Creation of new jobs	1			1					1		1	1			1	1
		A14	Reduction of unnecessary hospitalizations / transport	1		1	1	1	1	1	1	1		1	1	1	1	1	1
		A15	Medical time optimization				1	1		1					1	1		1	

After the identification of expected values by stakeholders, we searched for indicators to measure the creation of these values (see in table 2 below). The rows of the first matrix OCSM1 (*A01* to *A15*) are now the columns of the second one OCSM2. If the indicator influences the value creation, the intersection is "1".

For instance, for the creation of value "better control of health spending" (A11), we consider the indicators "service access cost" (I01) and "yearly average rate of medical expenses per patient" (I07). Another example: the indicators "number of telemedicine jobs" (I09) and "yearly telemedicine turnover in France" (I10) influence the value creation "activity development" (A12).

Table 2. Indicators for Expected Value Measurement

OCSM N°2 : Indicators for Expected Value Measurement

		STAKEHOLDERS' EXPECTED VALUES														
		QUALITY OF CARE								SOCIAL		ECONOMIC				
		A01	A02	A03	A04	A05	A06	A07	A08	A09	A10	A11	A12	A13	A14	A15
INDICATORS		Improving access to healthcare	Improving colaboration among HP	Shared medical decision	Work delegation	Creation/ Evolution of jobs and medical practices	Improving continous healthcare	Transport risk reduction	Data security and safety	Improving medical knowledge	Improving the quality of life	Better control of health spendings	Activity develop ment	Creation of new jobs	Reduction of unnecessary hospitalizations / transport	Medical time optimization
I01	Service access cost	1										1				
I02	Advanced Encryption Standard (AES) encryption								1							
I03	No of emergency visits related to avoidable inter-institutions transportation per year		1													
I04	Average number of physicians consulted for a treatment decision			1												
I05	Availibility rate of specialised medical time						1			1						
I06	Mortality rate related to inter-institutions transportation							1			1					
I07	Yearly average rate of medical expenses per patient											1				
I08	No of transports per patient per year							1							1	
I09	No of telmedecine jobs					1							1	1		
I10	Yearly French telemedecine turnover												1			
I11	The rate between the time spent with patients and total time activity				1		1									1

The next step is the design of EclairAge process. Based on the previous tables, we model the geriatric tele-expertise service taking into account the value creation expected by EclairAge stakeholders (OSCM1) and the indicators quantifying the value creation (OSCM2). The workflow described below (figure 1) was validated by the stakeholders involved (the geriatrician, EHPAD's medical staff, SAMU and SOS Doctors representatives, patients).

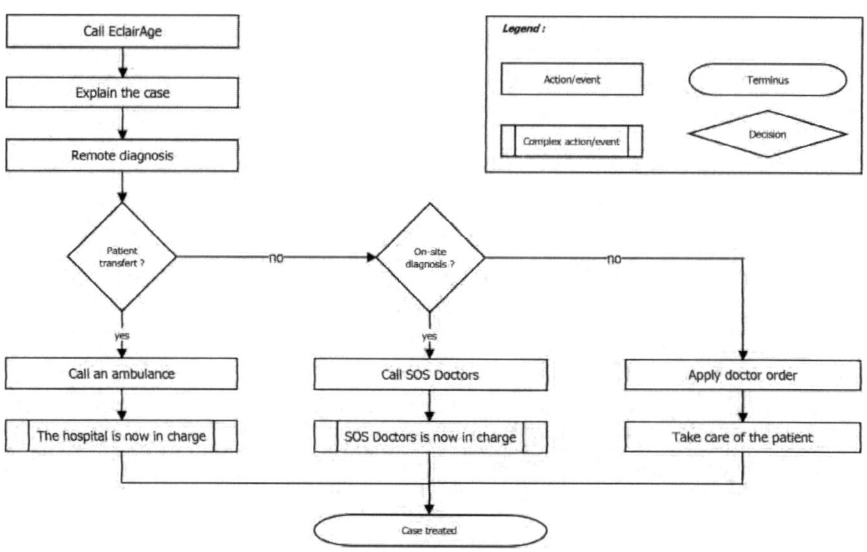

Figure 1. EclairAge Workflow

In case of an alarming situation, the nurse contacts EHPAD's coordinator physician or the patient's attending physician. If available, the case is treated. If not, the nurse calls the emergency service or EclairAge.

Two main benefits arise from the utilization of EclairAge geriatric hotline. First, based on the immediate transmission of EclairAge Emergency Files (containing the complete medical record of all EHPAD' residents), the on-call geriatrician is able to make personalized medical decisions to treat the case avoiding transfers to the hospital or SAMU.

In the case of a confirmed emergency, EclairAge can initiate the intervention of SOS Doctor (private emergency service). As a result, a physician was sent to the caller EHPAD. Therefore, the case was treated without transferring the resident to the hospital. Second, when transfers are unavoidable, the pre-diagnosis performed by EclairAge's geriatrician and the transmission of resident's emergency file may shorten the waiting time at the hospital reception desk.

In the case of a differed emergency, the on-call geriatrician can schedule a hospitalization and help find a hospital bed in geriatrics, saving time and improving residents' reassurance.

Summing up, EclairAge process was designed to improve access to healthcare, in particular to specialist consultation (geriatrician), to reduce transportation risks, to avoid unnecessary transfers, to improve collaboration among HP, to help shared medical decision, to avoid unnecessary hospitalizations by "filtering" patients, to save everyone's time (medical time optimization).

In order to quantify all these gains, we specify quantitative and qualitative indicators like EclairAge access cost, number of emergency visits related to avoidable inter-institutions transportation per year, mortality rate related to inter-institutions transportation, yearly average rate of medical expenses per patient etc.

All five experimenters EHPAD implemented EclairAge process (described in figure 1). Based on this process, the next matrix OCSM3 is Activity Analysis for Value Creation (table 3). We describe the main EclairAge activities (AC01 to AC10) in order to create the expected values for all stakeholders. If the indicator allows the measurement of the activity then the intersection between activities and indicators is "1". For instance, to measure the activity "intervention in EHPAD" we use the indicators: service access cost, number of emergency visits related to avoidable inter-institutions transportation per year and the yearly average rate of medical expenses per patient.

Table 3. Activity Analysis for Value Creation

OCSM N°3 : Activity Analysis for Value Creation										
		INDICATORS								
		I01	I02	I03	I04	I05	I07	I08	I09
		Service access cost	AES encryption	No of emergency visits related to avoidable inter-institutions transportation per year	Average no of physicians consulted for a treatment decision	Availibility rate of specialised medical time	Yearly average rate of medical expenses per patient	No of transports per patient per year	No of telmedecine jobs
ACTIVITIES	AC01 Call EclairAge	1			1	1			
	AC02 Medical Decision				1				
	AC03 Call an ambulance			1			1	1	
	AC04 Plan an emergency visit								1
	AC05 Medical advice								1
	AC06 Call SOS Doctors								
	AC07 Send in the geriatric mobile team / Prepare the intervention						1		1
	AC08 Intervention in EHPAD	1		1			1		
	AC09 Report of decisions and outcomes								1
	AC10 Fill in patient's file	1	1						1

Table 4 illustrates the last matrix of OCSM linking activities to resources. If the activity consumes resources, their intersection is "1". For example, to perform the activity "call EclairAge", human resources (EHPAD's nurse, EclairAge geriatrician) and material resources (a telephone and subscription, a touchpad and EclairAge software) are necessary. To plan an emergency visit, the coordination responsible needs EclairAge software installed on a computer.

Table 4. Resource Identification to Perform the Activities

OCSM N°4 : Resource Identification to Perform Activities

			ACTIVITIES									
			AC01	AC02	AC03	AC04	AC05	AC06	AC07	AC08	AC09	AC10
			Call EclairAge	Medical Decision	Call an ambulance	Plan an emergency visit	Medical advice	Call SOS Doctors	Send in the geriatric mobile team/Prepare the intervention	EHPAD Intervention	Report of decisions and outcomes	Fill in patient's file
RESOURCES	R01	EHPAD medical staff	1	1	1		1	1		1		1
	R02	Geriatrician	1	1						1		
	R03	Coordination responsible				1			1		1	
	R04	Computer				1			1		1	1
	R05	Telephone	1	1								
	R06	Phone subscription	1	1								
	R07	EclairAge software	1	1	1	1		1	1	1	1	1
	R08	Touchpad	1	1	1			1	1	1		
	R09	Nurse								1	1	

The advantage of using Multiple Domain Matrix (MDM) is to represent and visualise the links between resources, activities, indicators, value creation and stakeholders. We can multiply the matrix of OCSM, as the number of rows of the first matrix equals the number of columns of the second matrix. In the case of numerous rows and columns, calculating each indicator is fastidious. Hence, identifying key indicators helps the evaluation of expected values for all stakeholders. The matrix product of OCSM2 and OCSM1 shows *the importance of indicators by stakeholder*. In our case, *I09* number of telemedicine jobs and *I11* the rate between time spent with patients and total time activity are the most important indicators for a majority of stakeholders. We can explain the results by the following: by implementing EclairAge, job creation in telemedicine is expected. Patients living in EHPAD are complaining that the nurse's average time per patient per day is 11 minutes. By delegating some tasks to the tele-expertise, EclairAge service, patients and healthcare professionals expect better use of the medical time spent by patient.

Furthermore, by multiplying matrix OCSM3 and OCSM2 we find the *key activities that respond to the majority of expected values*. Here, activity *AC01* (call EclairAge), *AC08* (intervention in EHPAD) and *AC10* (fill in patient's file) covers a wide range of expected values. Therefore, by improving only this key activity, it will improve the whole system.

In addition, we can facilitate some activities to better satisfy expected values considered of high priority. We can therefore allocate more resources or of better quality to key activities. We can use OCSM matrix as a support to decision making by changing or suppressing non-core activities. Indeed, a time consuming activity or an activity that only answers to few expected values is inefficient.

4. CONCLUSION

Our research framework aims at defining the system boundaries of eHealth innovations. It uses complex system-modelling tools based on value analysis, suitable for the economic evaluation of projects. Organisation Complex System Modelling (OCSM) enables the design of new process based on the value creation expected by the system's stakeholders. Applying Multiple Domain Matrix (MDM) to OCSM helps us represent the resources used, by which process, for the creation of which values. Applied to EclairAge case study, this methodology helped us design the new process while tackling with the complexity of organizational change.

Further work is needed to exploit the process designed and to compare *a-posteriori* the expected values with the completed values.

REFERENCES

Bergmo, T., 2009. Can economic evaluation in telemedicine be trusted? A systematic review of the literature. *In Cost Effectiveness and Resource Allocation*, 7, 18.

Chanal, V., 2011. *Business Models dans l'innovation : pratiques et méthodes*. Presses Universitaires de Grenoble.

Chase, J., 2001. Value creation in the product development process. Massachusetts Institute of Technology.

Chesbrough, H., Enkel, E., Gassmann, O., 2009. Open R&D and open innovation: exploring the phenomenon. *In R&D Management*,Vol. 39, pp. 311-316.

Davalos, M., French, M., Burdick, A., and Simmons, S., 2009. Economic Evaluation of Telemedicine: Review of the Literature and Research Guidelines for Benefit–Cost Analysis. *In Telemedicine and e-Health*, 15, pp. 933-948.

Jean, C., Duong, T., Cardinal, J., Jankovic, M., Bocquet J., and Espinoza, P., 2016. Sharing economic value between the stakeholders of a telehealth project: methodological issues?, *In Eur. Res. Telemed.*, vol. 5, pp. 37–44.

Hahn, G., Kuhn, H., 2011. Value-based performance and risk management in supply chains: A robust optimization approach. *In International Journal of Production Economics*.

Lasbordes, P., 2009. La télésanté: un nouvel atout au service de notre bien-être - Un plan quinquennal éco-responsable pour le déploiement de la télésanté en France.

Osterwalder, A. and Pigneur, Y., 2010. *Business Model Generation*. John Wiley & Sons.

Schindler, A. 2009. Vers la multi-performance des organisations : conception et pilotage par les valeurs du centre de recherche intégré MIRCen du CEA. Doctoral thesis, Laboratoire Génie Industriel, Ecole centrale Paris.

A NOVEL WEB APPLICATION (FOODCOACH) TO MONITOR AND IMPROVE EATING BEHAVIORS: PILOT RANDOMIZED CONTROLLED TRIAL

Diana Trojaniello, Francesca Sacchitelli, Elena Massa, Pietro Vismara and Alberto Sanna
Ospedale San Raffaele, Via Olgettina 60, Milan, Italy

ABSTRACT

Nowadays, overweight and obesity are considered a global public and health problem. The use of ICT in weight management programs has been reported in improved long-term adherence and nutritional habits. Few studies validated ICT based approaches versus standard paper-based approaches. This study aimed at investigating the efficacy and acceptability of a novel Web App, i.e. FoodCoach, to guide and monitor dietary behaviours in an healthy sample population. A two arms randomized controlled trial study has been designed. Fourty healthy subjects have been included in two groups: the first group (control) obtained a standard paper-based not-personalized dietary intervention; the second group (experimental) obtained a novel Web-based personalized dietary intervention through the FoodCoach Web App. Both the groups were remotely followed for 40 days by expert dieticians, the first through telephone and emails, the second trough the Web App. At the start and the end of the 40-days experimental period, physical appointments with the dieticians and research team at the hospital premises have been organized. The results showed a significant weight loss, Body Mass Index and Fat Mass percentage reductions in the experimental group participants, while no differences in the control group subjects have been found. The acceptability of the proposed Web App was explored by means of questionnaire and engagement metrics (frequency of usage): a good level of engagement over the 40 days and high scores in terms of usability and user experience of the participants have been showed.

KEYWORDS

Nutritional Platform, Mediterranean Diet, Web Platform, E-Health, Remote Monitoring, User Experience

1. INTRODUCTION

Overweight and obesity are a global public health problem. About 2 billion of people worldwide are overweight and one third of them obese (Seidell and Halberstadt, 2015). In Italy, the Organisation for Economic Co-operation and Development (OECD) reports about 1 in 10 people is obese and OECD projections indicate that overweight rates will increase by a further 5% within ten years (*OECD statistics*, 2018). Wrong eating behaviors and a lack of physical activity are considered the primary causes.

Nowadays the gold standard treatment in clinical practice for overweight and obesity management consists of: a) a balanced diet prescribed by a nutrition expert (e.g. dietologist, dietitian); and b) a proper diet-daily assessment performed by both the dietitian and the patient itself, i.e. self-assessment (Naska, Lagiou and Lagiou, 2017). Several tools, such as 24-h dietary recall, Food Frequency Questionnaire FFQ (retrospective) and daily diet-diaries (prospective), have been recommended for dietary assessment. In particular, paper diet-diaries are commonly advocated as the gold standard tool for patient's eating behavior monitoring: guidelines recommend that patients provide a detailed account of timing, type and structure of everything eaten or drunk for at least three consecutive days (Burke *et al.*, 2005). Studies showed that completing food diaries, in combination with dietary advice, is associated with weight loss (Kong *et al.*, 2012). However, the paper diet-diaries are characterized by a low adherence (Glanz *et al.*, 2006; Knudsen *et al.*, 2011; Gondolf *et al.*, 2012) mainly due to the burden associated to a) the food intake monitoring through a not automated tool and b) the memory effort requested. Therefore, the need of more engaging tools to improve diet diary filling usage and diet adherence is widespread.

Recent innovations in information and communication technology (ICT) offer new opportunities to overcome such limitations in diet management by providing new solutions for dietary assessment (Personal

Digital Assistant, PDA, or Web-based diaries) and, at the same time, allowing experts to follow patients located in remote areas. Indeed the use of ICT tools in daily life activities is widespread among the population and, in this context, the diffusion of ICT tools, including mobile Health Applications (mHealth Apps), aimed to assist, inform and guide users in health issues management is promising (Ozdalga, Ozdalga and Ahuja, 2012). Therefore a growing interest has been registered in the use of ICT technologies to handle diet interventions: nutrition Apps for mobile devices (e.g., PDA, smartphones, tablets) are becoming increasingly accessible (Gilmore *et al.*, 2014) and can assist the patients with the difficult task of intake recording for dietary assessment and self-monitoring (Lieffers and Hanning, 2012). The majority of commercially available nutritional mApps allow users to retrieve information on foods, track their food consumption and receive pills to lose weight or to improve nutritional behaviours. Indeed the use of ICT tools for personalized diet interventions has been found to improve adherence in diet-diary filling, increased awareness of the actual eating habits, help subjects in following dietary prescriptions and allow experts to remotely access to the patient's nutritional data (Probst and Tapsell, 2005). A recent study found that nutritional mApp use is also linked to significant short-term weight loss and that this weight loss is associated with adherence (Jacobs, Radnitz and Hildebrandt, 2017). On the other hand, nutritional mApps require users to actively engage with it in order to benefit and reach the personal objectives. Despite the proliferation of nutritional apps and their apparent potential in dietary interventions, there is currently limited evidence on the experience of using these ICT tool and how participants perceive the comparison between novel and traditional methods of dietary assessment (Carter *et al.*, 2013; Jimoh *et al.*, 2018). Indeed, the validation of the proposed ICT tools in terms of effectiveness, adherence and user engagement related to the dietary intervention against standard methodologies is desired. In addition, a stand-alone intervention (e.g. without the expert support), as that commonly proposed by the commercially available ICT tools, could be inappropriate (e.g. the patient could follow an unbalanced diet) and not effective.

We developed a responsive Web App for desktop and mobile devices which offers a Personalized Nutritional Plan (PNP) to guide users towards a healthy diet and behaviors, called FoodCoach (FC), by integrating Care-receiver (i.e. the patient) and Expert Care-giver (i.e. the nutrition expert) assessment and fostering Care-receiver reflective learning and self-management. The new developed tool allows the Care-receiver to: a) access the PNP established by the nutrition expert; b) track the food and beverages consumption; c) set periodical appointments (weigh yourself, appointment with the experts); d) self-monitor and track the progresses; e) access educational material provided by the experts; f) chat with the experts. At the same time, the proposed tool allows the Expert Care-giver to profile the Care-receiver, to check the automatically generated PNP, to validate the automatically generated dietary prescriptions and to remotely monitoring the Care-receiver progresses. The FC Web App has been benchmarked against commercially available systems, such as "A better plan (Microsoft HealthVault)", "Nutrino" and "SparkPeolple" , and contains a) a large, detailed food database (i.e. "DataFood") which includes the INRAN Italian database, the french CIQUAL database, and the american USDA one; b) weekly/daily personalized plan by Expert Care-giver; c) recipes for every suggested food including food pictures, nutritional details, ingredients and the descriptions; and d) the possibility to receive, in a quasi-real time manner, dietitian feedbacks and advices. These features have been added to engage users with the Web App in a real-life setting.

The current study aims to evaluate the efficacy, ease of use, acceptability, user engagement and perceived effectiveness of FC platform in providing a personalized nutritional intervention, for the diet management of a healthy population sample compared with a more traditional paper-based approach, through a two-arms randomized controlled pilot.

2. MATERIALS AND METHODS

2.1 The FoodCoach Platform

The FC platform can be described as a food recommendation system that uses a recipe dataset and annotations to recommend menus. It is a responsive web application for desktop and mobile devices and offers to patients a PNP prescribed by an expert dietitian, based on metabolic needs and preferences. It is based on a Front-Office portal addressed to patients and on a Back-Office one addressed to dietitians. Dietitians insert patients personal data (e.g. age, weight, BMI, nutritional habits, etc.), thus allowing the platform to properly compute the PNP and dietary suggestions. The FC platform (Back-Office) is characterized by the following sections:

- The *Personalized Nutrition Plan* (*PNP*) section through which the user accesses the PNP and consult the suggested meals everyday. The user may also choose between three different substitutions for any single meal. In the PNP are suggested some recipes;
- The *Food Diary* section which allows the user to daily record meals consumed, also being able to specify the quantity, the place and the time. It is also possible to indicate a missing meal and the motivation. In addition, it is also possible to track the daily water intake (*Water filling*);
- The *Food Frequency Questionnaire (FFQ)* section in which the user auto-fills a questionnaire to rebuild its eating habits, based on the frequency of consumption of a list of foods.
- The *Statistics* section allows the user to monitor the PNP adherence downloading reports that show progresses and body changes;
- The *Educational Material* section allows the user to check informative material related to eating behaviours suitable for a healthy and active lifestyle, such as information concerning nutrition;
- The *Weight Ritual* section allows the user to weekly update the body weight. The user also receives a weekly message as reminder for the body weight measurement. In this way the personal statistics remain updated and the clinical expert is able to monitor user's progresses;
- The *Recipes* section in which the user checks and chooses recipes whose ingredients correspond to the foods provided by the PNP for the suggested meals on the basis of the user's nutritional demand.
- The *Chat* section with the dietitian allows the user to get in touch with the dietitian directly, asking questions and doubts and getting clarifications.

In Figure 1 the landing page of the FC Web App has been showed.

Figure 1. FC landing page: the responsiveness of the proposed Web App allows the users to access it through both PC and mobile devices (i.e. tablet, smartphone)

2.2 Experimental Protocol

The research has been conducted in Milan (Italy) at the San Raffaele Hospital (HSR) premises, from March 2017 to June 2017. The experimental protocol was approved by the HSR Ethical Committee (Protocol 653704-1 approved on 10/11/2018) and informed consent forms have been edited in accordance to the Declaration of Helsinki.

2.2.1 Study Design and Randomisation

The trial design was a single blind controlled parallel 2-arm randomized pilot intervention trial (see Figure 2) undertaken between March and June 2017. Eligible participants were randomised in a 1:1 ratio to either "experimental" (n=20) or "control" group (n=20). The randomisation sequence was generated by a person unrelated to the project. As a pilot trial, the primary outcomes were the users' engagement, acceptance and adherence to the intervention (frequency of use). Secondary outcomes were anthropometric measures. Once included in the study, participants (n=40) have been invited at HSR facilities to attend a first baseline assessment. During the ***baseline assessment***, the participants were assigned to one of the groups (experimental or control) according to the randomization list. Weight, height, Body Mass Index, Fat Mass percentage and body circumferences have been measured during a standard nutritional visit. Baseline self-completed questionnaire was also provided to collect information on demographics and food consumption habits, i.e. Food Frequency Questionnaire (FFQ). For the experimental group participants, the dietician created a FC user profile along with a PNP. Credentials were sent by email to the participant to access the FC platform.

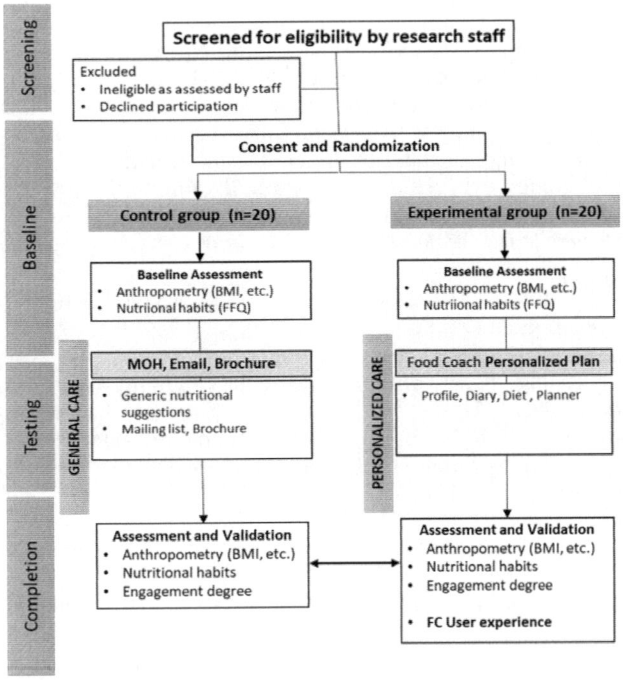

Figure 2. Flow of participants through a randomized 2 arms pilot trial of the FC Web App (N=40)

Participants were then instructed for the ***experimental period*** activities according to the belonging group:

(a) *Control Group* (CG): 40-days "not-personalized" intervention consisting in (a) to follow the nutritional guidelines (INRAN guidelines) provided by the nutritionist twice a week via email; b) to fill the paper diet-diary daily; c) to register their weight weekly and insert the information in the paper diet-diary.

(b) *Experimental Group* (EG): 40-days "personalized" intervention consisting in (a) to follow the *PNP* daily provided by FC; b) to fill the FC *food diary* daily; c) to register their weight weekly and insert the information in the *weight ritual* section.

Participants enrolled in the EG were then introduced by the research team to the FC platform usage to assess the usability of the platform. At the end of the baseline assessment, EG participants were asked to fill the System Usability Scale (SUS) checklist to rank the FC usability.

After the 40-days experimental period, both CG and EG participants underwent to a second appointment at HSR facilities (***final assessment***) where received a second nutritional visit. Weight, height, Body Mass Index, Fat Mass percentage and body circumferences have been measured again. Final self-completed questionnaire was also provided, i.e. FFQ. At the end of the final assessment, EG participants were asked to fill questionnaire about user engagement and user experience to assess the FC user experience (UX).

2.2.2 Recruitment Strategy

Participants were recruited among HSR workers and relatives by advertising through email, intranet and brochures. Advertising material encouraged participants to contact the research team for the preliminary screening. The eligibility criteria were: absence of metabolic diseases, food allergies or any pathology directly related to nutrition; a value of Body Mass Index (BMI) higher than 29.9; aged 18 to 65 years; willing to commit the necessary time and effort to the study; not pregnant, breast-feeding or planning a pregnancy; no involvement in other weight management program; able to read and write in Italian; able to access Internet; willing to be randomized to 1 of the two groups.

A total of 355 subjects were interested in participating to this research. From them, 89 were screened to fulfill the inclusion criteria and 40 subjects in total were enrolled. 49 have been excluded due to the presence of food allergies, metabolic diseases, and for intestinal, liver and kidney diseases.

2.2.3 Interventions

The activities took place both at HSR premises and at participants home. For each participant, two visits at the hospital, one at the beginning and one at the end of the study period (40 days), were scheduled. The experimental period lasted 40 days in order to ensure an adequate range of time allowing the subjects to adhere to the nutritional plan suggested both in personalized (EG) and not personalized (CG) intervention. The subjects were visited at the start and at the end of the experimental period by a registered dietitian. During the 40 days, the subjects were asked to carefully follow the given nutritional suggestions, both personalized and not personalized. At the end of the experimental period, they were invited at HSR to undergo a second nutritional visit and to be evaluated in terms of user engagement degree and UX.

2.2.4 Assessment Tools

a) ***Nutritional assessment*** (both CG, EG)
   - *Anthropometric Indices*: BMI, Fat Mass percentage (%FM) and Weight (W) were considered to evaluate body changes in between the start and the end of the experimentation.

b) ***Usability assessment*** (only EG)
   - *System Usability Scale (SUS)* (Brooke, 1996) a simple, ten-item Likert scale giving a global view of subjective assessments of usability. It is generally used after the respondent has had an opportunity to use the system being evaluated.

c) ***User engagement assessment*** (only EG)
   - *Frequency of use (FoU):* the rate of usage (weekly and overall) of the FC functionalities: Food Diary (and Water filling), Weight Ritual, Chat.
   - *User Engagement Questionnaire* (O'Brien and Toms, 2010): a multidimensional scale to measure user engagement based on six attributes of user engagement (Perceived Usability, Aesthetics, Focused Attention, Felt Involvement, Novelty, and Endurability)

d) ***User experience assessment*** (only EG)
   - *Standardized User Experience Percentile Rank Questionnaire (SUPR-Q)* (Sauro, 2015): 8 item questionnaire for measuring the quality of the website UX. It is a 5-point rating scale to measure perceptions of Usability, Trust, Credibility, Appearance and Loyalty for Websites.

2.3 Data Analysis

- ***Personalized intervention efficacy*** (both CG, EG)

First of all, a Shapiro test for data normality check has been done on the anthropometric indices. Mean (m) and standard deviation (sd) of the anthropometric indices collected during the first (W_{IN}, BMI_{IN}, $\%FM_{IN}$) and second (W_{OUT}, BMI_{OUT}, $\%FM_{OUT}$) visits have been computed for each group. A T-test between IN and OUT measures for each group (CG, EG) has been done to verify if significant differences were present after the 40-days studies in one or both the two groups.

- ***Usability, User Engagement and Experience*** (only EG)

Usability - According to the level of agreement selected for each statement of the SUS questionnaire, a SUS score has been computed for each participant. According to (Bangor, Kortum and Miller, 2009) depending on whether the reported SUS score was greater or smaller than 68, the FC platform was defined usable or not.

User Engagement - The FoU has been computed for each FC section (Food Diary and Water filling, Weight Ritual, Chat) for each participant as the percentage of no use, less of 50% and more than 50% usage. In addition, the weekly usage for the experimental period (6 weeks) has been computed. The user engagement has been also explored by means of a specific questionnaire, i.e. the *user engagement questionnaire*: for each questionnaire dimension (Perceived Usability, Aesthetics, Focused Attention, Felt Involvement, Novelty, and Endurability) mean values and standard deviations have been computed among the EG participants.

User experience (UX): the user experience has been explored by means of the *SUPR-Q*: for each questionnaire dimension (Usability, Trust, Credibility, Appearance and Loyalty) mean values and standard deviations have been computed among the EG participants.

3. RESULTS

In the following, the main results of the study have been provided.

3.1 Population Characteristics

Subjects' characteristics (Age, Gender, Occupation) are reported in Table I. Among the fourty subjects recruited for the study (20 in EG, 20 in CG), 4 subjects from the CG and 2 subjects from the EG dropped out. Characteristics and data of drop-out subjects have not been included in the following table.

Table 1. Characteristics of the participants enrolled in the 2-arms pilot

	CG (n=16)	**EG (n=18)**
Age [m (sd)]	43,7 (9,4)	39 (11)
Gender (female) [%]	37,5%	61,1%
Occupation, employee [%]	18,7%	33,3%

3.2 Personalized Intervention Efficacy

Anthropometric indices at the baseline (W_{IN}, BMI_{IN}, $\%FM_{IN}$) and final (W_{OUT}, BMI_{OUT}, $\%FM_{OUT}$) assessments have been reported in Table 2 along with the results of the T-test performed for each antropometric index by comparing the baseline and the final measure for both the CG and EG.

Table 2. Baseline (IN) and final (OUT) anthropometric indices and the p-value obtained for each IN-OUT comparison

	CG			**EG**		
	m	*sd*	*p-value*	*m*	*sd*	*p-value*
W_{IN}	74,5	12,7	ns	72,6	11,3	<0.01
W_{OUT}	74,0	12,6		71,1	11,2	
BMI_{IN}	25,4	3,0	ns	24,8	2,7	<0.01
BMI_{OUT}	25,2	3,1		24,2	2,5	
$\%FM_{IN}$	25,7	7,0	ns	25,1	7,5	<0.05
$\%FM_{OUT}$	25,2	7,1		24,0	7,3	

Significative differences ($p<0.05$) were found for all the anthropometric indices IN-OUT in the EG, while no differences have been found in the CG data.

3.3 Usability, User Engagement and User Experience

3.3.1 Usability

The SUS scores obtained in the EG group are the following: the 89% of participants rated it as usable (SUS>68) while the remaining 11% rated it unsuable (SUS≤ 68).

3.3.2 User Engagement

In Figure 3, Figure 4, Figure 5, Figure 6 both the overall and weekly FoU for each FC section have been reported. The results of the User Engagement questionnaire have been reported in Figure 7.

3.3.3 User Experience

The Figure 8 shows the SUPR-Q questionnaire results averaged on EG participants.

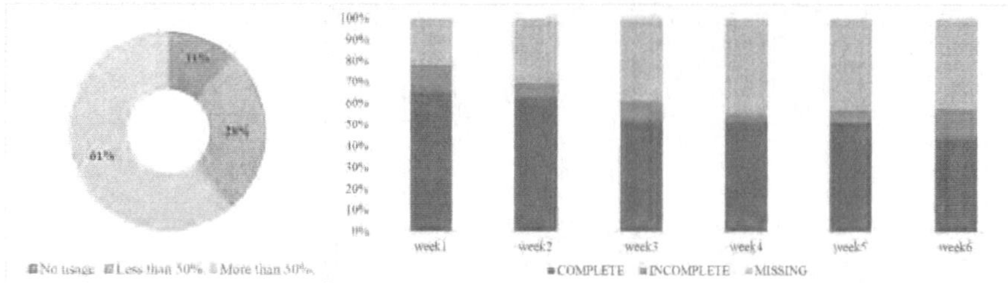

Figure 3. (a) Diary overall frequency of usage; (b) Diary weekly frequency of usage

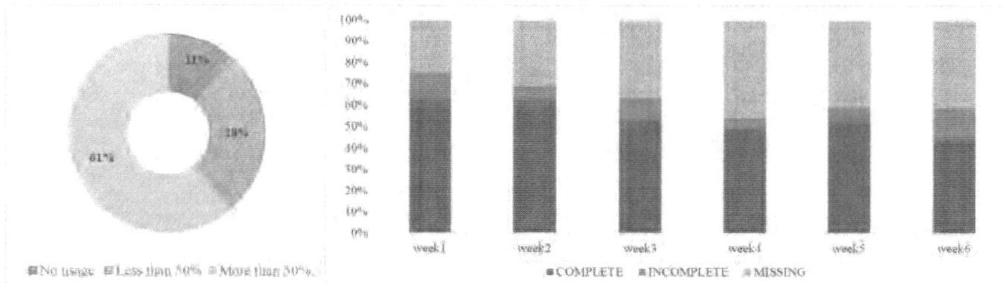

Figure 4. (a) Water filling overall frequency of usage; (b) Water filling weekly frequency of usage

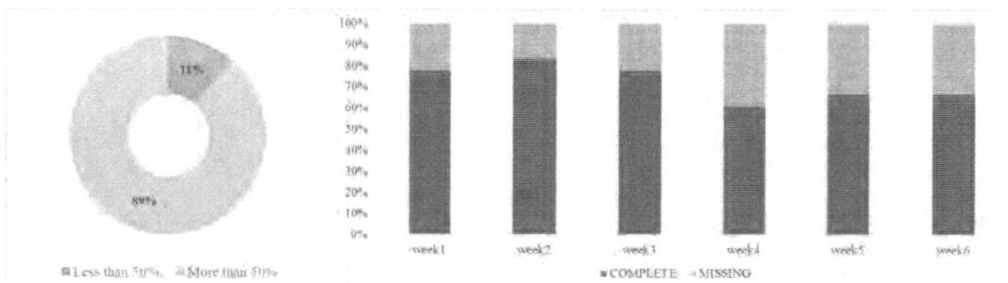

Figure 5. (a) Weight ritual overall frequency of usage; (b) Weight ritual weekly frequency of usage

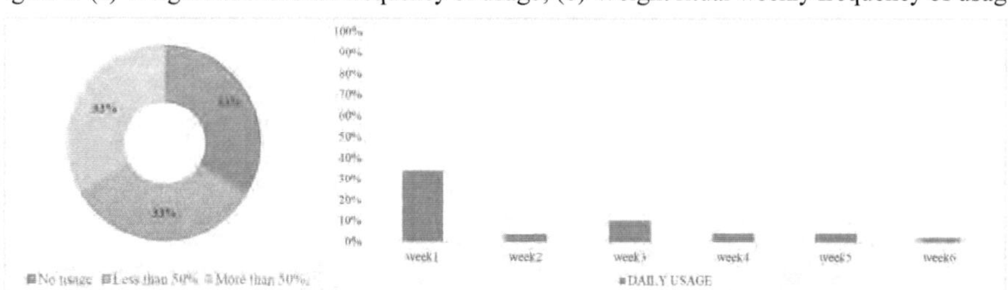

Figure 6. (a) Chat overall frequency of usage; (b) Chat daily frequency of usage per week

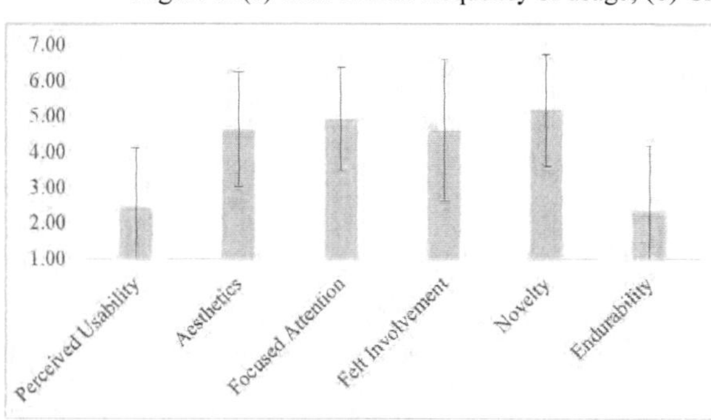

Figure 7. User engagement scores averaged on EG participants

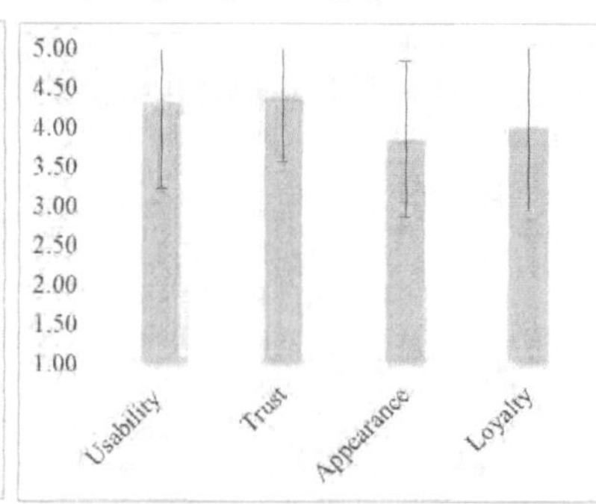

Figure 8. SUPR-Q scores averaged on EG participants

4. DISCUSSION AND CONCLUSION

The current study showed that the FC Web App could be a feasible and acceptable instrument for monitoring and improving eating habits in the sample population. In terms of recruitment, about 9 times (n=355) the target number of subjects (n=40) showed interest in partecipating in the study.

E-mail communication was the most successful recruitment strategy. Overall, the trial suffered of 15% attrition rate: among the 6 subjects who abandoned the trial, 2 belonged to the EG and 4 to the CG, i.e. 50% more subjects dropped out in CG. Drop out may reflect a person's dissatisfaction with the intervention and unequal dropout among groups is likely to be intervention-related. Attrition rate in the EG (10%) seems to be lower with respect to other studies specifically based on Web-based interventions (Neve *et al.*, 2010; Carter *et al.*, 2013) where an attrition rate of 20% has been found. Although the pilot trial was not statistically powered to detect a difference in weight change among the groups, it provided some data on effect sizes within each group. In fact, CG showed a mean weight reduction of 0.5±0.1 Kg, while the EG showed a mean weight reduction of 1.5±0.1 Kg. The same happened for the BMI and Fat Mass Percentage: a reduction of 0.2±0.1 (BMI) and 0.5±0.1% (FM) characterized the CG while the EG was characterized by a reduction of 0.6±0.2 (BMI) and 1.1±0.2% (FM). The t-test carried out to compare the IN-OUT data for each group showed a statistical difference only in the EG, for all the antropometric indices (W, BMI, FM%). For the EG, the SUS score showed an high perceived usability of the platform, i.e. only 2 over 18 subjects reported a SUS score lower than 68. After the 40-days trial, the user engagement of the EG participants has been evaluated both by analysing the actual frequency of usage of the various FC sections and by collecting subjective feedbacks on the overall usage over time from the participants. Both the results showed that a high, both objective and subjective, engagement rate. The "Diary" and "Water filling" sections have been used, at least one time, by 16 over 18 participants (89%). The "Weight ritual" section has been used at least once by all the participants and by 16 over 18 (89%) for more than 3 weeks. The "Chat" section has been used at least one by 12 participants (66%) while only 33% used the function for more than 50% of the trial period. The frequency of usage of the various sections showed that all the FC functionalities have been used at least one time by the majority of the subjects. The rate over the weeks has been also analysized to understand if any engagement loss trend was present: only the "Chat" section showed a significant higher usage during the first week (34%) with respect to the last one (2%). This result is quite expected since the main doubts or questions are likely to arise in the first period more than the last one. Regarding the Diary, Water filling and Weight ritual no specific usage trends have been observed during the trial period with exception to an overall (complete + incomplete pages) FoU reduction from the first to the last week of 10% for the Diary (from 68% to 58%) and of 6% for the Water filling (from 75% to 59%) sections. Engagement data obtained through the questionnaire analysis showed high levels of novelty, focused attention and felt involvment. Instead, the perceived usability after 40-days of usage showed an overall low score: this means that the daily usage of FC has been perceived less usable than expected (i.e. comparing with the pre-usage SUS). Finally, the high scores obtained in the UX questionnaire showed an overall acceptable UX after the 40-days of usage.

The present study showed some limitations: the sample group was limited; the FC Web App was a prototype and participants reported that they encountered bugs, thus sometimes affecting their engagement; participants knew what interventions were available and when included in the CG the engagement was lower.

In conclusion, this preliminary study showed a good level acceptance and efficacy of the FC Web platform with respect to a not-personalized standard diet intervention. As future studies, the adherence to the suggested personalized (PNP) and not personalized (INRAN guidelines) intervention by analysing the food diaries (both paper based and Web based) data will be explored. In addition, a comparison analysis of the eating behaviors as collected through the FFQ_{IN} and FFQ_{OUT} will be performed for both the groups to understand the compliance to the Mediterranean Diet, before and after the intervention.

ACKNOWLEDGEMENT

The research leading to these results has received funding from the European Union's Horizon 2020 research and innovation programme under grant agreement n° 653704 – OPERANDO.

REFERENCES

Bangor, A., Kortum, P. and Miller, J. (2009) 'Determining what individual SUS scores mean: Adding an adjective rating scale', *Journal of usability studies*, 4(3), pp. 114–123. doi: 66.39.39.113.

Brooke, J. (1996) 'SUS-A Quick and Dirty Usability Scale'. doi: 10.1002/hbm.20701.

Burke, L. E., Warziski, M., Starrett, T., Choo, J., Music, E., Sereika, S., Stark, S. and Sevick, M. A. (2005) 'Self-monitoring dietary intake: current and future practices.', *Journal of renal nutrition : the official journal of the Council on Renal Nutrition of the National Kidney Foundation*. United States, 15(3), pp. 281–290.

Carter, M. C., Burley, V. J., Nykjaer, C. and Cade, J. E. (2013) 'Adherence to a smartphone application for weight loss compared to website and paper diary: pilot randomized controlled trial', *Journal of medical Internet research*. JMIR Publications Inc., 15(4), pp. e32–e32. doi: 10.2196/jmir.2283.

Gilmore, L. A., Duhé, A. F., Frost, E. A. and Redman, L. M. (2014) 'The technology boom: a new era in obesity management', *Journal of diabetes science and technology*. SAGE Publications, 8(3), pp. 596–608. doi: 10.1177/1932296814525189.

Glanz, K., Murphy, S., Moylan, J., Evensen, D. and Curb, J. D. (2006) 'Improving dietary self-monitoring and adherence with hand-held computers: a pilot study.', *American journal of health promotion : AJHP*. United States, 20(3), pp. 165–170. doi: 10.4278/0890-1171-20.3.165.

Gondolf, U. H., Tetens, I., Hills, A. P., Michaelsen, K. F. and Trolle, E. (2012) 'Validation of a pre-coded food record for infants and young children.', *European journal of clinical nutrition*. England, 66(1), pp. 91–96. doi: 10.1038/ejcn.2011.133.

Jacobs, S., Radnitz, C. and Hildebrandt, T. (2017) 'Adherence as a predictor of weight loss in a commonly used smartphone application.', *Obesity research & clinical practice*. Netherlands, 11(2), pp. 206–214. doi: 10.1016/j.orcp.2016.05.001.

Jimoh, F., Lund, E. K., Harvey, L. J., Frost, C., Lay, W. J., Roe, M. A., Berry, R. and Finglas, P. M. (2018) 'Comparing Diet and Exercise Monitoring Using Smartphone App and Paper Diary: A Two-Phase Intervention Study', *JMIR mHealth and uHealth*. JMIR Publications, 6(1), pp. e17–e17. doi: 10.2196/mhealth.7702.

Knudsen, V. K., Gille, M.-B., Nielsen, T. H., Christensen, T., Fagt, S. and Biltoft-Jensen, A. (2011) 'Relative validity of the pre-coded food diary used in the Danish National Survey of Diet and Physical Activity.', *Public health nutrition*. England, 14(12), pp. 2110–2116. doi: 10.1017/S1368980011001650.

Kong, A., Beresford, S. A. A., Alfano, C. M., Foster-Schubert, K. E., Neuhouser, M. L., Johnson, D. B., Duggan, C., Wang, C.-Y., Xiao, L., Jeffery, R. W., Bain, C. E. and McTiernan, A. (2012) 'Self-monitoring and eating-related behaviors are associated with 12-month weight loss in postmenopausal overweight-to-obese women.', *Journal of the Academy of Nutrition and Dietetics*. United States, 112(9), pp. 1428–1435. doi: 10.1016/j.jand.2012.05.014.

Lieffers, J. R. L. and Hanning, R. M. (2012) 'Dietary assessment and self-monitoring with nutrition applications for mobile devices.', *Canadian journal of dietetic practice and research : a publication of Dietitians of Canada = Revue canadienne de la pratique et de la recherche en dietetique : une publication des Dietetistes du Canada*. Canada, 73(3), pp. e253-60. doi: 10.3148/73.3.2012.e253.

Naska, A., Lagiou, A. and Lagiou, P. (2017) 'Dietary assessment methods in epidemiological research: current state of the art and future prospects', *F1000Research*. F1000Research, 6, p. 926. doi: 10.12688/f1000research.10703.1.

Neve, M., Morgan, P. J., Jones, P. R. and Collins, C. E. (2010) 'Effectiveness of web-based interventions in achieving weight loss and weight loss maintenance in overweight and obese adults: a systematic review with meta-analysis.', *Obesity reviews : an official journal of the International Association for the Study of Obesity*. England, 11(4), pp. 306–321. doi: 10.1111/j.1467-789X.2009.00646.x.

O'Brien, H. L. and Toms, E. G. (2010) 'The development and evaluation of a survey to measure user engagement in e-commerce environments', *Journal of the American Society for Information Science & Technology*, 61(1), pp. 50–69. doi: 10.1002/asi.21229.1.

OECD statistics (2018). Available at: http://www.oecd.org/els/health-systems/obesityandtheeconomicsofpreventionfitnotfat-italykeyfacts.htm (Accessed: 17 January 2019).

Ozdalga, E., Ozdalga, A. and Ahuja, N. (2012) 'The smartphone in medicine: a review of current and potential use among physicians and students.', *Journal of medical Internet research*. Canada, 14(5), p. e128. doi: 10.2196/jmir.1994.

Probst, Y. C. and Tapsell, L. C. (2005) 'Overview of computerized dietary assessment programs for research and practice in nutrition education.', *Journal of nutrition education and behavior*. United States, 37(1), pp. 20–26.

Sauro, J. (2015) 'SUPR-Q: A Comprehensive Measure of the Quality of the Website User Experience', *Journal of Usability Studies*, 10(2), pp. 68–86. Available at: http://www.upassoc.org.

Seidell, J. C. and Halberstadt, J. (2015) 'The global burden of obesity and the challenges of prevention.', *Annals of nutrition & metabolism*. Switzerlanworld, 66 Suppl 2, pp. 7–12. doi: 10.1159/000375143.

"SHARING IS CARING": DESIGNING A VALUE-SENSITIVE MHEALTH PLATFORM FOR SHARING TYPE 1 DIABETES MANAGEMENT WITHIN FAMILIES

Elettra Oleari[1], Clara Pozzi[2], Francesco Sardu[2], Francesca Sacchitelli[2], Virginia Sanchini[2] and Alberto Sanna[1]

[1]*Ospedale San Raffaele, Milan, Italy*

[2]*Fondazione Centro San Raffaele, Milan, Italy*

ABSTRACT

Within a mobile-technology enhanced educational framework, this study investigates needs and expectations of children (aged 8-14) with type 1 diabetes, and their parents, to define the functionalities of *PAL Inform,* a web-based monitoring module. The aim is to design it as a tool for parents to be informed on the educational progresses, as well on diabetes-related data, of their children without violating their values (privacy, trust, autonomy). Through a set of co-design activities, carried out with 39 children and 45 parents, it was possible to define which kind of information parents would monitor and children (especially pre-adolescents), in turn, would share, without compromising family dynamics: i.e. glycaemic values and trends, insulin doses, carbohydrates intake, generic emotional status, educational progresses on diabetes-related knowledge. Leveraging on these insights, a set of functional requirements was elicited for the future monitoring module implementation.

KEYWORDS

Type 1 Diabetes, Children, Self-Management, mHealth, Focus Group, Role Playing

1. INTRODUCTION

Type 1 Diabetes Mellitus (T1DM) is the most pervasive endocrine-metabolic condition in childhood, with more than 28,000 newly diagnosed worldwide, from 0 to 19 years old, in 2017 (International Diabetes Federation 2017). Diabetes self-management plays a crucial role for young patients in order to maintain health-related quality of life and to avoid long-term complications, like heart and blood vessel diseases or nerves, kidneys and eyes damages. However, it is not an easy goal to attain for the younger ones, since it requires long-term motivation and perseverance (Hoey et al. 2001). Moreover, the onset of T1DM raises new challenges for the entire family as, on the one hand, many new notions have to be learnt, and, on the other, the whole family has to find a way for restoring an internal balance, adapting to the diagnosis and new daily routines (Marshall et al. 2009; Jaser 2009). The final aim of this process is the gradual acquisition of autonomy by the side of children and/or adolescents in the management of their condition, including blood glucose monitoring, insulin administration and carbohydrates (CHO) counting (Schilling et al. 2002).

Autonomous management of diabetes in children/adolescents is strongly affected by a wide range of personal and environmental factors, such as their developmental stage and the level of support and care provided by the reference adults, especially their parents (Naranjo 2014; Scott 2013; Wu et al. 2014). In fact, several studies show that a shared responsibility of T1DM within the family is consistently associated with better adherence and health functioning across adolescence (Helgeson 2008; Wysocki 2009). Moreover, parental involvement in diabetes management reveals to be essential to the maintenance of psychological health, glycaemic control, and the development of self-care behaviours in young patients (Anderson 1997; Anderson and Wolpert 2004; Berg et al. 2008). In this view, children are expected to be more responsible towards diabetes care when they reach pre-adolescence; however, at this age, adherence and glycometabolic control decrease. Transition into adolescence, in fact, is characterized by oscillations between dependence on parents and desire of independence, with feelings of increased self-reliance and autonomy (Steinberg and Silverberg 1986). A chronic condition like diabetes creates a situation of dependence precisely in the phase in which adolescents feel the greater pressure to define themselves as separate individuals (Coupey and Cohen 1984). To avoid such complications, current treatment standards foster the importance of continuous parental involvement in T1DM managing process, ranging from support during childhood to supervision during adolescence, through intervention programs, that aim to reinforce parental disease acceptance, paternal involvement, family warmth, cohesion and supportive attitudes (Berg et al. 2008; Young et al. 2014).

In this context, technology revealed to play an important role in supporting self-management of chronic conditions such as T1DM (Franklin et al. 2008; Holtz and Lauckner 2012). Amongst them, mobile health (mHealth) applications have been shown as promising tools, thanks to their easy access and wide usage, particularly by young people, potentially capable of improving patients' and families education (Cafazzo et al. 2012; Calvo et al. 2014; Frøisland et al. 2012) and enhancing communication with care providers in effective ways (Holtz et al. 2017). On the wake of these evidences, the PAL project (grant agreement n° 643783) aims to develop a Personal Assistant for healthy Lifestyle (PAL), a System that helps children/pre-adolescents, parents and Health Care Professionals (HCPs) to advance T1DM self-management, so as to establish an adequate shared patient-family-caregiver responsibility before adolescence.

The present work aims to empirically investigate needs and expectations of families dealing with T1DM, in order to detect characteristics, contents and functionalities of a value-sensitive interface, called *PAL Inform*. This represents a monitoring and informing tool *for parents* to improve their knowledge on children's self-management, while fostering a positive attitude towards T1DM shared experiences and communication, without violating children's private sphere. Drawing from these premises, a set of co-design methodologies, i.e. Focus Group and Role Playing sessions, have been put in place. Data collected have been then analyzed and formalized into functional specifications to implement *PAL Inform* in such a way as to satisfy all the end-users desiderata and make it actually usable in daily lives. The study has been designed and conducted by a pool of experts in the fields of psychology, mHealth technologies, service design and bioethics in order to have comprehensive scientific basis on which to structure the research. The paper is organised as follows: *Section 2* is devoted to the presentation of the PAL Educational Framework; *Section 3* presents the experimental procedure and *Section 4* study results. Finally, the discussion reflects on some of the insights gained during the study and presents an anticipation of the future developments of this work.

2. PAL SYSTEM EDUCATIONAL FRAMEWORK

The PAL System is structured as a multicomponent platform connected to a central knowledge-based Artificial Intelligence (AI) (Figure 1). PAL technologies for children/pre-adolescents are a suite of mHealth games linked to a conversational agent, both embodied (the PAL robot) and virtual (its avatar). Therefore, children can play with the PAL robot in contexts like hospitals, and with its avatar at home using the *MyPAL application for tablets*, which comprises a quiz, various mini-games, and an interactive diary where to input data regarding T1DM (e.g. glucose values, insulin doses, carbohydrates' intake) and other activities (e.g. sports and emotions). Thanks to these interactions, children are supposed to acquire knowledge and to be motivated in pursuing correct behaviors, so as to better manage their conditions (Peters et al 2019).

The educational contents for children/pre-adolescents are created and monitored by HCPs, who set personalized T1DM self-management learning goals through the *PAL Control* interface (Peters et al 2017). These goals, strictly linked with daily-life situations and tuned on specific age ranges, are supposed to be pursued in time by children through the activities and games proposed by *MyPAL app* and the PAL agent. For example, to autonomously participate to a birthday party it is important to be able to count CHO; therefore PAL will propose to the child quiz questions and levels of the sorting game on this topic (Figure 1).

The third component, which represents the subject of this contribution, is named *PAL Inform.* It is meant to be a browser interface allowing parents to have a dedicated access, in order to see the PAL-mediated educational progresses of their children and data regarding their T1DM management retrieved by the System (e.g. glycaemia values inserted in the diary of *MyPAL app*).

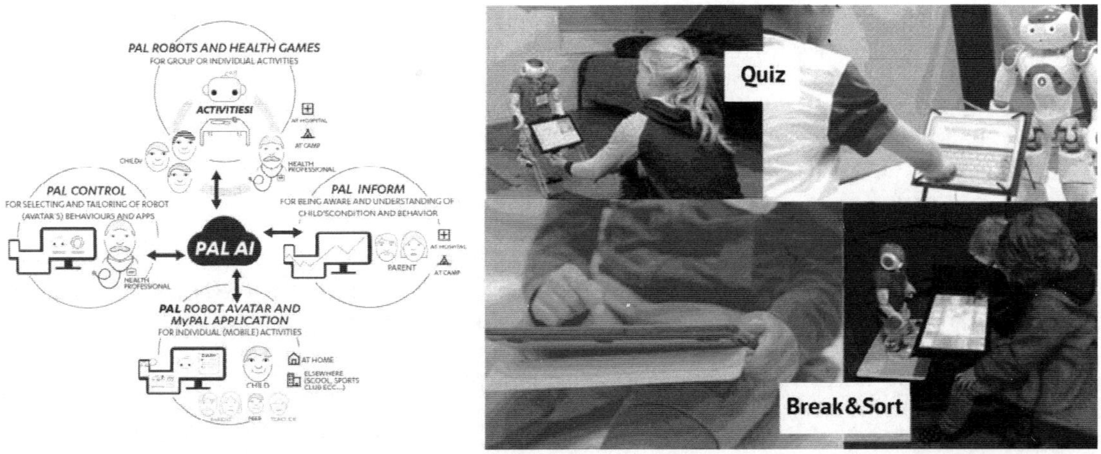

Figure 1. On the left: the PAL technological components, with the details of their specific end-users and application settings; on the right: children playing educational activities with the PAL robot and its avatar

3. DESIGNING PAL INFORM: AN USER-CENTERED INVESTIGATION

A dedicated user-centered investigation has been devised in order to derive coherent, and respectful of the various points of view (value-sensitive), functional specifications for *PAL Inform*. The study presents the following endpoints: *(i)* to understand which are, according to parents, the determining factors (e.g.: knowledge, understanding, trust) able to offer their children a support, through a shared responsible process, in achieving an autonomous management of their disease; *(ii)* to explore parents' needs and expectations towards a technological tool, able to provide them with a certain level of monitoring regarding different aspects of their children's T1DM management; *(iii)* to understand children's opinions on *PAL Inform* and what kind of information they would be willing to share without considering it as an intrusion.

3.1 Materials and Method

A set of formal co-design methods has been outlined to create a proper and engaging research environment:

1. *Structured Questionnaires* are quantitative research methods, made up by direct and close questions. For the present contribution, a specific Questionnaire[1] was drawn for the participating parents to explore: *(i)* their experiences with day-life management of T1DM, with particular attention to the shared responsibilities on diabetes tasks – see Q1, Q2, *(ii)* the perceived level of autonomy of their children and parents' feelings about it – see Q3, Q4, Q10, *(iii)* the principal parental worries linked to diabetes – see Q6 to Q9 and *(iv)* their desires and expectations regarding *Pal Inform* – see Q11 to 16. The questionnaire takes inspiration from already existing clinical instruments (Lewin et al. 2009; Streisand et al. 2005) and the researchers' clinical experience with T1DM young patients and, since it was not already a published or validated tool, it has been internally validated in its contents and structure by a pool of experts (2 psychologists and 1 pediatric diabetologist) of San Raffaele Hospital, before its use.
2. *Role Play* is an experiential activity originating from the Classical Psychodrama (Castro et al. 1993; Moreno 1973; Robinson 2004), which provides participants with an active experience simulating a real-life situation, characterized by certain social roles. In this case, the aim was to make children/pre-adolescents assuming both their own role and their parents' one regarding T1DM management and education, in order to find possible solutions about *PAL Inform* and subsequently compare the gained insights with those derived by the (real) adults participating;

[1] For the full questionnaire see: http://www.pal4u.eu/wp-content/uploads/Structured-Questionnaire-for-Parents.pdf

3. *Focus Group* (Morgan 1998; Migliorini and Rania 2001) is a qualitative research technique, where a group of people is invited to discuss on a theme. Here, this was used to derive parents' needs, issues and desiderata regarding a technology-enhanced monitoring of their children T1DM management.

This research protocol has been approved by the Ethics Committee of San Raffaele Hospital (Milan, Italy), protocol number PAL-643783-5, in the session of the 13/07/2017.

3.2 Measures

Based on the exploratory nature of the study, an inductive qualitative content analysis was performed to derive the key areas of interest for the PAL end-users (Krippendorff 2004). *Frequency of occurrence* of questionnaire's and templates' responses, *participants' verbal responses* and discussions from the audio/video recordings of the activities, as well as *researcher's notes* during their execution, were taken into account to identify the key themes emerged in the investigation.

3.3 Participants

The activities took place during two educational Camps for children/ pre-adolescents with T1DM: the former held in August 2017 for pre-adolescents aged 11 to 14 y.o. (without their parents), the latter in October 2017 for children aged 8 to 10 y.o. with their families. To take part to the activities, participants had to sign a specific informed consent (in case of the minors both the parents or legal tutors had to sign it), and had to be native language speakers. Only the parents compiling the on-line format of the Structured Questionnaire were excused from signing the informed consent as data collection in this case was anonymous (see Section 3.4).

In total, 39 of 44 pre-adolescents (mean age= 12.73 y.o, std.dev= 0.79) attending the August camp and 33 amongst their parents, as well as 6 families participating at the October camp (i.e. all the 12 parents) participated to the study. Notably, researchers chose to explore only expectations and perceptions of the pre-adolescents group regarding *PAL Inform*, leaving aside those of the younger ones. The rationale behind this choice is that, as already explained, preadolescence is a critical developmental phase generally characterized by changes in relationships with adults, which influences, and is influenced, by the management of a chronic disease (Foster et al. 2016). Children, instead, are more parents' dependent on T1DM management.

3.4 Procedure

At the diabetes camps, dedicated rooms were set up with all the needed materials to carry out the activities, which were supervised by a moderator and by a psychologist, and their audio/video recordings.

Structured Questionnaire for parents - parents of both the age groups were involved: those of the younger children filled in the questionnaire before starting the Focus Group session, while those of the older ones were provided an online version of the same questionnaire and had the chance to complete it anonymously at home while their children were at the August camp.

Role Play sessions, for pre-adolescents of 11 to 14 y.o. at the August camp - participants were randomly assigned to one of two groups (from 3 up to 5 components each) on the basis of the role they were supposed to interpret: the "parents" group and the "children" group. Participants were asked to think about a possible future platform (*PAL Inform*) that would help families with children of their age to manage T1DM and learn about it. The two groups imagined which functions they would like to find on this tool and for which purposes, thinking to real situations of everyday-lives, helped by a set of "situation card", like for example a school trip or sports. After discussing on these, each group wrote all their considerations down on dedicated templates, which are meant to help them focusing on the following aspects: the kind of information they would like (or not) to share, the reasons of those choices and the corresponding priority orders.

Focus Group activity for parents of children of 8 to 10 y.o. at the October camp - the 12 parents were invited to discuss their needs and wishes about *PAL Inform*. The parents had at disposal, to elicit the discussion, the same templates used by children during the Role Playing.

4. RESULTS

We first analyzed the Structured Questionnaire for parents to characterize the parental participant's population and gain a first preliminary overview of: *(i)* the main needs and worries of parents regarding their children's T1DM management and *(ii)* expectations about *PAL Inform.* Then, all the dialogs in the audio/video recordings of the Role Play and Focus Group were transcribed, in order to obtain a database of contents compatible with the other materials collected in the activities (i.e. templates formats and researchers' notes) and suitable to be studied by an open coding process (Flick 2009). All the raw data were then reviewed and annotated by the research team to identify recurrent themes, from which grouping data into broader categories expressing the same concepts, focusing the attention on how to properly share data on T1DM management and on T1DM education within the family. These themes were therefore compared with the preliminary insights of the Structured Questionnaire so as to obtain a final list of key topics to be translated into functional requirements for the *PAL Inform* future implementation (see Section 5).

4.1 Characterization of the Parental Pool

Data collected from the Questionnaire revealed a parental attitude towards the care of their children in line with the level of autonomy expected by the two age groups (Q1 to Q3). The younger children share the responsibility of the major T1DM tasks, i.e. glycemic checks and insulin injections, with their parents: 10 out of 12 parents (83%) stated that these tasks are performed alternately by themselves and their children. Pre-adolescents, instead, are more independent and usually manage on their own the glycaemia (24/33 cases, 73% of the pool) and the injections (23/33 cases, 70%). The major parental concerns, in both the groups, are linked to possible children's diabetes-related malaises far from their direct control, like severe hypoglycemic episodes away from home – e.g. Q5: 9/12 and 14/33 cases, Q8: 6/12 and 26/33 cases, Q9: 11/12 and 17/33 cases for children and pre-adolescents respectively. However, the idea that their children will improve in autonomy towards T1DM as time passes (Q4), makes all the parents confident and is reported as a stimulus for reassuring their children on their own self-management capabilities. Parents would like, through *PAL Inform*, to globally monitor T1DM self-management of their children, especially (Q14): blood sugar levels – 12/12 and 23/33 cases, CHO intake – 6/12 and 8/33 cases and insulin dosages – 4/12 and 11/33 cases for children and pre-adolescents respectively. Moreover, on this basis, parents would like to interact if necessary with their children on T1DM related topics, e.g. decide together the insulin dosage – Q15: 7/12 and 19/33 cases. The need for monitoring T1DM-related information is thought to be mainly directed at the time spent at school, Q13: 9/12 and 14/33 cases, or sport/free activities, Q13: 8/12 cases.

4.2 Role Play Sessions for Pre-Adolescents of 11 to 14 y.o

Overall, 5 Role Play sessions (i.e. 10 groups of children, 5 impersonating the parents and 5 themselves) took place during the August camp. The collected insights are grouped in two clusters, which are presented in the following with a fine grained description of the main sub-topics emerged:

Sharing data regarding T1DM management – to elicit the discussion among the participants, the templates covered the entire range of information which are usually pertinent to T1DM self-management and could be retrieved through the use of the *MyPAL* diary in real-life situations: *(i)* data on glycemic control (e.g. timing and values of the glycaemic checks), *(ii)* data on meals and CHO counting, *(iii)* data on the timing and entity of insulin dosages, *(iv)* data on physical activity and other events, *(v)* data on emotional status. Both the "parents" and "children" groups, agreed on the importance of sharing information regarding glycemic values (10/10 groups), CHO counting (5/5 "parents", 4/5 "children") and insulin doses (5/5 "parents", 4/5 "children"). An aspect that was not enthusiastically accepted by the "children" group (2/5 groups), and in any case did not find a great interest either by the "parents" group (2/5 groups), is the possibility to share the number and details of meals away from home. Differently, with respect to additional information on sports, all the "children" groups were willing to share it in its entirety, while the "parents" groups showed interest only for type and duration of the exercises. Lastly, most of the "parents" were not willing to be informed on emotion-related data (1/5 groups) and "children" were prone to share only an overview of their feelings, like a "mood of the day" emoticon (4/5 groups), but not personal information written within the dedicated *MyPAL* diary section.

Sharing the educational progresses on T1DM – the activity also focused on the possibility of sharing the T1DM educational path envisioned by the PAL System and the T1DM personalized goal achievements set by HCPs. There was a clear positive attitude in the "children" group towards the sharing of these kinds of data (4/5 groups); the "parents" groups, instead, were split between those who were afraid of being intrusive (3/5 groups) and those who stressed the importance of understanding any improvements towards autonomy (2/5).

4.3 Focus Group Activity for Parents of Children Aged 8 to 10 y.o

As for the Role Playing activity, the discussion held between the 12 parents focused on the following topics:

Sharing data regarding T1DM management – through *PAL Inform*, parents would like to monitor glycemic values (12 out of 12) and corresponding timing (10/12 cases), as long as the insulin doses (10/12 cases) and the amount of CHO intake per meal (12/12 cases). In addition to those, parents would like to monitor, in relation to the corresponding glycemic trends, the duration of eventual sport activities (8/12 cases). Interestingly, knowing the emotional status of their children, derived from the MyPAL diary, is perceived as an added value to the daily therapy management (10/12 cases). From an extract of the parents' discussion: "... to understand how they [emotions] could influence [my child's] glycaemia and help him in managing the therapy".

Sharing the educational progresses on T1DM – all the parents showed an interest in being aware of the progresses in the T1DM educational path of their children, as well as in the fulfilment of their goals, in order to help them; if needed (2/12 cases), they praise and reward them whenever they accomplish a new goal (6/12 cases) and understand their actual improvements towards autonomy (4/12 cases).

An overview of the all the findings of the *PAL Inform* co-design investigation is reported in Table 1.

Table 1. Overview of the results obtained from the *PAL Inform* investigation, with the details of the T1DM-related topics under analysis, where Yes or No indicates the willingness (or not) to share/know the specific type of data

		QUESTIONNAIRE		ROLE PLAY		FOCUS GROUP
		Parents (8-10 y.o)	Parents (11-14 y.o)	"Children"	"Parents"	Parents (8-10 y.o)
Glycaemia	Glycaemic Values	Yes	Yes	Yes	Yes	Yes
	Number of insertion	No	No	No	No	No
Insulin	Insulin doses	Yes	Yes	Yes	Yes	Yes
Nutrition	CHO counting	Yes	No	Yes	Yes	Yes
	Meal details	/	/	No	No	No
Activities	Sport	Yes	Yes	Yes	Yes	Yes
Emotional Status	Overview (emoticons)	No	No	Yes	Yes	Yes
	Details (notes)	No	No	No	No	No
T1DM education	Goal achievement	/	/	Yes	Yes/No	Yes
	Overall progress	/	/	Yes	Yes/No	Yes

5. DISCUSSION AND FUTURE PERSPECTIVES

The present study aims to explore and collect parents' and their children's needs, values and expectations regarding *Pal Inform*, a browser module for monitoring data on T1DM management and education within the family. A suite of co-design activities was exerted to understand which kind of information may have been smoothly shared and, on the contrary, which ones may have represented a source of tension between children (especially the pre-adolescent ones) and their parents, thus making *PAL Inform* counter-productive.

Despite an overall positive parental perception of children's autonomy in both the age groups, a generalized need of control emerged. Indeed, all the parents expressed their wish to monitor their children, especially during those hours in which they are not with them, such as at school and during free/sport time. Inasmuch as they are concerned of possible diabetes-related critical episodes, such as hypoglycemic crises, blood sugar levels represent the parameter the parents care mostly. In this regard, it is interesting to note that both children and parents figure *PAL Inform* as an useful tool for sharing data related to T1DM therapy management, i.e. glycaemia values and insulin doses, as supposedly able to provide a mutual reassurance. The nutritional domain, instead, represents an area in which children are reluctant to share too much details with their parents. While carbohydrates counting is considered as a parameter worthwhile to be shared, pre-adolescents appear unwilling to disclose the details of their daily food intake. This result is, however, in line with the literature, as food transgressions are quite common during this age (Kelly et al. 2005). Regarding data on the emotional status, children want to share only superficial data, with a refusal towards more details. This is observed in parents as well, thus showing respect for the privacy of their children together with an attitude of non-intrusiveness. Parents and children, moreover, demonstrate a common interest in sharing information regarding diabetes-knowledge acquisition and progresses, and the related motivations are all connected to the empowerment of trust between family members: "...so our parents could trust us more" [citation from one of the children groups during the Role Play session], "...so I can realize his level of awareness and help him in case of need" [citation from one of the parents during the Focus Group"].

To conclude, on the wake of these results, Table 2 reports the functional requirements elicited for the upcoming implementation of *PAL Inform*. Once the module will be up and running, an user experience evaluation is going to be put in place in order to verify its effectiveness and reliability in real-life usage.

Table 2. List of PAL Inform Functional Requirements

Requirement number	Description	Example
1	PAL Inform has to allow parents to visualize children's glycaemic data	• Time of the glycaemic test • Glycaemia value
2	PAL Inform has to show when glycaemia goes above/ below certain thresholds	Setting personalized hypo/ hyper-glycaemia values
3	PAL Inform has to display children's glycaemic trend in time	Customizable data visualization: daily, weekly, monthly
4	PAL Inform has to display each dose of insulin performed	• Time of the injection • Amount of insulin
5	PAL Inform has to display the number of CHO assumed for each meal	• Time • Quantity
6	PAL Inform has to display the total of CHO assumed in a day	
7	PAL Inform has to display the sport activity/ies performed	• Time • Type of activity • Last glycemic value before sport
8	PAL Inform has to display a summary of the emotional status but NOT its details	• Type of emoticon
9	PAL Inform has to notify when the child achieves a T1DM goal	• Name and Level of the goal • When it has been achieved • Brief description of the goal
10	PAL Inform has to display meaningful aggregated data on a customizable timeline	• Glycemic Value • Insulin Bolus • CHO intakes
11	PAL Inform has to enable parents to send their children messages	• Discuss on the dosage of insulin

ACKNOWLEDGEMENT

This work is funded by the EU Horizon PAL project (grant nr. 643783). We wish to thank all the participating families and the "Sostegno70- insieme ai ragazzi diabetici" patients' association for their support.

REFERENCES

Anderson BJ, Ho J, Brackett J et al. ,1997. Parental involvement in diabetes management tasks: relationships to blood glucose monitoring adherence and metabolic control in young adolescents with insulin-dependent diabetes mellitus. *Journal of Pediatrics,* 130(2): 257–65.

Anderson BJ and Wolpert HA, 2004. A developmental perspective on the challenges of diabetes education and care during the young adult period. *Patient Education and Counseling* 53(3): 347–352.

Berg CA, Butler JM, Osborn P, et al., 2008. Role of parental monitoring in understanding the benefits of parental acceptance on adolescent adherence and metabolic control of type 1 diabetes. *Diabetes Care* 31(4):678–83.

Cafazzo J, Casselman M, Hamming N, et al., 2012. Design of an mHealth App for the Self-management of Adolescent Type 1 Diabetes: A Pilot Study. *Journal of Medical Internet Research* 14(3): e70.

Calvo RA, Pardo a, Zhao Y, et al., 2014. TransitionMate: A Mobile Application for Chronic Illness Transition Support. *Workshop on Supporting Autonomy in Technology Design*, pp.1–4.

Castro B, Castro D, and Poznanski R, 1993. Use of role-playing in child psychotherapy and psychopedagogy. *Annales Médico-Psychologiques* 151(7): 535–7.

Coupey SM and Cohen MI, 1984. Special considerations for the health care of adolescents with chronic illness. *Pediatric Clinics of North America* 31:211–219.

Flick, U, 2009. *An Introduction to Qualitative Research*. SAGE. ISBN 9781446241318.

Foster C, Bellando J and Wang Y-CA, 2016. Diabetes Control and Adherence in Adolescence. *Pediatric Annals* 45(9): e327–e331.

Franklin VL, Greene A, Waller A et al., 2008. Patients' engagement with "Sweet Talk" - a text messaging support system for young people with diabetes. *Journal of Medical Internet Research* 10(2):e20.

Frøisland DH, Eirik Å and Finn S, 2012. Improving Diabetes Care for Young People With Type 1 Diabetes Through Visual Learning on Mobile Phones: Mixed-Methods Study. *Journal of Medical Internet Research* 14(4):e111.

Helgeson VS, Reynolds KA, Siminerio Let al., 2008. Parent and adolescent distribution of responsibility for diabetes selfcare: links to health outcomes. *Journal of Paediatric Psychology* 33(5):497–508.

Hoey H, Aanstoot H-J, Chiarelli F, et al., 2001. Good Metabolic Control Is Associated With Better Quality of Life in 2,101 Adolescents With Type 1 Diabetes. *Diabetes Care* 24(11): 1923–8.

Holtz B and Lauckner C, 2012. Diabetes management via mobile phones: a systematic review. *Telemedicine journal and e-health* 18(3):175-184.

Holtz BE, Murray KM, Hershey DD et al., 2017. The design and development of MyT1DHero: A mobile app for adolescents with type 1 diabetes and their parents. *Journal of Telemedicine and Telecare* 0(0): 1–9.

International Diabetes Federation, 2017. *Diabetes Atlas Eighth edition 2017*.

Jaser S, 2009. Family Interaction in Pediatric Diabetes. *Current Diabetes Reports* 11(6): 480–485.

Kelly SD, Howe CJ, Hndler JP, et al., 2005. Disordered eating behaviors in youth with type 1 diabetes. *Diabetes Educator* 31(4): 572–583.

Krippendorff K, 2004. *Content Analysis: An Introduction to Its Methodology*. Sage Publications.

Lewin AB, LaGreca AM, Geffken GR, et al, 2009. Validity and reliability of an adolescent and parent rating scale of type 1 diabetes adherence behaviors: the Self-Care Inventory (SCI). *J Pediatr Psychol*. 34(9):999-1007.

Marshall M, Carter B, Rose K, et al, 2009. Living with type 1 diabetes: Perceptions of children and their parents. *Journal of Clinical Nursing* 18(12): pp.1703–1710.

Migliorini L and Rania N, 2001. I focus group: uno strumento per la ricerca qualitativa. *Animazione Sociale*, pp.82–88.

Moreno J, 1973. *Psicodramma e vita*. Rizzoli Editore.

Morgan D, 1998. The Focus Group Guidebook. Available at: http://sk.sagepub.com/books/the-focus-group-guidebook.

Naranjo D, Mulvaney S, McGrath M et al., 2014. Predictors of self-management in pediatric type 1 diabetes: individual, family, systemic, and technologic influences. *Current Diabetes Reports* 14:544.

Peters, R. et al, 2017. Guidelines for Treebased Learning Goal Structuring. *Proceedings of the 22Nd International Conference on Intelligent User Interfaces (IUI '17)*. ACM, New York, NY, USA, pp. 401–405.

Peters, R. et al, 2019. Usability of the PAL Objectives Dashboard for Children's Diabetes Self-Management Education. *Proceedings of the 5th International Conference on e-Society, e-Learning and e-Technologies*, Vienna, Austria.

Robinson DT, 2004. *The SAGE Encyclopedia of Social Science Research Methods*.

Schilling LS, Grey M and Knafl KA, 2002. The concept of self-management of type 1 diabetes in children and adolescents: an evolutionary concept analysis. *Journal of Advanced Nursing* 37(1):87–99.

Scott LK, 2013. Developmental Mastery of Diabetes-Related Tasks in Children. *Nursing Clinics of North America* 48(2): 329–342.

Steinberg L and Silverberg SB, 1986. The vicissitudes of autonomy in early adolescence. *Child Development* 57, 841–851.

Streisand R, Swift E, Wickmark T, et al, 2005. Pediatric Parenting Stress Among Parents of Children with Type 1 Diabetes: The Role of Self-Efficacy, Responsibility, and Fear, *Journal of Pediatric Psychology*, 30(6): 513–521

Wu YP, Rausch J, Rohan M, et al, 2014. Autonomy support and responsibility-sharing predict blood glucose monitoring frequency among youth with diabetes. *Health psychology: official journal of the Division of Health Psychology*, American Psychological Association, 33(10): 1224–1231.

Wysocki T, Nansel TR, HolmbeckGN, et al., 2009. Collaborative involvement of primary and secondary caregivers: associations with youths' diabetes outcomes. *Journal of Pediatric Psychology*. 34(8):869–81.

Young MT, Lord JH, Patel NJ et al., 2014. Good cop, bad cop: quality of parental involvement in Type 1 Diabetes management in youth. *Current Diabetes Reports* 14(11): 546.

SUPPORTING RELAXATION: SLOWBREATH FOR PEOPLE WITH MYALGIC ENCEPHALOMYELITIS

Jorun Børsting
Department of Informatics, University of Oslo, Oslo, Norway

ABSTRACT

This paper describes the design process resulting in a cushion for relaxation called SlowBreath. The cushion was made with the intent to support people with Myalgic Encephalomyelitis (ME) in illness self-management. In particular, the focus was on energy balancing (activities interlaced with relaxation and rest) that people with ME need to master. When SlowBreath was finished to the extent that it could be regarded as a research product, five people diagnosed with ME participated in exploring the fit of SlowBreath in their lives, as well as the experiences that it generated. In this paper, the importance of design for minimal interaction for this particular user group is highlighted. The findings from the study are articulated as a set of implications for the design of everyday digital artefacts, aiming to scaffold self-management for people with ME.

KEYWORDS

HCI, e-Health, Relaxation, Myalgic Encephalomyelitis, Self-Management, Research Products

1. INTRODUCTION

Myalgic Encephalomyelitis (ME) still has no scientifically established cause and subsequently, no curative treatment. It is established, however, that ME is a chronic, multisystem illness that severely affects the life quality of people diagnosed with the illness (Hvidberg et al., 2015). ME is classified in four primary categories, ranging from very severe (completely bedridden), to mild (reduced in function by at least 50% over a long time) (Carruthers et al., 2011). The illness is also considered to be common, with studies indicating prevalence as high as 3-5% by some definitions of the illness (Líndal et al., 2002). Post-Exertional Malaise (PEM) is a key symptom, present in all people with ME. A cluster of other symptoms is also usually present. The most frequent ones are hypersensitivity to external sensory inputs (touch, smell, taste, visual, or auditory), fatigue, sleep disturbance, muscular pain, nausea, stomachaches, headaches, immune dysfunction and impaired concentration and short-term memory. PEM causes intense exhaustion, and an increase in the intensity of other symptoms after mental or physical exertion. The exertion needed to trigger PEM is different from person to person and sometimes varies with the same person on a day to day basis. For some, it might be caused by a small amount of exercise. For others, everyday activities like doing the laundry, or sitting upright, can trigger the PEM. While struggling with these fluctuating symptoms and health issues, self-management strategies seem to be, at present, the best option. *Energy balancing* (Lacourt et al., 2018) is one commonly used self-management strategy which involves balancing periods of rest and activity to avoid overexertion (Goudsmit et al., 2012).

Only recently, has technology design for people with ME received attention within the HCI community (Culén et al., 2019). People with ME are found to present a particularly interesting technology design challenge, based on how they need to minimize their interactions, both with technology and other people. Furthermore, exploring how technology can support self-management strategies, such as energy balancing or increasing wellbeing, becomes relevant.

This paper explores whether relaxation, or meditation, facilitated by digital artefacts such as SlowBreath, see Figure 1, can be a component in energy balancing and self-management of ME. SlowBreath gently guides and supports users at moments when relaxation and de-stressing are important. The aim of the work presented in this paper, was not to establish practices of regular relaxation or meditation by using SlowBreath. Instead, the focus was on identifying design implications for digital artefacts that can scaffold experiences of relaxation and meditation, by exploring form, computational materials, the materiality of artefacts, and bodily experiences that they facilitate. In addition, the paper contributes by addressing methodological challenges of doing research for and with vulnerable people and proposing an alternative to traditional participatory approaches.

Figure 1. The designed artefact, SlowBreath, aims to scaffold relaxation and meditation practices

2. RELATED WORK

In this section, related work is divided into two main themes: technology design and ME and research on relaxation and meditation in HCI.

2.1 Technology Design and ME

As mentioned in the introduction, designing technologies for people with ME has only recently gained attention within the field of HCI. Culén et al., (2019) made the observation that aspects of rich communication and minimal interaction (Janlert and Stolterman, 2017; Kaye, 2006) are relevant for technology design for people with ME, who require minimal physical, cognitive, and sensory exertion. Furthermore, a framework consisting of a set of strategies for designing technologies that support relatedness (while involving minimal interactivity and engagements) was proposed. The strategies were *physicalness*, *expressivity*, *awareness*, *balance*, *extension-of-self*, *coolness*, and *acts-of-care*. The strategies were developed based on a study with a *research product*. In that study, the research product was a networked robot, placed in the everyday lives of nine adolescents with ME to explore the extent to which the robot mitigates the sense of social isolation. The findings indicated that the research product was highly effective at improving the social relatedness of participants, while adhering to the need to reduce interactivity and engagements to a minimum. Research products are to be understood as described by Odom et al. (2016), as artefacts having the interrelated attributes *Inquiry-driven*, *Finish*, *Fit* and *Independent*. *Inquiry-driven* implies that experiences with the artifact drive the research inquiry. The *Finish* highlights the actuality of the design artifact, i.e., what it is, rather than what it might become. The *Fit* relates to the lived experiences with the artifact. Notably, the artifact should be perceived as neither too strange nor too familiar, for the artifact to gain an actual role in the context of users' lives. Lastly, the *Independent* implies that a research product can be used in everyday life without interventions from the researcher.

The research products used in this study was instrumental for gaining better insights on how the product would actually be used and to what extent it would fulfill its intent. This is important in general, but in particular when the research involves vulnerable users. Similar ideas did also guide the present work.

2.2 Research on Meditation in HCI

The rapid pace of modern life, has increased the interest in ancient practices of mindfulness and meditation as ways to cope with high levels of stress, thus improving the health and wellbeing. Here, the term mindful meditation is used to describe the process of taking time to slow down and focus on the present moment (Kabat-Zinn, 1990). Prior research on mindfulness meditation has shown that it can produce a variety of health benefits (Brown et al., 2015). For beginners in mindfulness meditation, breathing exercises are often the best way to start. Either human voice or meditative music guides the breathing. It is difficult for novice meditators to remain focused during meditation. Therefore, it frequently takes months of regular practice to develop a state of flow in the meditation (Cochrane et al., 2018).

Currently, in HCI there is a growing focus on mindfulness, wellbeing, and technology (Calvo and Peters, 2014; Niksirat et al., 2017). Several studies within HCI have investigated how to design technologies for mindfulness meditation to increase wellbeing. Most common approaches provide guided meditation through various senses like touch, audio, and vision (Niksirat et al., 2017). Several studies also focus on the user's bodily experiences (Bergström and Jonsson, 2016; Höök et al., 2016) as a guide for the design of meditation technologies. Others, provide visual feedback in real time, displaying bodily states based on data from wearable bio-sensors (like breathing or heartbeat) to encourage mindfulness and developing meditation strategies to increase wellbeing (Sanches et al., 2010).

Prpa et al., (2018) performed a research through design study focusing on breath and mindfulness meditation. They explored how people become aware of breath through embodied interaction, through the design of an immersive virtual environment with sounds to facilitate breath awareness as part of meditation. They found that participants' own accounts of their experience can describe their breathing data. Further, they state that their study: *"revealed the process of becoming aware of breathing in VE and demonstrated the importance of a sense of agency, understanding of control and possible subtle impact of the environment on breathing patterns without users being aware of it"* (ibid., p.80).

The somaesthetics, in general, focuses on research in HCI that engages with the body (Shusterman, 2012). Two particular examples are relevant to mention. The Soma Carpet and Breathing Light are research artefacts designed with a somaesthetic approach to meditative bodily introspection (Höök et al., 2016; Ståhl et al., 2016). The Soma Carpet provides heat stimuli to subtly guide the attention of people interacting with the artefact to different body parts. Furthermore, the Soma Carpet is described to support guided meditation, body scanning, Feldenkrais-sessions, or to facilitate relaxation (Ståhl et al., 2016). With a similar aim, The Breathing Light lamp creates an ambient light dimming connected to data from breath. The Breathing Light also uses sounds, like the sound of waves or bird songs, to facilitate the meditative experience. The studies revealed that the Soma Carpet and the Breathing Light are successful in generating a feeling of relaxation, softer movements, and increased awareness of breath (Ståhl et al., 2016).

This research aims to reduce interactions even further, focusing on the very subtle vibrational guidance of the breath and warmth towards relaxation.

3. FRAMING THE RESEARCH

In this study, research products were used to perform inquiries into how technology can support the self-management of ME. The inquiry was performed through material explorations in the everyday lives of five people with ME. Due to the nature of explorations, the research was qualitative.

3.1 The Method

The main method was in-depth interviews with participants. The author traveled to the homes of the participants for an interview, and left SlowBreath in participants' homes for at least a week. During this period, participants were to use SlowBreath freely and on their own terms. The elicitation diary was used to collect data on the use. The participants were given notebooks for the purpose. Whenever the participant was ready to return the artefact, the second visit, and the second interview took place. The involvement of the researcher in the field deployment was minimalized, to lower the strain of participating in the study for the ME ill participant, as described in (Culén et al., 2019). The data from the research product deployment was analyzed using content analysis with a-priory coding (Lazar, 2010).

3.2 Participants

The five participants in the study all have mild to moderate ME diagnosis. Four participants are female, and one is male. In the rest of the paper, pseudonyms are used to preserve their privacy. Hedda and Mathea had a rather similar living situation, both live in an apartment on their own. Elin lives with her fiancé. Astrid lives with her husband and two children and has had the ME for fourteen years, through the largest part of her children's childhood. Henrik lives with his parents.

3.3 Ethics and Dissemination

A strong focus on research ethics and how to involve participants in the research while reducing negative health implications due to the energy needed to participate in the study was an important ongoing concern. The study was also reported to and approved by the national ethical review board.

4. THE DESIGN OF SLOWBREATH

The SlowBreath did not emerge through the ideation process, or in an instant. Rather, through four years of research with technology for people with ME, the common concern with how to relax and know when to do it emerged as important. All people with ME who crossed my research trajectory struggled with energy balancing, and almost all had limited experience with meditation. Thus, the design focused gradually on making a device that gently supports users in relaxation or mindfulness meditation. Through the vibration, the user was given a guided breathing experience in time intervals lasting four-seconds. Both the duration of the breath and the quality of vibration could be, in theory, endlessly manipulated until the 'right' combination for a participant was found. However, two particular patterns were explored with participants. The first one was a four-seconds breath pattern, inspired by traditional mindfulness meditation practices. The second pattern was giving repetitive pulsing vibrations to increase mindfulness. The heat functionality was intended to be calming and relaxing. The combined use of heat and vibration aimed to give participants the 'ideal' combination to reach the relaxed, meditative state.

In line with (Mazé and Redström, 2005), this research explored a plethora of combinations and dynamics of the materials, forms, and computational expressions to create the SlowBreath, Figure 1. In what follows, excerpts from the initial interviews with users provide for a richer understanding of the condition. In the light of this, the choice of the shape, computational materials, and other materials are discussed.

4.1 Energy Balancing and Stress

When asked about what living with ME implies, Elin said: *"People often don't understand that having ME is so much more than just being tired, the symptoms and PEM are so much more complex than just being tired".* All participants discussed the PEM, and how it was triggered for them. They all wished they had an explanation (medical) for what it is that they have. Elin addresses the enigma of PEM: *"A lot of effort goes into trying to find patterns that could give insights into where my limits are in regards to overuse of energy, or what it is that exactly makes me worse".*

All participants had sleep difficulties, both in regards to poor sleep quality, and non-restorative sleep that affected their energy levels. Two of the participants had developed a strategy of looking at data about sleep quality in the morning, as one of the indicators of their energy level on a particular day. Hedda got the sleep quality data from her Fitbit, while Astrid got it from her CPAP machine, that she uses for her sleep apnea.

However, self-monitoring the illness through logging and self-reporting can be difficult. Elin pointed out that she was on a self-management course for the people with ME, where she was to log her energy levels extensively. The tracking was not working for her: *"The tracking was misplaced, since I can't trust my feelings at the moment as indicators of where my limit goes. Because if something is too much, I do not feel that at the moment. I can feel good doing something, but the next day I find out that it was all too much what I did the day before".* She further explains: *"It feels very varying what I can do and not do. I can do something one day and endure it fine, but the next day I do the same thing and get really sick".* In addition, the act of tracking extensively was making her sicker.

The participants describe the enigma connected to the cause of PEM as a huge challenge, especially since it affects their social lives. When their condition is so unpredictable, it makes planning (desired) social activities difficult. All participants admitted that they sometimes do activities that they know will cause PEM, but they went ahead because they felt like doing something important or meaningful, no matter the cost. However, for all participants, such overuse of energy resulted in weeks or even months of having increased illness symptoms and often being entirely bedbound during that time.

Hedda draws attention to another common problem of self-blame: *"Everybody has their window of opportunity where there are small things that they can do to affect their daily health. However, it is also a part of the illness to get really sick without having done anything wrong in regards to self-management and energy balancing. Sometimes, the illness simply gets worse. It is important at that time to remember that if you get worse and worse, you should not blame yourself. What we know for sure, is that energy balancing can often help people with ME. It comes without any risks of worsened health for the individual. Since there is no cure for the illness yet, energy balancing is the only 'treatment' we have now that can have an effect".* Henrik addresses his concern regarding the increased focus on energy balancing: *"I can become scared of over-exertion and what happens in my body if I go beyond my limit. It is very important to focus on energy balancing, but not to the point where it becomes scary to do anything".* He further elaborated that being scared resulted in a feeling of being stressed and provided an example of how technology for self-management of ME can be stressful: *"If you are, for example, going to have something to remind you to rest, it must provide reminders in a way that doesn't make you stressed".*

From all of the above, it is difficult for people with ME to not feel the stress of living with the condition. In fact, the states of anxiety, confusion, even panic are frequent. Ways to relax then can be a part of those small daily things that can be done by ways of self-help.

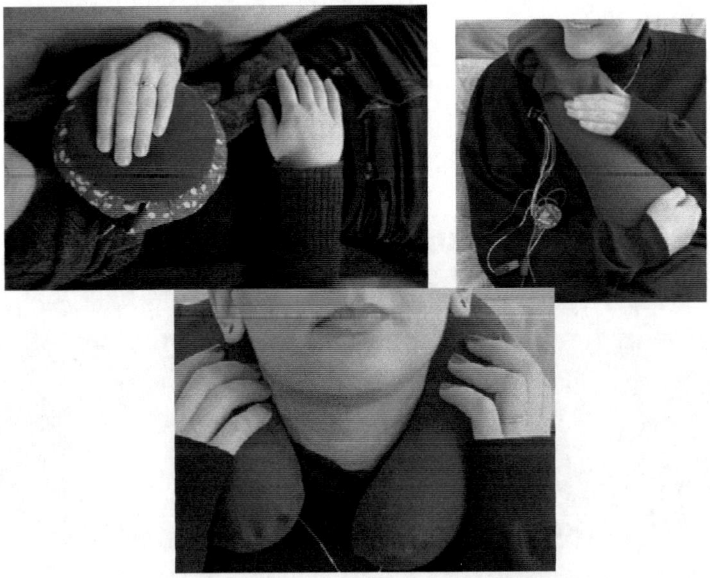

Figure 2. (left) How Hedda used the circular shaped SlowBreath for the first time. (middle) Elin trying the rectangle shaped meditation cushion. (right) Elin trying the U-shaped cushion

4.2 Three Shapes to Explore Bodily Experiences of Participants

As mentioned, there were developed three distinct shapes, in the following, the forms are presented, in addition to the selected computation and materials. Also, an account is given of how participants interacting with them during interviews perceived these shapes. The three distinct shapes were used to explore materials, forms, and shapes through bodily experiences of artefacts similar to *soma shapes and tools* (Höök, 2018).

4.2.1 Form

Several shapes and sizes were used (see Figure 2) and addressed during interviews with the participants to explore how the form shaped the properties of the meditation artefact. There were three main shapes:

i. Long and rectangle shaped, it was weight down by its filling that was buckwheat.
ii. U-shaped cushion, very lightweight filled with small Styrofoam balls.
iii. Circular shape filled with poly-fil.

All the different meditation cushions had reused fillings, and the outer material was ecological cotton. While all three shapes had different qualities that in some way were relevant for generating meditative experiences, the circular shape was found to be more configurable during the design process, and was thus selected to be used as research products placed in the everyday lives of participants for a little over a week of use. However, the other shapes and different size variations of the circular shape was used during interviews, to further inquire into the bodily experiences of all the developed shapes.

4.2.2 Computation

SlowBreath was programmed in Arduino and Python, it contains an Adafruit FLORA microcontroller, two electric heating pads, five vibrating mini discs, and one haptic motor driver. Through changing the code controlling the vibration as part of the material exploration, it opens up for investigating how the temporal form and additional properties of SlowBreath were manifested through interacting and using the artefact.

4.2.3 Material

SlowBreath was sewn of ecologic cotton fabric and filled with repurposed poly-fil. Through the exploration of materials bringing the fabric closer to something natural by choosing cotton over synthetic materials became important. The material was a key aspect of shaping the feeling of holding and touching SlowBreath and experiencing the vibration and heat provided by the computation.

5. FINDINGS

The findings are presented from the bodily experiences with the three different shapes and lived experiences with the research product.

5.1 The Bodily Experiences of the Three Shapes

In this section, the findings regarding participants perceptions of the three shapes are provided.

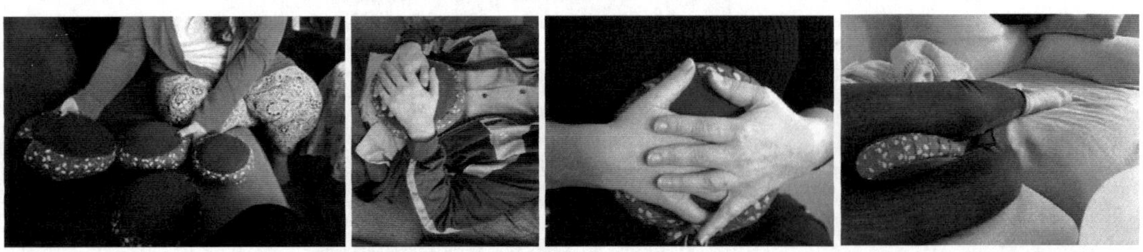

Figure 3. (left) Mathea exploring different sizes of SlowBreath, (middle left) Henrik trying the guided breathing, (middle right) Astrid using the meditation vibration functionality (right) Hedda using the heating functionality

5.1.1 Rectangle Shaped Meditation Cushion

The rectangle shape reminded the participants of the heat pads that can be heated in the microwave oven pointing out that the cushion had similar shape and weight. All participants liked the weight of the rectangle shaped cushion. Elin commented: *"I like the feeling of the added weight given by this cushion. For me, the weight is calming"*. Hedda similarly commented: *"I like the heaviness. For example, I also think that a large, heavy duvet is more relaxing than a thin blanket when sleeping. With the added weight, it provides conditions for relaxation in a good way"*.

5.1.2 U-shaped Meditation Cushion

The u-shaped cushion gave clear associations to traditional neck cushions, which also restricted the participants imagined ways of using this meditation cushion. Elin addressed this by stating: *"I cannot see myself using it in other areas than the neck, I would feel quite stupid by embracing or cuddling something that looks like a neck cushion. The other cushions can be used in many different ways more naturally than the neck cushion one"*. However, the participants did like to use the heat functionality at their neck area, as it was described as a body part that was often cold.

5.1.3 The Circular Shape of SlowBreath

The notion that the circular shape is more versatile surfaced during the research explorations in the design process and was validated through interviews with participants. Elin's initial comment regarding SlowBreath was: *"It is so lightweight and can be versatile in use"*. She immediately placed SlowBreath at her chest, with the heat towards her body and placed her hand on the vibration. She stated: *"It feels good to have the cushion on my chest and the added heat is great"*. She also commented that the meditation cushion would be suitable for placing on her abdomen to visually follow her breathing during the meditation. When first introduced to SlowBreath, both Hedda and Astrid laid down and placed it on their chests. In contrast, Henrik and Mathea both placed it on their abdomen. However, all participants were actively exploring different placements of SlowBreath during the first interview.

Scaling up and down the size of the cushion was also explored. A total of five different sizes of SlowBreath were made, as can be seen in Figure 3 (left). The participants interacted with the different sizes of cushions. It became evident that perceived affordances of the artefacts varied based on size. Hedda and Elin perceived the smallest cushion as toy-like, as Elin articulated: *"the smallest meditation cushion feels like a kind of teddy-bear. Intuitively, I just want to lift it and cuddle it."* The largest meditation cushion was seen as an ordinary cushion that invited to be used as a head rest, or placed in-between one's legs for added comfort (relevant for those who are in laying down position a lot). However, the sizes in-between these two were found to be more ambiguous as Hedda elaborated: *"I think it loses its teddy-bear feeling already at one size bigger. The smallest is fun and cute, the biggest is clearly a cushion. But the ones in the middle are a little unrecognizable. They are a mysterious thing that one does not know what is"*. The ambiguous aspect of the middle-sized cushions also sparked curiosity, as Astrid pointed to: *"I kind of like the middle-sized version, somehow it makes me curious about what it can do for me, and I like that feeling"*.

5.2 Lived Experiences with the Research Product

The mid-size, circular shaped cushion was used in the field study. The aim of the study was to allow the participants to experience SlowBreath in their everyday lives and for relaxation, as part of energy balancing. The closing interview with participants contained rich descriptions about their individual, lived experiences with the cushion. The analysis of interviews focused on the a-priory categories: 1) bodily experiences 2) relevance of the cushion for energy balancing, 3) tension between meditation and relaxation, and 4) feedback on the design with suggestions for re-design, which are used bellow to organize the findings.

5.2.1 The Bodily Experiences

The vibrational functionality was liked by all participants. Some needed to get used to it, as Elin describes: *"The vibration functionality to guide the breathing was comfortable, but it took some time to get into it. I like it"*. After testing SlowBreath in their homes, Elin and Hedda, as first users of the cushion, did raise some issues. For Hedda, the mindfulness breathing was new, and she found it challenging. She would have liked to keep the cushion and try more, but with different vibration patterns. She described her difficulty: *"I tried to breathe with the vibration, but I got so out of breath and just wanted to breathe normally. So, it was hard and tiring to follow the guided breathing"*. Elin found the vibration and the heat to be too intense for her to facilitate a relaxing and meditative experience. Even though Hedda and Elin did not have positive experiences with the implemented functionality, both expressed that the potential of the cushion is great and that with lower intensity of the vibration and heat it could have been quite beneficial toward relaxation. The vibration and the heat then were adjusted, impacting very positively the experiences with the artefact for the remaining three participants. This also showed how even minor alterations to the computation materials had a significant impact on bodily experiences.

Astrid's experience with SlowBreath was very positive: *"I have used the guided breathing the most, it was easier to find a rhythm in my breath with it. I experienced that the four seconds vibrations calmed down my breathing also. I usually find it difficult to calm my thoughts, but this too was easier with SlowBreath. I could relax every time I used it."*

"It felt very comfortable to just lie down and feel the vibration", Mathea pointed out, elaborating on her experience of using SlowBreath: *"When using SlowBreath, I noticed that I draw in the breath with the first vibration and then at the moment when I start breathing out the next vibration comes naturally. I experienced that I developed a nice rhythm in tune with SlowBreath, the intervals felt really nice. And that experience shows how the name* [SlowBreath] *reflects exactly what it does"*.

Henrik described his experiences with SlowBreath as positive – on most days. On days when the illness symptoms were increased, the vibration was still perceived as too strong.

5.2.2 Potential of Being Incorporated in Practices of Energy Balancing

Hedda, although she thought that the cushion could be very useful, had some reservations for her own use. She had been sick for so long that she had developed her own practices of resting as part of energy balancing. Thus, she was not sure if she would be incorporating SlowBreath in her life as a regular practice of relaxing.

Henrik's conclusion was similar: *"Since I've been sick for a while, to go and rest is so natural for me. I somehow know exactly what to do, and how to control my breathing. If I would now start using SlowBreath, it will be a bit distracting. But then, if I had SlowBreath earlier in the course of my illness, it would have been ideal."* He also pointed out that he still finds it challenging to relax through breathing exercises when he is not at home, like when traveling, or sitting at the doctor's waiting room. He saw the potential for himself to use SlowBreath in such situations.

Mathea, Elin, and Astrid had not resolved challenges related to relaxation practices, bringing forward the fact that they are very easily distracted. Mathea found SlowBreath successful in helping her keep the focus on the breath, thus, leading to less distraction. She would have liked to incorporate SlowBreath in her current practices of energy balancing. Also, Astrid reported that SlowBreath was very successful in helping her meditate and experience fewer distractions, making her regular practice easier and better.

All participants meant that SlowBreath had a potential to help develop practices of relaxing, and energy balancing, in the early onset of the ME illness. Using SlowBreath in the course of this study, made all participants reflect on how their own practices of energy balancing evolved since the onset of the illness.

5.2.3 Resting vs. Mindfulness Meditation

Prior to the engagement in this study, an inquiry on meditation was done with the ME organization. It revealed that there are two fractions: people who are positive towards meditation, and the ones that are negative. Hedda addressed this directly: *"I know that the ME community is very divided when it comes to meditation, about half practice meditation and finds it helpful, while the rest prefer to rest without meditation"*.

The participants in the study also had different personal opinions about meditation, especially prior to using SlowBreath. Elin, Astrid, and Henrik like meditation, while Hedda and Mathea do not find mediation useful and were critical towards its role in energy balancing. Mathea had a strong reaction towards framing SlowBreath as an aid for meditation. She would have preferred if this activity was framed as resting: *"I react to the use of the word meditation. When I think about meditation, I think more about the east and finding Zen and such stuff, which does not suit me. So, for me, it is more a relaxation aid than a meditation aid"*.

5.2.4 The Re-design of SlowBreath

The participants overall liked the simplicity and the minimal interactions (choice of vibration and level of heat). They liked being able to remove all lights, close their eyes, lay in a comfortable position, and wear custom molded earplugs that removed all sound while using meditation devices, in order to minimize sensory engagements. When they were asked to compare SlowBreath to a range of existing solutions with audio or visual feedback, as described in Section 2.2, all participants said that the haptic feedback is strongly preferred to any other type of feedback.

Participants expressed the desire for more control over parameters (vibration modes and heat in more granular increments), as represented by what Elisabeth said: *"I would have liked to be able to change the length of the vibrational interval to six seconds instead of four. While using SlowBreath, I experienced that my breathing became so calm that longer intervals would be better. A button to adjust the strength and the depth of the vibration would be nice to have, so that one could do the adjustments with the eyes closed".* Mathea, Henrik, and Astrid also pointed out that their reaction to haptic feedback depended on the state of the illness and the intensity of symptoms. For that too, adjustments would be important. Mathea expressed this common opinion as follows: *"I have noticed that regarding the vibrations, I sometimes thought that it was a little too strong, and sometimes that it was a little too weak. So, it would have been nice if I could adjust the strength of the vibrations".*

In addition to control over the parameters mentioned above, the participants would have appreciated having a timer. As it was, when using SlowBreath all participants set the alarm for 20-30 minutes. Mathea suggested the timer to have at least to options, a short 10-minute session, and a longer, 20 minutes one. The short option was perceived as a low threshold that was actually achievable for all.

Astrid and Mathea proposed an alternate placement of the vibration discs and the heat pads. Elisabeth explained: *"I liked the feeling of the heat very much. So, I would always place the heat down on my stomach, and then, I would have wanted to have the vibration into my body, too. With SlowBreath as is, when the heat is placed down, the vibration side faced up. I had to use my hands to feel the vibration, but the hands are often something you should mentally disconnect from during meditation".*

Inspired by thinking of how SlowBreath could help him practice energy balancing outside the home, Henrik came with the following suggestion: *"I would have liked the cushion shaped like a fish. With a size that fits perfectly into my hands. It would be nice with a silicone surface, or something else that is easy to clean. I would then not have to worry about bringing home the bacteria from outside. If some kind of transparent silicone was used, then one could also choose whether you wanted vibration or a dimmed light to guide the breathing. The light alternative is suitable because vibration sometimes becomes a bit too disruptive. But also, the vibration feels really good. I could then just sit and have the fish inside my hand and use the vibration to calm down".*

Other comments on the shape included: *"The biggest round cushion gives the feeling of a real cushion, which for me resonates with an object made for a different use. If it was even slightly bigger, it could be seen as a sofa pillow, and as such, be very accessible. That is positive. However, people have different styles they prefer regarding their pillows",* Elisabeth said. A re-design into a shape that fitted commercial sized cushion covers was perceived as a good option for Hedda: *"It would be good if you can use your own cushion covers, then you have all the possibilities, and the style of the meditation cushion could be guided by a personal taste. It would be easily accessible since it is just there, on your couch. Nice with something you do not have to use the energy to get, or tidy after the use".* As a possible negative aspect of this, she reflected, was that it could be easily forgotten – while in plain sight. Discussing the option of making the meditation cushion in teddy-bear shape, Hedda said: *"An advantage with the teddy-bear shape is the possibility to personalize it. It could be given a name, and one could say 'I have to go to Teddy now, and rest a bit with him, coming back soon'".* The personalization could be a fun thing to do that could make it easier and perhaps more socially acceptable to take a break when in the company with others. Elin reflects along these lines: *"In social situations, it can be very difficult to say that one needs a break. For example, if I am caught in an intense personal conversation, it can feel rude to break out of it. But the breaks for us with ME are necessary, or else one can easily sit for too long, risking becoming sicker. There is not a good way to tell others that you are in a bad shape".*

All participants said that the aesthetics of the cushion was very important. As they spend a lot of time at home, every detail is extra important. Hedda articulated this point as follows: *"Whether it is a cushion or a bear, I would have appreciated if it was designed with simplicity, and perceived as pretty. Then it is not an ugly thing that is at odds with the rest of the apartment when you bring it home".*

6. DISCUSSION

In this section, the three main implications of this work for designing technology for self-management of ME, and for further research within HCI are discussed: 1) SlowBreath as part of energy balancing, 2) implications for design for relaxation for users with ME, and 3) The research approach.

6.1 SlowBreath as Part of Energy Balancing

The results of the study confirmed that people with ME are a user group with very different needs regarding the design of digital artefacts, including meditation cushions such as SlowBreath. The participants' condition demands minimal engagement and minimal interactions, in line with findings in (Culén et al., 2019). Working with them clearly demonstrated that over-burdening any of the five senses is not beneficial, and can trigger the hypersensitivity to external sensory inputs. The haptic feedback was found to be suitable, and in combination with heat, it was particularly helpful for people with ME, who often experience symptoms connected to abnormalities in the body temperature regulation (Carruthers et al., 2011). The findings indicate that self-management of ME through energy balancing can be helped with SlowBreath.

The findings also show that further improvements are necessary should SlowBreath be developed into a product. Most notably, the ability to easily (preferably with closed eyes) regulate the strength of the heath and vibration, the length of the vibrational cycle, as well as the length of a session would all be desirable features towards effective energy balancing.

6.2 Meditation and ME, Implications for Design

One of the insights was that, despite the small number of participants, the stand on meditation and ME reflected the pattern of the larger ME community – some participants were considering meditation as beneficial from the start, while others had reservations, and preferred to avoid the use of the word meditation all together (opting for describing the activities involved as the rest or relaxation). Apart from the terminology, the participants agree that the following implications regarding design for relaxation are important:

– The artefact for rest or meditation has to be carefully matched with their actual lives and needs that vary from day to day.

– The vibration and heat have to be adjustable and changeable during the experience of resting or meditating.

– A careful inquiry into how to support the development of sustained, everyday practices of using the artefact for meditation should be conducted, also from the perspective of those who do not like to call this practice meditation.

– In the same way as the computational materials need to be adjustable and controllable by participants, its shape and material components (fabrics) should match the style of the surroundings and hold whatever aesthetic standards the participant demands. Seeing how the actual home interior differed profoundly between participants, yet they all wanted the cushion to match their own interior, the ability to personalize the materials is particularly important when designing everyday artefacts for people with ME.

– The accessibility of the artefact within the home, is another important aspect when designing meditation artefacts for people with ME. If the artefact is an integral part of the home décor, and does not exert extra efforts to take it out, use it, or tidy after use, if it can facilitate resting at any time without taking extra energy (implying that it is accessible wherever rest and meditation take place), then it is beneficial for users with ME, and lowers the threshold for use.

– Reminders to use the artefact should be considered when designing meditation artefacts for people with ME. Forgetting to take breaks is one of the main barriers to energy balancing. Setting up mobile phone reminders is a strategy that the participants already rely upon in their everyday lives, to overcome the cognitive symptoms associated with ME (challenges with memory and concentration). Thus, providing reminders is an important step towards facilitating the habitual use of SlowBreath as a self-management strategy for energy balancing.

6.3 The Research Approach

Even though explorations were initially done by the author as a first-person research (Lucero et al., 2018), it was essential to deploy the SlowBreath in the field and with real users to get a proper indication of how well this product would work for people with ME. The insight from this study is the same as described in (Culén et al., 2019) – the research products placed in everyday lives offered the possibility for participants to use the research product as if it was a finished product, and to experience the artifact on their own terms. Of particular importance was that the use does not trigger the PEM. Furthermore, doing research with SlowBreath placed at people's homes gave the possibility to explore the bodily experiences of people with ME in a different way than traditional participatory approaches (where the design process is a co-design, but the artefacts are not always ready to be used at people's homes, in contrast to first-person design and lived experiences by users).

6.4 Study Limitations

The length of the study presented a limitation. Seeing how long it could take to develop the skill of meditation though mindfulness breathing and to develop and sustain practices of using SlowBreath as part of energy balancing, a lot longer study would have been needed. For in-depth insights on practices of meditation using SlowBreath by novice meditators and the potential health benefits would require a different time scale.

Another limitation is the small number of participants. This is frequently the case for similar qualitative studies in HCI, especially when the research involves vulnerable users.

Last, the interview data was coded and analyzed by the author only. To increase the validity of the analysis, as well as discuss the relevance of a-priory categories chosen, a collaboration with another researcher would be in place. However, SlowBreath is the last of several technology design explorations for people with ME done by the author during the last years. Thus, the author has relied on the extensive experience in the research domain that became the foundation for the approach taken.

7. CONCLUSION

This paper presented the design and deployment of a device called SlowBreath. SlowBreath was made with the intent to support people with ME in self-management through energy balancing. Unlike existing meditation aids that employ combined elements of audio, haptic and visual feedback to guide the user in achieving a meditative state, SlowBreath takes the alternate approach relying on minimal interactions and haptic feedback through vibration and heat to avoid PEM due to sensory overload.

Overall, all five participants who took part in the study preferred the haptic feedback over the other devices available that used combinations of haptic, audio and visual feedback. The findings also highlight the importance of exploring the combinations and dynamics of the materials, form and computational expressions when designing for technologically-mediated meditative experiences.

ACKNOWLEDGEMENT

Thanks are due to the Norwegian ME organization for helping with the participant recruitment process, and providing helpful advice anytime it was needed. I would also like to thank the participants who generously shared their time, and provided valuable input and feedback on their experiences of having SlowBreath in their lives for the purpose of this study.

REFERENCES

Bergström, I., Jonsson, M., 2016. Sarka: Sonification and Somaesthetic Appreciation Design. *Proceedings of the 3rd International Symposium on Movement and Computing - MOCO '16*. Thessaloniki, GA, Greece, pp. 1–8.

Brown, K. W., Creswell, J. D., Ryan, R. M., 2015. *Handbook of mindfulness: theory, research, and practice*. The Guilford Press, New York.

Calvo, R. A., Peters, D., 2014. *Positive computing: technology for wellbeing and human potential.* The platform studies series. The MIT Press, Cambridge, Massachusetts, USA.

Carruthers, B. M., van de Sande, M. I., De Meirleir, K. L., Klimas, N. G., Broderick, G., Mitchell, T., et al., 2011. Myalgic encephalomyelitis: International Consensus Criteria. *Journal of internal medicine*, Vol. 270, No. 4, pp. 327-338.

Cochrane, K. A., Loke, L., de Bérigny, C., Campbell, A., 2018. Sounds in the moment: designing an interactive EEG nature soundscape for novice mindfulness meditators. *Proceedings of the 30th Australian Conference on Computer-Human Interaction - OzCHI '18.* Melbourne, Australia, pp. 298–302.

Culén, A. L., Børsting, J., Odom, W., 2019. Mediating Relatedness for Adolescents with ME: Reducing Isolation through Minimal Interactions with a Robot Avatar. *Proceedings of the 2019 on Designing Interactive Systems Conference 2019 - DIS '19*, ACM, San Diego, USA. Doi:10.1145/3322276.3322319

Goudsmit, E. M., Nijs, J., Jason, L. A., Wallman, K. E., 2012. Pacing as a strategy to improve energy management in myalgic encephalomyelitis/chronic fatigue syndrome: a consensus document. *Disability and rehabilitation*, Vol. 34, No. 13, pp. 1140–1147.

Hvidberg, M. F., Brinth, L. S., Olesen, A. V., Petersen, K. D., Ehlers, L., 2015. The Health-Related Quality of Life for Patients with Myalgic Encephalomyelitis/Chronic Fatigue Syndrome (ME/CFS). *PLOS ONE*, Vol. 10, No. 7, pp.1-16.

Höök, K., Jonsson, M. P., Ståhl, A., Mercurio, J., 2016. Somaesthetic Appreciation Design. *Proceedings of the 2016 CHI Conference on Human Factors in Computing Systems - CHI '16.* ACM Press, California, USA, pp. 3131–3142.

Höök, K., 2018. *Designing with the body: somaesthetic interaction design*. The MIT Press, Cambridge, Massachusetts, USA.

Janlert, L.-E., Stolterman, E., 2017. The Meaning of Interactivity—Some Proposals for Definitions and Measures. *Human–Computer Interaction*, Vol. 32, No. 3, pp. 103–138.

Kabat-Zinn, J., 1990. *Full catastrophe living: using the wisdom of your body and mind to face stress, pain, and illness.* Delacorte Press, New York, NY. USA.

Kaye, J. J., 2006. I Just Clicked to Say I Love You: Rich Evaluations of Minimal Communication. *Extended Abstracts on Human Factors in Computing Systems CHI'06*. New York, NY, USA, pp. 363–368.

Lacourt, T. E., Vichaya, E. G., Chiu, G. S., Dantzer, R., Heijnen, C. J., 2018. The High Costs of Low-Grade Inflammation: Persistent Fatigue as a Consequence of Reduced Cellular-Energy Availability and Non-adaptive Energy Expenditure. *Frontiers in behavioral neuroscience*, Vol. 12, No. 78, pp. 1-20.

Lazar, J., Feng, J. H., Hochheiser, H., 2010. *Research methods in human computer interaction*, John Wiley & Sons, NJ, USA.

Líndal, E., Stefánsson, J. G., Bergmann, S., 2002. The prevalence of chronic fatigue syndrome in Iceland - A national comparison by gender drawing on four different criteria. *Nordic journal of psychiatry*, Vol. 56, No. 4, pp. 273–277.

Lucero, A., Desjardins, A., Neustaedter, C., Höök, K., Hassenzahl, M., Cecchinato, M.E., 2018. A Sample of One: First-Person Research Methods in HCI, *Proceedings of the 2019 ACM Conference Companion Publication on Designing Interactive Systems.* ACM.

Mazé, R., Redström, J., 2005. Form and the computational object. *Digital Creativity*. Vol. 16, No. 1, pp. 7–18.

Niksirat, K. S., Silpasuwanchai, C., Ahmed, M. M. H., Cheng, P., Ren, X., 2017. A Framework for Interactive Mindfulness Meditation Using Attention-Regulation Process. *Proceedings of the 2017 CHI Conference on Human Factors in Computing Systems - CHI '17.* Denver, Colorado, USA, pp. 2672–2684.

Odom, W., Wakkary, R., Lim, Y., Desjardins, A., Hengeveld, B., Banks, R., 2016. From Research Prototype to Research Product. *Proceedings of the 2016 CHI Conference on Human Factors in Computing Systems, CHI '16. ACM.* New York, NY, USA, pp. 2549–2561.

Prpa, M., Tatar, K., Françoise, J., Riecke, B., Schiphorst, T., Pasquier, P., 2018. Attending to Breath: Exploring How the Cues in a Virtual Environment Guide the Attention to Breath and Shape the Quality of Experience to Support Mindfulness. *Proceedings of the 2018 on Designing Interactive Systems Conference 2018 - DIS '18.* Hong Kong, China, pp. 71–84.

Sanches, P., Höök, K., Vaara, E., Weymann, C., Bylund, M., Ferreira, P., et al., 2010. Mind the body!: designing a mobile stress management application encouraging personal reflection. *Proceedings of the 8th ACM Conference on Designing Interactive Systems - DIS '10.* Aarhus, Denmark, pp. 47-56.

Shusterman, R., 2012. *Thinking through the Body: Essays in Somaesthetics*. Cambridge University Press, Cambridge, United Kingdom.

Ståhl, A., Jonsson, M., Mercurio, J., Karlsson, A., Höök, K., Johnson, E.-C. B., 2016. The Soma Mat and Breathing Light. *Proceedings of the 2016 CHI Conference Extended Abstracts on Human Factors in Computing Systems - CHI EA '16*, ACM Press, Santa Clara, California, USA, pp. 305–308.

USING ICT IN OCCUPATIONAL STRESS MANAGEMENT: THE VIEWS AND EXPERIENCES OF SRI LANKAN STRESS EXPERTS

Manoja Weerasekara[1], Åsa Smedberg[2] and Hélène Sandmark[2]
[1]*School of computing, NSBM Green University Town, Pitipana, Homagama, Sri Lanka*
[2]*The Department of Computer and Systems Sciences, Stockholm University, Kista, Sweden*

ABSTRACT

ICT supported health intervention is a phenomenon that has grown stronger in recent years. Although eHealth and mHealth care are an expanding area of ICT, the development has not been as significant as in other fields. Successful implementation of eHealth or e-mental health applications requires close collaboration between healthcare practitioners and system developers. Complex and multilayered factors determine healthcare practitioners' intention to use ICT in their practice. This study aims to understand the Sri Lankan stress experts' experience and perception of using ICT tools in their stress management practices. Eight stress experts in Sri Lanka were interviewed. Through thematic analysis, we identified three main themes and multiple subthemes. Results show that practitioners used only a limited set of internet resources in their stress management practices and preferred face-to-face consultations over distance/online communication. Overall, they appreciated the immediate benefits of ICT interventions while highlighting disutilities like the risk of misleading information and insufficient access to resources. Moreover, clients were regarded as empowered through increased access to new and more information. They were also encouraged to make their own decisions to a greater extent. However, there were also objections from practitioners who were reluctant to accept the client's ability to make decisions through self-education. This study can possibly provide a better understanding of the impact of ICT on stress management practices and knowledge to design and develop ICT tools for stress management.

KEYWORDS

eHealth, e-Mental Health, Health Intervention, Stress Management, Sri Lanka: Stress Expert

1. INTRODUCTION

Stress is a global phenomenon and a major healthcare challenge in the 21st century. Almost everyone is exposed to a wide range of pressures in both personal and work life. Some individuals cope very well with these pressures. On the other hand, some individuals find it difficult to manage the situation, and that is when individuals fall into the "stress." Occupational stress is a branch of stress which is associated with one's work environment or job conditions. The imbalance of occupational "demands" and individual's "supply" will cause occupational stress (Mustafa et al., 2015); where high pressure, unrealistic goals and expectations within the occupation do not match with an employee's knowledge, skills and capacity. It is commonly believed that stress to a certain level is acceptable and considered positive, whereas the excessive amount of stress is deemed to be negative or potentially harmful to both physical and mental wellbeing (Lazarus, 1993). Nowadays, stress-related disorders are a significant public health matter in many countries and many organizations. The analysis of Global Employee Assistance Programs conducted across Asia, Europe, Africa, North America and South America, covering more than 100,000 employees, revealed a significant increase in stress reactions, anxiety and depression (Debnam, 2016).

Technology has become an integral part of people's everyday functioning, changing their lifestyle rapidly. People are willing to receive quality services effectively and efficiently through information and communication technology (ICT) enabled processes. Not only with the areas of business, education and transport, now there is a rising demand also for ICT supported healthcare delivery (Burney, 2010). Healthcare services, with eHealth and mHealth approaches, offer enormous potential for effective information exchange, reduction of costs and improvement of healthcare access. The term eHealth is referred as "health services and information delivered or enhanced through the Internet and related technologies" (Eysenbach, 2001, p. 2), whereas mobile health, or mHealth, is defined as the "use of mobile computing and communication technologies in health care and public health" (Free et al., 2010, p. 1). ICT supported health interventions provide avenues for a collaborative environment where patients and healthcare professionals connected virtually enable more personalised and customised services (Burney, 2010). Besides the ICT tools for collaboration between healthcare providers and patients, there are also online support groups allowing patients to gain a more autonomous role (Barrett, 2005). With online support groups, more patient participation and control can be achieved (Barak, Boniel-Nissim & Suler, 2008; Barrett, 2005). Through the interaction with others, patients and other people with health concerns can learn about what resources to access, develop their ability to make personal decisions and to think critically (Barak et al., 2008). They get support to develop skills and abilities together with a more positive image of themselves leading to 'patient empowerment.'

ICT plays a significant role in identifying and overcoming stress-related disorders. It can help to improve the efficiency of interventions and make it possible to reach a broader audience regardless of time and geographical barriers (Cohen, Kessler & Gordon, 1995). In the same way, ICT interventions allow the user to access information in a quick, easy and confidential manner (McKenna & Bargh, 2000). Although eHealth and mHealth in mental healthcare are an expanding area, further development is needed (Schmidt & Wykes, 2012). ICT enabled tools in mental health care are underused (Ennis, Rose, Callard, Denis, & Wykes, 2011; Mclaren, 2005), and there is just little scientific evidence supporting investment in ICT-based psychosocial programs for resilience or prevention of posttraumatic stress symptoms, depression, or anxiety (Breslau & Engel, 2016). This may be due to the unique challenges faced in developing and applying/using such ICT tools in the health sector, including financial capacity; infrastructure availability; knowledge and access to such resources; leadership and governance issues; and security, social and technical issues (Ariani, Koesoema & Soegijoko, 2017).

Sri Lanka is a country with a national awareness on mental health concerns due to occupational stress. According to recent statistics at the Mental Health Unit of Sri Lanka's Health Ministry, one in five public sector employees suffers from stress, and this has become a real issue that needs to be addressed at the workplaces (ColomboPage, 2017). Sri Lanka is a small island lying southeast of India with a population of 21.02 million in 2019. It is the 58th most populous country in the world with multi-cultural and multi-ethnic identity. Sri Lanka is a developing country with an economically active population around 8.6 million in the first quarter 2017 of which 62.7% are males and 37.3% are females (Sri Lanka Labour Force Survey, 2017). Among them about 45% are engaged in the service sector, 28% in the industry sector and 27% in the agriculture sector. With the fast-paced industry and service environment, the competitive run to reach the next level, long working hours and lack of work-life balance have resulted in mental stress and mental disorders among workers (Rajgopal, 2010). Mental health care is provided through a combination of public, private and non-government services (De Silva, 2002). The psychiatric treatment service system is comprised of a network of hospitals at national, provincial and district levels. Primary care is managed by the district hospitals, the peripheral units and the rural hospitals, with referrals to higher institutions when necessary. Secondary care is provided by the base hospitals and, increasingly, by psychiatric inpatient units in district hospitals. Tertiary and sub-specialty inpatient psychiatric care is provided by the National Institute of Mental Health (World Health Organization, 2017). There is a small private psychiatric service sector. Psychiatrists working in the public sector may also carry out private practice outside the public working hours and provide approximately 50% of outpatient care (De Silva, 2002). Mental healthcare is also provided by Ayurvedic and other indigenous practitioners (De Silva, 2002). Mental Health ATLAS 2017 Member State Profile of Sri Lanka recorded that there are only 1,480 mental healthcare professionals in both government and non-government sector, which gives 7.14% mental healthcare workers per 100,000 population. Considering the current shortage of mental healthcare human resources, there has been particular interest in introducing innovations to deal with the situation (Minas, Mendis, & Hall, 2017).

Thus, this study specifically aims to understand the Sri Lankan stress experts' experience and perception of using ICT tools in their practice. The outcome of this research could feed-in as knowledge to design and development of ICT tools for stress management to yield optimum efficacy levels.

2. RESEARCH AIM AND OBJECTIVES

Based on the introduction, this study aims to explore the Sri Lankan stress experts' experience and perception of the use of ICT for stress management interventions. This will provide a holistic understanding of perceived benefits and challenges of using ICT in their practice for stress management in the Sri Lankan context. The specific objectives of this research are: (1) to explore the use of ICT in stress management practices, (2) to identify the types of technologies that the stress experts prefer to use, and (3) to explore the challenges and benefits the stress experts perceive while using different technologies in their practice.

3. LITERATURE REVIEW

3.1 Stress and Occupational Stress

The term "stress" evolves after the 1920s, and it is a form of the Middle English "destresse," derived via Old French from the Latin word 'stringere', "to draw tight" (Keil, 2004). In a medical context stress is a physical, mental or emotional factor that causes bodily or mental tension. Stress can be external coming from the environment, psychological, or social situations. Stress can initiate the "fight or flight" response, a complex reaction of neurologic and endocrinologic systems. Occupational stress is associated with one's work environment or job conditions. Studies exploring the impact of work-related stress on organizational outcome have discovered forms of stress behavior affecting competitiveness, productivity and the public image of the organization. For example, besides the impact on workers' physical wellbeing, a poor psychosocial working environment contributing to work-related stress can result in increased absenteeism and presenteeism, as well as reduced commitment, motivation and satisfaction, along with a higher rate of staff turnover and intention to quit (Quick, 2013). At a broader level, stress-related disorders impact the economic and social environment of both individuals and organizations (Narváez, Tobar, & López, 2014).

3.2 ICT in Healthcare

ICT acts as a catalyst agent of economic and social development and impacts across different sectors in society, including the healthcare sector. Use of ICT to deliver healthcare service has existed since 1999 (Della Mea, 2001) and has expanded rapidly over the years. ICT usage in healthcare mainly moves hand-in-hand with eHealth. eHealth technologies offer an enormous potential to advance healthcare communication, improve healthcare access and develop personalised medical service at a reduced cost (Ryu, 2010). For the current study, the definition of ICT provided by Afolabi and Abidoye (2011) seems much relevant. They describe ICT as "the usage of electronic devices such as computers, telephones, internet and satellite systems to store, retrieve and disseminate information in the form of data, text image and others" (Afolabi & Abidoye, 2011, p. 2). This allows the researcher to focus on ICT in a holistic manner covering all the possible electronic, mobile and internet technologies use within the healthcare sector.

3.3 ICT Supported Occupational Stress Interventions

The use of ICT can help to improve the efficiency of interventions. It is possible to carry them out at a larger scale, regardless of time, place or group of people (Cohen, Kessler, & Gordon, 1995), and the user can access information in a quick, easy and confidential manner (McKenna & Bargh, 2000). Despite massive growth in the eHealth area, there is much to be studied to harness the electronic data for use in mental health services (Schmidt & Wykes, 2012) showing that e-mental healthcare is an area of potential improvements in treatment delivery, shared care and decision making. For successful implementation of ICT in healthcare, especially in developing countries, there is a need to consider existing and potential challenges (Ariani et al., 2017). These challenges include funding needs, infrastructure, knowledge and research evidence, leadership and governance, security and interoperability, and sociocultural and technological environments (Ariani et al., 2017). Healthcare providers' active participation is a vital aspect of the ICT supported interventions. Their positive and active involvement will help to accelerate the adoption and deployment of such interventions within society. Therefore it is essential to recognize their views and experiences as they will have their own unique set of motivations, attitudes, knowledge, and skills, resulting in a plethora of unique challenges for engaging in such interventions (Ariani et al., 2017).

4. METHODOLOGY

This study is exploratory and qualitative and employs a phenomenological approach to understand the stress experts' experience and perception of using ICT in their stress management practices. As Penner et al. (2008) note: "a phenomenological analysis does not aim to explain or discover causes"…it is "to clarify the meanings of phenomena from lived experiences" (p. 93). Thus, this approach strengthens the objective of the study. A combined convenience and purposeful criterion sampling method was used to select respondents. Five of the interviewees were identified from the list of counsellors available at the psychologymatters.asia (www.psychologymatters.asia) website and through major governmental and non-governmental counselling service websites, one was added through a known contact and the last two were recommended by two of the other interviewees. The selected interviewees include counsellors from different governmental and non-governmental organizations (NGO). Two counsellors were affiliated with well-known government organizations, five counsellors were from leading non-governmental organizations and one was from a private organization. There were seven interviews done with the eight participants; one of the interviews was a group interview with two of the participants. The interviews were conducted in the period January 2019 to February 2019. This purposive selection of participants provides an opportunity to receive different views and ideas from different perspectives to enrich the understanding of an experience (Polkinghorne, 2005). Demographic data of the interviewees is detailed below in Table1.

Table 1. Respondent Demographic Data

No:	Gender	Age Range	Education Qualifications	Experience	Affiliation
I01	Female	50-55	Bachelor of Medicine	Less than 5 years	Government Organization
I02	Female	50-55	Not revealed	More than 10 years	Government approved charity Organization
I03	Female	45-50	Postgraduate Diploma in Counselling	More than 20 years	Non governmental organization
I04	Male	60-65	Diploma in Public Health-Royal Society of Health-England	More than 15 years	Non profit non governmental organization
I05	Male	50-55	Postgraduate Diploam in Counselling	More than 15 years	Non profit non governmental organization
I06	Male	50-55	Postgraduate Diploma in Counselling	More than 15 years	Private organization
I07	Female	50-55	Diploma in Psychology and Diploma in Counselling	More than 25 years	Non governmental organization
I08	Female	30-35	Diploma in Counselling and Pshychotherapist	Less than 5 years	Non governmental organization

Data were collected through semi-structured interviews, which allowed the researcher to focus on specific themes while adhering to the conversational style of interviewing (Polkinghorne, 2005). All interviews took place in face-to-face mode based on the respondent's availability and preference. The first author made sure that the interviews were conducted in settings where the participants felt comfortable and convenient. The interviews were conducted using English or in the participant's native language (Sinhala), and each interview took 20-30min. Interview questions focused on background data and current stress management practices, how the interviewees used ICT in their stress management practices and other counselling activities and their experience and perception of ICT supported stress management interventions. All the interviews were audio recorded and transcribed to written text for analysis purpose. The verbatim transcribed interviews were analysed using inductive thematic analysis. Thematic analysis is the process of identifying, analysing, and reporting patterns from responses or data which allow the researcher to organise and describe the data in detail (Braun & Clarke, 2006).

5. ETHICAL CONSIDERATIONS

This section of the article discusses the ethical considerations, the choices made to conduct an ethically sound investigation for the research approach. Before the data collection, each respondent received information about the study and how the collected data would be handled during and after the study, together with a request for their written consent to proceed. The interviews were recorded digitally, and each recorded data identified through a serial number guaranteed the anonymity of the respondents to anyone outside the study. The first author took ownership of the collection and responsibility for protecting it from unauthorized accesses. When interviews were done in other languages than English proper translation mechanism was followed involving professional translators to assure the accuracy and non-biases. Both translations and original transcriptions were checked by the first author, who read through the text and verified and reconciled it with the audio recording. The translations and transcribed text were verified with the respondents to assure that their original input is taken for the analysis purposes.

6. RESULTS

As a result of the coding of the participants' responses, three major themes emerged. In this section, the themes and their subthemes (see Table 2) are explained in detail. The descriptions are supported with direct quotations from the participants.

6.1 Theme 01: Counsellors' Positive Experience of ICT Interventions in Stress Management Practices

The theme "Counsellors' positive experience of ICT interventions in stress management practices" elaborates the various types of ICTs used in the stress counselling process and their purposes and advantages as perceived by the Sri Lankan counsellors. The theme consists of several subthemes.

6.1.1 Provides Rich Sources of Resources

The "Provides rich sources of resources" subtheme examines the types of ICT tools accessed and their usage during the stress counselling process. It was evident from the interviews that participants accessed a wide range of online materials to facilitate their stress counselling process. They used audio clips, video clips, guided imaginations, online assessment tools and other online tutorial for self-education purposes and to facilitate their stress management practices, for example simple relaxation activities. This enabled participants to use standard materials published from various professional organizations around the globe. They often mentioned that these materials improved their counselling process and that they were appreciated by their clients. This did not only apply to their stress related practices but also their other counselling approaches. Following extracts confirm this viewpoint:

I05: "We are also trying to find IT tools in our therapies. We online search for therapeutic music, we get therapeutic music from internet, we use such music for relaxation techniques, for laughing therapy we use video clips from the internet. There are many other therapies available online. First, we check whether they are appropriate to them."

I06: "So now annually and even daily basis all the diagnosis systems and other details for diseases and the disorders are changing so that now the previous year, whatever the things that we found as symptoms for a disorder may not be relevant, it's updated or changed with new information. Especially it is done by APA and other associations in the world. So, the information technology is a useful tool to refer to those associations' websites and their new methods to identify the new diagnosis methods."

Though there were objections from governmental counsellors, other participants believed that these online resources provided an opportunity for clients to seek information and educate themselves on their current situation.

I06: "Yes the point is now at this moment the world is very wide and that means many details and information available in the world. As a client, he can get to know about his disorder or the psychiatric problems or personal problems referring that information."

6.1.2 Creates Reliable Communication Channel between Client and Counsellor

The subtheme "Creates reliable communication channel between client and counsellor" consists of two parts. Firstly, ICT as a tool to reach clients. ICT tools allowed counsellors to access larger audience in an efficient manner. Participants mentioned that they used mobile and office telephone lines to reach their clients and answer client queries. They also mentioned that they used email to communicate with their clients. In some private organizations they used Skype to reach clients who were unable to visit the counsellor's place. Secondly, ICT to stay in touch with their clients. Participants referred to their use of ICT as a reliable mode of communication to increase their availability to clients. As stated by the participants, ICT devices allowed them to work from office and provide immediate advice to clients and to closely monitor their progression on assigned activities. ICT enabled communication channels also supported practitioners to reach clients who were reluctant to attend counselling sessions. These secure and confidential online approaches helped counsellors to build initial rapport and continue with further investigations. This is mentioned in following statements:

I04: "We have a hotline as online counselling. We can use the phone to start the initial conversation."

I03: "We give homework activities to our clients if we can make them online, we can see whether they're doing them and monitor their progress and advise them so it's convenient to client."

I06: "We need to keep contact with our clients more frequently at the time we need to pass messages. If we make this automated even when counsellor is busy this will continue the dialogue with the client. Then client will not get disappointed and the process also will continue. This is really good for clients with suicidal thoughts and also when clients are at critical and emotional situations."

6.1.3 Provides Efficient Ways of Sharing Information

This subtheme elaborates on how ICT can be used to share information among the public and clients. Participants also indicated that ICT allowed them to save time by eliminating travel to disseminate information as they could share their information through ICT systems such as websites, social media and e-mail. These ICT systems enabled wide distribution of information in much user-friendly manner. This supported them to publish materials and service information easily. It was also said that clients perceived this positively as they could receive information without physically attending the premises. The followings illustrate how they used ICT to share information:

I03: "On the one hand we are using our web site and it contains individual/group counselling programs. You can find a hotline so they can speak to us. There are many queries come from email asking for various information. We use Facebook to conduct campaign and publish information about our workshops."

I07: "We do have a web site and next we have a Facebook page, so we put all the required information there, we conduct email campaigns with an institute who does email campaign. We send emails to our counsellors' network and they pass the message to their friends and clients."

I02: "At the moment we are trying to develop our website and Facebook page to do all chat and increase the awareness."

6.1.4 Enables Improvements to Existing Traditional Processes

Last subtheme of theme 1 discusses how ICT can be used as a tool to enhance traditional stress management activities. Participants discussed how ICT allowed them to establish and practice more comprehensive activities within their traditional stress management activities. These ICT enabled forms allowed them to be more interactive with clients and efficiently uplift client's mental conditions. One participant, a counsellor from a government institute mentioned that ICT enabled them to restructure traditional practices adding technological exposure to create more avenues for clients to engage. As an example, one participant mentioned that instead of using traditional activities now they are practicing more interactive activities like karaoke, movie discussions, typing and online games with their clients. They have accessed YouTube, online music portals and multimedia projectors for these activities. The following illustrate their efforts and ideas on how to involve ICT during stress management practices:

I05: "On Sunday, we all watch films using multimedia projectors, after that we discuss the film. We use Karaoke app to play background music, and they sing along. They loved these activities. Likewise, we use a computer, a machine, to change people."

I06: "I believe that these computer-based techniques are more productive than our traditional approaches. We can develop customized programs for them, but there's only less focus or enthusiasm for such initiatives so if you can build up such programs, of course, we can use them at our centers."

Practitioners mentioned that they used ICT supported interventions for clients with severe mental illnesses and not only with the stress management activities. It is well explained with the following quote. (Note: Angoda is the location for National Institute of Mental Health).

I05: "If we look at Angoda mental hospital, still they ask these people to make envelopes, covers, etc. When a person comes with a good family background and social status, they tend to think "I was asked to do all these because I'm not mentally healthy" and they will come up with negative attitudes. They will reject these activities. But they prefer IT based techniques over the traditional approaches. It gives them self-pride."

Subsequently, theme 2 elucidates the counsellors' perceived negative experience of ICT interventions in stress management practices.

6.2 Theme 02: Counsellors' Negative Experience of ICT Interventions in Stress Management Practices

The theme "Counsellors' negative experience of ICT interventions in stress management practices" particularizes the counsellors' challenges encountered in the use of ICT and their adverse opinion concerning the ICT usage in stress counselling practices. This theme consists of four subthemes.

6.2.1 Lack of Face-to-Face Communication

This subtheme addresses the participants' negative viewpoints on using ICT for stress counselling practices. Participants appreciated the value of face-to-face communication in their counselling process. They explained that during the counselling process they meet with the client in a private and confidential setting to explore the problem client is having. They emphasized that this live conversation involved active listening, observing body language and facial expressions. They believed that other forms of communication settings could not allow this kind of rich communication. It is well evident through the following:

I03: "In counselling we should go one –to – one and I must see client's emotions, body language, change of facial expressions, whether they get angry while speaking, is he in sad mood or has a thought on commit suicide. This can be done only when we are seeing the client so we can hear what they hide from words. If we connect in other ways, we cannot get such hidden information."

I06: "When we do counselling, we need to keep eye contact and be emotional and interactive. So, if we ask them to use computer and look at web sites, that does not match with the counselling procedure. Counselling procedure is not like that. It is always a live conversation to sort out the problem [...]."

6.2.2 Unavailability of Required Resources

This subtheme examines the difficulties identified by the participants while searching for required content. They mentioned that sometimes available content is outdated, incorrect or biased. Since there is no universal standard followed by the content developers, counsellors believed that they need to be more careful when retrieving online content. They also mentioned that some important and more standardized contents are not freely accessible and required to be purchased. This is well evident through:

I06: "Actually sometimes when we refer or when access to some website content, they are not updated, that's one thing, the other thing is some content is not free, you need to pay."

I03: "Different people, organizations publish different materials but sometimes material that we need is not readily available. We need to verify content before we use."

6.2.3 Limited knowledge, resources and unfavorable experience

The subtheme "Limited knowledge, resources and unfavorable experience" elucidates how counsellors discuss how the technical knowledge and experience impact on the ICT usage. The participants discussed that they have limited knowledge and training on ICT supported interventions. They also mentioned that even though they have required skillset sometimes there are no adequate ICT resources within their work premises. Participants further elaborated on their own negative personal experience of ICT in general. This is mentioned in these responses:

I05: "We have only eight computers, but we have 50 patients sometimes, it's a limitation."

I04: "Today we can see lots of people are on the train, but they are using their mobile phones and they don't want to talk to anyone. When they are crossing the road, they are using. I don't know what's going on with them. This is a problem. This has made many accidents as well. People are addicted to these mobile devices. You know, nowadays these devices computer and mobile phone is a problem for youngsters. Anyway, I think these computers and mobiles should be limited. Especially, it should not be a frequent thing for these stressed people. If they really want, they can access them once in a while."

6.2.4 Risk of Misleading Information

The last subtheme "Risk of misleading information" addresses how ambiguous material appears online could negatively affect the stress counselling process. The stress experts mentioned that this has twofold effect; firstly, it can mislead counsellors when they use it for self-education and secondly it can negatively impact on clients. Such false information and unsuitable content can rapidly spread on the Internet and reach larger audience and leave them at a vulnerable position. The participant also mentioned that when clients seek for information on their own, they can be misled or misguided while accessing inappropriate or irrelevant content. In such situations counsellors found it difficult to manage clients as they were reluctant to listen to the counsellors' advises. This is mentioned in following responses:

I03: "Some clients study themselves and they do not listen to us and they try to talk what they know. In that situation it is very difficult to move with them and bring them to the correct track. It can happen. Client doesn't know what exactly he needs to do. Sometimes information available online may be not suitable and relevant to that client, if he is using them it will make their situation more complex."

I04: "Different people publish different thoughts online, but we need to make sure the validity of the contents. With wrong information client might get wrong decisions."

I06: "If people share only good document or good tools or good pictures or ideas over Facebook/email/G+/twitter then social media is good. Negative part of the social media is it can harm people. Because social media is accessed by large number of people, within a minute you can pass wrong idea among the society. Yes. You know, uneducated people and those people who like to mislead other people also have access to these platforms. That is the problem. Now actually in our country most of the time we see how people misuse social media. So that's why I don't recommend always the social media for the counselling part or the development of mind of the people."

6.3 Theme 03: The Perceived Challenges of Counselling Practice

The final theme "The perceived challenges of counselling practice" elaborates problems identified in the counselling process in general and how ICT can be used to overcome such complications. This theme comprises of two subthemes: "Breaking the barrier" and "Fight or fly – a societal challenge".

6.3.1 Breaking the Barrier

This subtheme discusses clients' unwillingness to seek professional support and how this can be alleviated through ICT. The participants strongly believed that only a few percentages of people seek for professional support in stress conditions while a large group who need support is left unobserved. They further elaborated on that people are not willing to open up and share their thoughts with an outsider due to many reasons. This involves their busy lifestyle and experienced stigmatizing attitudes. Participants explained that ICT could be used to create positive social awareness about mental illnesses and motivate people to seek support. They further mentioned that social media (e.g. Facebook) allows people across communities to organize various online campaigns and local events to create awareness on mental health and support individuals. Following quotes show the barriers identified:

I02: "Even sometimes most of these people they don't care about stress. Stress management is not much addressed on the worker level context. It's mostly the management level problem."

I03: "First thing is we have to get people to talk and we have to say it's okay to talk. It's okay to talk. It's not a thing to be shy about, and talk about your issues, to open up, so that message has to go into our society."

I04: "Some people don't like to tell their issues. Just they come with a certain thing, but they may be having a different issue, but they say something different when they come to us."

6.3.2 Fight or Fly – a Societal Challenge

The last subtheme concerns the cultural aspect of dealing with stress. Participants mentioned that misbeliefs and strong social stigmatizing attitudes in the society are major barriers to mental health support seeking behavior. Living in a society with negative opinion on people with mental illness may further avoid individuals seeking for professional support. They further elaborated that this public stigma on mental illness can be reshaped with proper social awareness campaign.

I02: "Our society is not that open they want to bottle it up because of stigma around mental health issues. It is a vigorous thing because you have to break this barrier, the mental stigma around mental health issues. It's great still even among other rich people, knowledgeable, educated people, everyone."

I03: "We are getting emails mentioning their problem and they ask whether their conversations are confidential, and they ask whether they can take the counselling without physically attending to the premises. They are shy and afraid, and they want to meet counsellors, but they will not come."

Table 2. Summary of themes and subthemes

Participants' views	Subthemes	Associated meaning
Theme 01: Counsellors' positive experience of ICT interventions in stress management practices		
ICT as a positive tool because it	Provides rich sources of resources	Allow counsellors to access more standardized online resources to support their counselling process
	Creates reliable communication channel between client and counsellor	Allow counsellors to easily reach and stay in touch with larger audience
	Provides efficient ways of sharing information	Facilitate a number of standardized and easy channels for sharing of information
	Enables improvements to existing traditional processes	Provide alternatives to enhance the traditional counselling approaches
Theme 02: Counsellors' negative experience of ICT interventions in stress management practices		
ICT as a negative tool due to	Lack of face-to-face communication	How ICT interventions might diminish the traditional face-to-face conversational approach
	Unavailability of required resources	Discusses on problems of unavailability of required materials online
	Limited knowledge, resources and unfavorable experience	Elaborates on how counsellor's knowledge and experience impact on ICT usage
	Risk of misleading information	Discusses on false information and unsuitable content availability
Theme 03: The perceived challenges of counselling practice		
ICT as a possible tool	Breaking the barrier	Discusses on clients' unwillingness to seek professional support and how this can be alleviated through ICT
	Fight or fly – a societal challenge	Elaborates on the cultural aspect of dealing with stress

7. DISCUSSION

The general objective of this study was to explore the Sri Lankan stress experts' views and experiences of ICT supported stress management interventions. The three main themes and ten subthemes extracted from the interviews provided valuable insights into the usage and role of ICT in stress counselling. The findings did not only highlight counsellors' negative and positive experiences of ICT usage but also pointed out the role that ICT plays to break the mental health stigma.

One key finding from this study was that the counsellors' overall positive experience of using ICT outweighed the negative ones. The interviews uncovered a variety of ICT tools used in the stress counselling practices. The counsellors mainly used websites and Facebook to disseminate information. They also used YouTube to play videos and audio during therapies. Further, counsellors elaborated that ICT supported interventions improved dissemination of health information, enabled remote consultation, facilitated collaboration and cooperation among clients, improved the efficiency of traditional stress management activities, supported more effective self-learning through access to information and facilitated public discourse and dialogue about mental health. These findings perfectly match with the results of a study carried out by Healthlink Worldwide (2007) focusing on the health sector of developing countries.

The study also revealed challenges and drawbacks of using ICT in stress management practices. The major constraint that the counsellors have identified is that ICT interventions could decrease the face-to-face communication. The most common way of arranging stress interventions can be described as confidential, individual, face-to-face counselling with a trained professional (Eisen, Allen, Bollash & Pescatello, 2008). Sri Lankan stress experts also adhered to this model and highly valued the importance of having face-to face conversation to build rapport with the client. This is a crucial point to consider when developing ICT supported stress interventions. The system developers need to make sure that ICT tools support similar functionality and capture all elements of the conversation. Next, counsellors were constrained by the limited access to useful interventions and validity of the online content. Ward et al. (2008) also highlighted similar issues identified around the flexibility of the ICT systems and whether they were 'fit for purpose', along with the confidence and experience of the users. Such unfavorable experiences could have resulted in negative opinions on ICT supported interventions in their work life. Therefore, it is important to consider participants' characteristics and practices in the intervention design.

During the interviews different arguments emerged on client empowerment and about improving health literacy through self-education. Empowerment in health care generally refers to the process that allows an individual or a community to gain the knowledge, skills and attitude needed to make choices about their care (World Health Organization, 2017). Research studies also show that active and positive participation in online communities and social media allows people to learn about different worldviews and positive stress-relief strategies that can spread to local context. While there were objections from government attached counsellors in our study, others suggested that ICT could empower their clients by providing opportunity for the clients to educate themselves, improve their health literacy and to make healthcare decisions.

The stigma of mental illness is common and defines a process involving discrimination, stereotypes and prejudice (Rusch, Angermeyer, & Corrigan, 2005). Mental health stigmatizing attitudes continue to be prominent in Sri Lankan as well. Persons experiencing mental health problems are subject to discriminatory practices, including decreased employment options (Consumer Action Network for Mental Health, 2015). Mental health stigma has been reported to affect health-seeking behaviors (Fernando, Pedersen & Weerackody, 2010) and quality of treatment received (Fernando, Deane & McLeod, 2010). As in the general population, health professionals also frequently have negative attitudes. A study showed that the majority of Sri Lankan doctors and medical students believed patients with mental illness are dangerous, unpredictable and difficult to communicate with, and that they perceive the world differently (Liyanage, Rowel, Hewawitharana, Silva & Dissanayake, 2018).

Despite the lack of comprehensive research evidence to date, ICTs hold potential in addressing the challenges of mental health care (Breslau & Engel, 2016). Similarly, the result of the study presented in this paper provides several promising insights for challenges identified in the area of stress interventions, and regarding mental health interventions in general, in the Sri Lankan context. Our study contributes to the area by exploring the Sri Lankan counsellors and stress practitioners' perceptions on ICT supported stress interventions. Thus, this study will lay an initiation for fruitful discussions on the role of ICT in stress management interventions.

8. CONCLUSION

Over time, the research literature and experience with ICTs will grow, and consensus will form on which ICT applications are of greatest benefit to patients and clients who are in need of stress management. Thus, the results of this study could contribute to the already existing knowledge base, giving a better understanding of the possibilities and difficulties related to using eHealth solutions in clinical stress management practices in Sri Lanka. It also clarifies what technologies are more commonly utilized in mental health and highlight how ICT could help to better target mental health problems. The findings could also contribute to future design and development of ICT tools for stress management in Sri Lanka. It is also relevant to investigate clients' and employees' experiences and views on ICT supported stress management in future studies.

REFERENCES

Afolabi, A. F., & Abidoye, J. A. (2011). The Integration of Information and Communication Technology in Library Operations Towards Effective Library Services. In *Proceedings of the 1st International Technology, Education and Environment Conference* (pp. 620–628).

Ariani, A., Koesoema, A. P., & Soegijoko, S. (2017). *Innovative Healthcare Applications of ICT for Developing Countries*. https://doi.org/10.1007/978-3-319-55774-8

Barak, A., Boniel-Nissim, M., & Suler, J. (2008). Fostering empowerment in online support groups. *Computers in Human Behavior, 24*(5), 1867-1883. https://doi.org/10.1016/j.chb.2008.02.004

Barrett, M. J. (2005). Patient Self-Management Tools: An Overview Patient Self-Management Tools: An Overview CALIFORNIA HEALTHCARE FOUNDATION, (June).

Braun, V., & Clarke, V. (2006). Using thematic analysis in psychology. *Qualitative Research in Psychology*, *3*(2), 77–101. https://doi.org/10.1191/1478088706qp063oa

Breslau, J., & Engel, C. . (2016). Information and Communication Technologies in Behavioral Health A Literature Review with Recommendations for the Air Force. *Rand Health Quarterly*, *5*. Retrieved from https://www.ncbi.nlm.nih.gov/pmc/articles/PMC5158232/#!po=95.0000

Burney, S. M. A. (2010). Information and Communication Technology in Healthcare Management Systems: Prospects for Developing Countries. *International Journal of Computer Applications*, *4*(2), 27–32.

Cohen, S., Kessler, R., & Gordon, L. (1995). Strategies for measuring stress in studies of psychiatric and physical disorders. Measuring stress: A guide for health and social scientists. New York. Retrieved from http://www.psy.cmu.edu/~scohen/Cohen, S et al (1995).pdf

ColomboPage. (2017). ColomboPage News Desk, Sri Lanka. Retrieved May 3, 2019, from http://www.colombopage.com/archive_17B/Oct10_1507658756CH.php

De Silva, D. (2002). Psychiatric service delivery in an Asian country: the experience of Sri Lanka. *International Review of Psychiatry*, *14*(1), 66–70. https://doi.org/10.1080/09540260120114096

Debnam, D. (2016). Huge Rise in Global Employee Depression, Stress, and Anxiety Since 2012. Retrieved from https://www.huffingtonpost.com/dean-debnam/huge-rise-in-global-employee_b_8923252.html

Della Mea, V. (2001). What is e-health (2). The death of telemedicine? *Journal of Medical Internet Research*, *3*(2).

Ennis, L., Rose, D., Callard, F., Denis, M., & Wykes, T. (2011). Rapid progress or lengthy process? electronic personal health records in mental health. *BMC Psychiatry*, p. 117. https://doi.org/10.1186/1471-244X-11-117

Eysenbach, G. (2001). What is e-health? *Journal of Medical Internet Research*, *3*(2), E20–E20. https://doi.org/10.2196/jmir.3.2.e20

Fernando, S. M., Deane, F. P., & McLeod, H. J. (2010). Sri Lankan doctors' and medical undergraduates' attitudes towards mental illness. *Social Psychiatry and Psychiatric Epidemiology*, *45*(7), 733–739. https://doi.org/10.1007/s00127-009-0113-6

Fernando, S., Pedersen, D., & Weerackody, C. (2010, July). Mental health in Sri Lanka. *Lancet (London, England)*. England. https://doi.org/10.1016/S0140-6736(10)61072-0

Free, C., Phillips, G., Felix, L., Galli, L., Patel, V., & Edwards, P. (2010). The effectiveness of M-health technologies for improving health and health services: a systematic review protocol. *BMC Research Notes*, *3*(1), 250. https://doi.org/10.1186/1756-0500-3-250

Healthlink Worldwide. (2007). *Improving Health , Connecting People : The Role of ICTs in the Health Sector of Developing Countries A Framework Paper*. Retrieved from http://www.infodev.org/infodev-files/resource/InfodevDocuments_84.pdf

Keil, R. M. K. (2004). Coping and stress: a conceptual analysis. *Journal of Advanced Nursing*, *45*(6), 659–665. https://doi.org/10.1046/j.1365-2648.2003.02955.x

Lazarus, R. S. (1993). From Psychological Stress to the Emotions " A History of Changing Outlooks. *Annual Review of Psychology*, *44*, 1–21.

Liyanage, U., Rowel, W., Hewawitharana, U. H., Silva, V. De, & Dissanayake, D. M. T. (2018). Stigma due to their work in mental health , among mental health professionals in Sri Lanka. *Sri Lanka Journal of Psychiatry*, *9*(1), 4–8.

McKenna, K. Y. A., & Bargh, J. A. (2000). The Implications of the Internet for Personality and Social Psychology. *Personality and Social Psychology Review*, *4*(1), 57–75.

Mclaren, P. (2005). health care. *Journal of the Royal Society of MedicinE*, *98*, 232–234. Retrieved from https://www.ncbi.nlm.nih.gov/pmc/articles/PMC1129049/pdf/00980232.pdf

Minas, H., Mendis, J., & Hall, T. (2017). Mental Health System Development in Sri Lanka. https://doi.org/10.1007/978-1-4899-7999-5

Mustafa, M., Illzam, E. M., Muniandy, R. K., Hashmi, M. I., Sharifa, A. M., & Nang, M. K. (2015). Causes and Prevention of Occupational Stress. *IOSR Journal of Dental and Medical Sciences*, *14*(November 2015), 98--104. https://doi.org/10.9790/0853-1411898104

Narváez, S., Tobar, Á. M., & López, D. M. (2014). Systematic review of interventions supported by ICT for the prevention treatment of occupational stress. In *Studies in Health Technology and Informatics* (Vol. 200, pp. 71–80). https://doi.org/10.3233/978-1-61499-393-3-71

Penner, J. L., & McClement, S. E. (2008). Using Phenomenology to Examine the Experiences of Family Caregivers of Patients with Advanced Head and Neck Cancer: Reflections of a Novice Researcher. *International Journal of Qualitative Methods*, *7*(2), 92–101. https://doi.org/10.1177/160940690800700206

Polkinghorne, D. E. (2005). Language and meaning: Data collection in qualitative research. *Journal of Counseling Psychology*, *52*(2), 137–145. https://doi.org/10.1037/0022-0167.52.2.137

Quick, J. C. (2013). *Preventive stress management in organizations* (2nd ed.). Washington, District of Columbia: American Psychological Association.

Rajgopal, T. (2010). Mental well-being at the workplace. *Indian Journal of Occupational and Environmental Medicine*, *14*(3), 63–65. https://doi.org/10.4103/0019-5278.75691

Rusch, N., Angermeyer, M. C., & Corrigan, P. W. (2005). Mental illness stigma: concepts, consequences, and initiatives to reduce stigma. *European Psychiatry : The Journal of the Association of European Psychiatrists*, *20*(8), 529–539. https://doi.org/10.1016/j.eurpsy.2005.04.004

Ryu, S. (2010). History of Telemedicine: Evolution, Context, and Transformation. *Healthcare Informatics Research*, *16*(1), 65–66. https://doi.org/10.4258/hir.2010.16.1.65

Schmidt, U., & Wykes, T. (2012). E-mental health - a land of unlimited possibilities. *Mental Health Journal*, *21*(4), 327–31. https://doi.org/10.3109/09638237.2012.705930

Sri Lanka Labour Force Survey. (2017). Retrieved from http://www.statistics.gov.lk/samplesurvey/LFS_Q1_Bulletin_WEB_2017_final.pdf

Ward, R., Stevens, C., Brentnall, P., & Briddon, J. (2008). *The attitudes of health care staff to information technology: a comprehensive review of the research literature. Health information and libraries journal* (Vol. 25). England. https://doi.org/10.1111/j.1471-1842.2008.00777.x

World Health Organization. (2017). *Mental Health ATLAS 2017 Member State Profile Sri Lanka*. Retrieved from https://www.who.int/mental_health/evidence/atlas/profiles-2017/LKA.pdf?ua=1

A COMPREHENSIVE STUDY OF IOT FOR ALZHEIMER'S DISEASE

Roberto Omar Andrade, Sang Guun Yoo and María F. Cazares
Facultad de Ingeniería de Sistemas, Escuela Politécnica Nacional, Ecuador
Facultad de Ciencias Psicológicas - IDEIAGEOCA, Universidad Politécnica Salesiana, Ecuador

ABSTRACT

The aging process includes the degradation of physical and neurological functions in the people. Although the process of degradation of functions is normal, in some people due to certain risk factors the impact can be high. The appearance of chronic diseases in older adults, one of which is Alzheimer's with about 46.8 million affected worldwide, can affect the quality of life of the person. Alzheimer's at the moment is considered a disease without a cure, an early detection together with an adequate cognitive stimulation can help to reduce the impact for the older adults and their relatives. Ambient Assisted Living - AAL solutions can help the person maintain a quality life, independent and socially active.
The inclusion of ICT (Information Communication and Technology) in AAL solutions increase their functionalities and the higher number of deployments of IoT in recent years and its future projection has generated the interest of its application in different domains of society such as: transport, agriculture and health; however, IoT is a relatively new technology and its development is still in progress. The contribution of this work is present an overview about the use of IoT in patients with Alzheimer's, and make an analyze the use of IoT in the processes of cognitive stimulation considering the possible technical limitations of IoT.

KEYWORDS

IoT, Ambient Assisted Living, Alzheimer's, Cognitive Stimulation

1. INTRODUCTION

The concern of health organizations in relation to older adults is based on the population growth of approximately 1 billion of people for every ten years around the world, in the case of the group of people who are more than 65 years old, their growth is more accelerated with a rate of 3% per year that of young and middle age groups (ONU, 2018). Low birth rates and longer life expectancy are transforming the composition of the world population, with a marked transition towards a much older population structure. Life expectancy for older adults will be 72 in 2020 and this increase to 84 years byN2050 (ONU, 2019). The population of the European Union on January 1, 2017 was estimated at 511.5 million, with a percentage of older people of 19.4%, showing an increase of 0.2 percentage points compared to the previous year and an increase of 2.4 percentage points compared to 10 years earlier.

With the process of aging since of 65 years of age, the nervous system and the brain suffer certain degradation that can generate neurological problems (Kerchner, 2016), so is important identify the type of disease that can affect the health of older adult (NCSC, 2019). In the Table 1, we show the most common neurological disorders associated with aging according to World Health Organization (HWO, 2019).

Table 1. Neurological disease associated with aging

Diseases	*System*	*People–Affected*	*Aging process*
Neuropathy	Nervous system	30 millon	yes
Alzheimer's	Nervous system	46.8 millon	yes
Parkinson	Nervous system	7 millon	yes
Lateral Sclerosis	Nervous system	2.5 millon	yes
Stroke	Nervous system	15 millon	yes
Muscular Dystrophy	Nervous system	1 millon	yes

Alzheimer's disease will could affect to approximately 200,000,000 individuals by the year 2050, and an early diagnosis allows life planning, decision making, and treatment when therapies. The inclusion of ICTs in the medical field has allowed improve the processes of treatment and accompaniment of some diseases for maintain an adequate quality of life for patients (Econsultancy, 2019).

Older adults in some cases live alone in their homes, so they will require remote medical assistance or some mechanism that allow know their health by the medical staff or their family, but also keep the privacy and security of the sensible information (Trendmicro, 2019). This population and health context for the following twenty years drives research into works about proactive and preventive health monitoring assisted by technology in smart domestic environments. The deployment of systems of Ambient Assisted Living (AAL) uses ICTs in the daily life of one older adult persons to enable them to live independently and stay active and socially connected (Medrano, 2018). The market for remote patient control is expected to grow from 10.6 billion in 2012 to 21.2 billion in 2017 (Forbes, 2019). This social and economic growth needs revitalize work on proactive and preventive health monitoring assisted by technology in home environments (Tsirmpas, 2017).

2. AN OVERVIEW ABOUT ALZHEIMER'S

Alzheimer's is currently a disease without a cure, and its evolutionary process offers great challenges for patients and their families. The psychological and emotional ramifications of early diagnosis without adequate therapy and resources that allow lead an adequate quality of life can be a frustrating process for the person and relatives. Portellano (2015) reports that the time of evolution of Alzheimer's is 12 years and defines three stages; each stage presents their own alterations and symptoms. The first stage denominated preclinical, with one to three years of duration presents abnormal biomarkers and without or only subtle cognitive impairment. The second stage, mild cognitive impairment is defined as the presence of abnormal pathophysiological biomarkers and episodic memory impairment, this stage has two to ten years of duration. Finally, the third stage called dementia presents abnormal biomarkers, and clear cognitive and functional impairment, this stage has eight to twelve years of duration. In the Table 2 we show the alterations and symptoms associated with each stage.

According to Muñoz (2009), the development of daily activities of people that suffer from deficiency in cognitive processes require the use of various strategies to compensate these losses. For example; devices such as electronic diaries, provide support for immediate memory, for the organization of time and activities, also voice recordings, alarms, key searchers support this cognitive process.

Table 2. Symptoms stage of Alzheimer's disease (Portellano, 2005)

Functions	*First stage – Preclinical*	*Second Stage – Mild Cognitive*	*Third stage - Dementia*
Memory	Immediate memory and fixation memory. Progressive forgetting and inability to new learn.	Severe amnestic disorder, episodic memory. They remember more easily the oldest events, but not those that have happened recently	None
Language	Impoverishment of language: difficulty in ascribing names to familiar objects or persons	Initial aphasia	Transcortical or sensory aphasia. Alters the symbolic structure of language, echolalia and palilalia
Behavior	Mood changes and manifestations of apathy, sadness and irritability	Conduct progressively indifferent, apathetic and sometimes restless, hyperactive and hallucinating	Conduct loss of behavioral initiative, low tolerance to frustration, paranoid ideas and lack of adaptation in their behaviors, with changing mood and labile affectivity. Constructive apraxia and dress apraxia ideomotor and ideatory apraxia
Awareness	Anosognosia: reduction of initiative and lack of awareness of their disease	None	None
Executive functions	None	Difficulty formulating concepts or solving problems	None
Motor skills	None	None	Generalized afaso-apracto-agnosico: gatismo syndrome urinary and fecal incontinence, muscle serious deterioration that affects all areas, total dependence on the environment, stiffness and flexed posture.

3. IOT SOLUTIONS FOR ALZHEIMER'S

Deployment and projection of IoT solutions in different domains of society, almost 20 billion of IoT devices by year 2020 worldwide (Forbes, 2018), motivates to researchers and industries consider the use of IoT solutions in the medical field to improve AAL solutions (Mulero, 2018). United Kingdom, for example, implements a IoT-based geographic sensors network to maintain a record of the clinical status of patients and thus avoid that them have to make continuous visits to the hospital. In this study, is our motivation analyze the application of Internet of Things - IoT in the processes of the evolution of Alzheimer's disease.

The research methodology used in this study considers the following steps:

1. Define our research question. What is the use of IoT in those related to Alzheimer's disease?

2. Define the search criteria. "Alzheimer's and IoT", "Alzheimer's and Internet of Things", and "Dementia and IoT".

3. Define the inclusion criteria. We consider the papers presented in scientific bases recognized worldwide as Scopus, IEEE, ACM, and Web of Science, published in the last 4 years (2015-2019).

4. Define the primary studies. We consider a screening process of two rounds; first we analyze the papers collected based on the title and abstract through a blind review by the authors of this study. The second round is to perform a full text review of the papers collected. From the screening process carried out, to a total of 35 articles were determined.

5. Perform a qualitative analysis of selected papers that will allow us to answer our research question, identifying the use of IoT in Alzheimer's persons. Based on the qualitative analysis made, the applications are mainly focused on the monitoring of patients with Alzheimer's. In the Figure 1, we show density analysis of works related to Alzheimer's and IoT, they are focus on the use of sensors, RFID, and medical/mobile computing for the process of health monitoring and location tracking. There is a relationship between the level of monitoring and the level of dementia of the person.

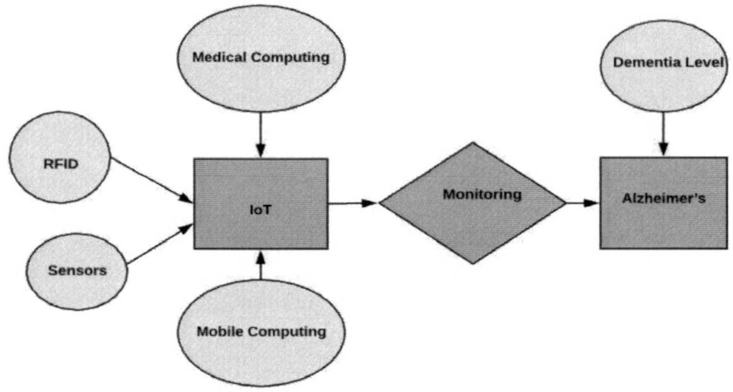

Figure 1. Density of works associated with Alzheimer's and IoT

In the Table 3, we show the IoT-based commercial projects that we have considered most relevant to the support for people with Alzheimer's with a description of their functionality.

Table 3. IoT projects for Alzheimer's disease

Project	***Description***	***References***
iCare: the patient with Alzheimer's must have a portable IoT device, which is the Apple Smartwatch. The second part is a download of the iOS application on the caregiver's iPhone that must be connected to the patient's clock	The application allows caregivers to track and find their loved ones, especially if they go out alone, and also shows the patient's heart rate to caregivers when they need it. It has another feature that is the list of reminders to schedule tasks, such as bathing, exercises, a patient's medications and medical appointments. In addition, the application has a section where caregivers can get information about Alzheimer's disease	(Aljehani, 2018)
City4Age Project	Research and innovation project funded by Horizon2020 with the objective of enabling age-friendly cities.	(EU, 2019)
Sphere Project	Install sensors inside the house of elderly people living alone. These sensors can detect the behavior by initial symptoms of dementia.	(Zhu, 2015)
Timeless	Application designed to help Alzheimer's patients recognize their family and friends through facial recognition.	(Timeless, 2019)
ROADMAP "Real-world results across the spectrum of Alzheimer's disease for better care: Multimodal data access platform".	Provide the basis for an integrated data environment and framework for real-world evidence (RWE in the Alzheimer's disease).	(Roadmap, 2019)
Tweri application: Alzheimer's Caregiving.	Application that uses GPS technology to detect Alzheimer's patients.	(Tweri, 2019)

Based on the comparison between the functionalities of the commercial solutions and the developments in the academic field, there is a difference in the objective of the proposals. In the academic field, IoT has been mostly used to develop monitoring solutions, while commercial solutions are mainly focused on cognitive reinforcement. An important gap to be consider in research field is find solutions for cognitive stimulation of Alzheimer's people, especially at the present time with the broad development of IoT, BigData and Machine Learning.

4. COGNITIVE STIMULATION USING IOT FOR ALZHEIMER´S

Alzheimer's is a disease without a cure but is possible make compensatory interventions to reduce the losses associated with cognitive processes, the Figure 2 shows some of the techniques that can be used for this objective. Regarding external aid the use of ICTs and specifically in related to this study IoT can be considered as a tool to improve its effectiveness. The combination of IoT with machine learning could be considered to improve the cognitive reinforcement in the related with metacognitive and specific skills training.

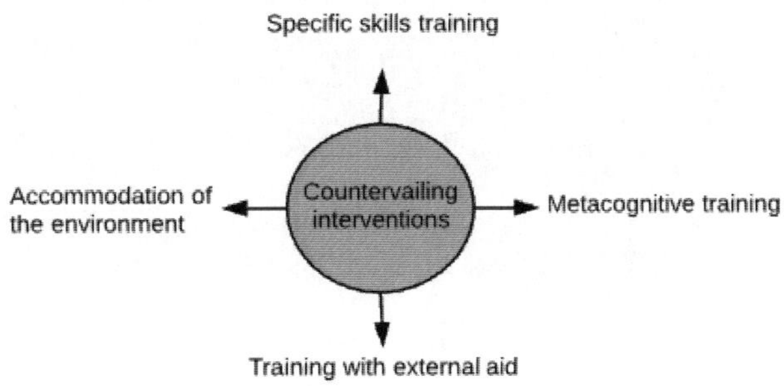

Figure 2. Interventions for cognitive stimulation

From the literature review in this study some points that we have considered relevant are the use of cognitive stimulation using fuzzy adaptive (Navarro, 2018), the use of wearable sensors for early detection (Roy, 2017) and the behavior analysis using multimodal sensing (Alvarez, 2018). At this point, our objective in this section is analyze based on the information gathered from the primary sources the general architecture that is being used for the use of IoT in Alzheimer's, for then identified how these solutions can add their functionality for cognitive reinforcement process. IoT solutions are mainly focus on determining the cognitive, physical and behavioral parameters of Alzheimer's patients in order to establish normal patterns of recognition of activity (Aitor, 2019). In the Figure 3, we present a flow about the interactions that allow establish the activity recogition in base our literature review. IoT solutions used for Alzheimer´s include a set of IoT devices in the phisical space in which the person interacts. In the Figure 4, we present an generic IoT solution that include camera for recognition of the people whom visit to older adult and the send the information of the visitant through of the voice service (Malavasi, 2017) also allow to send photographs stored in the IoT server related for stimulate their memory. The location of sensors will allow tracking of the person related to their daily activities, walking time patterns or detect if the person leaves the physical space. One of the limitations of using cellular networks is the cost of data use, the use of solutions such as LoRA, Sigfox or NB-IoT that are specially designed for IoT can be more efficient for these designs (Sinha,2017), although a possible limitation could be the bandwidth. Figure 5 show the three IoT networks for communication between older adult and care services or family identify in base on literature review.

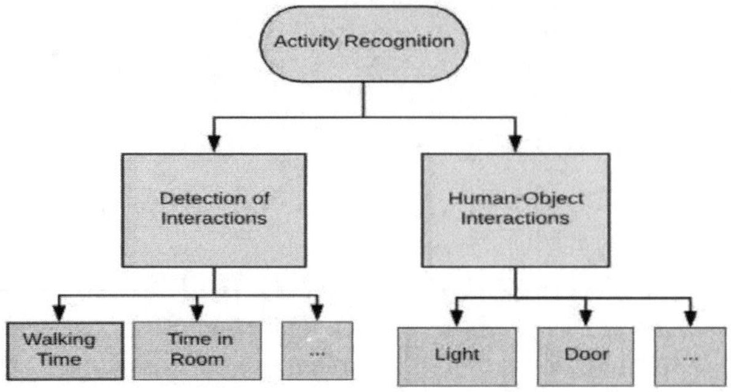

Figure 3. Activity recognition for IoT Alzheimer's solution

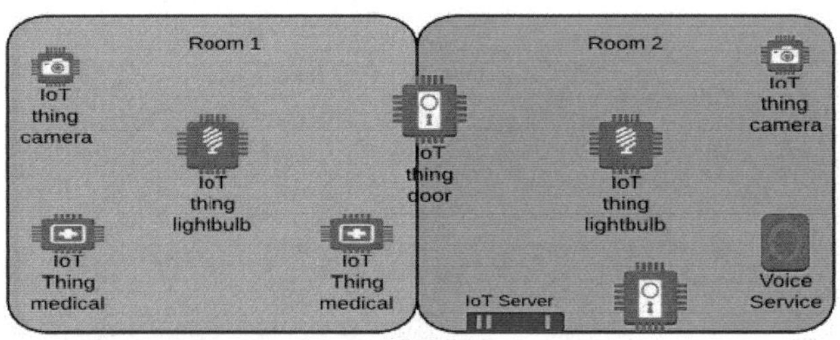

Figure 4. Activity recognition for IoT Alzheimer's design

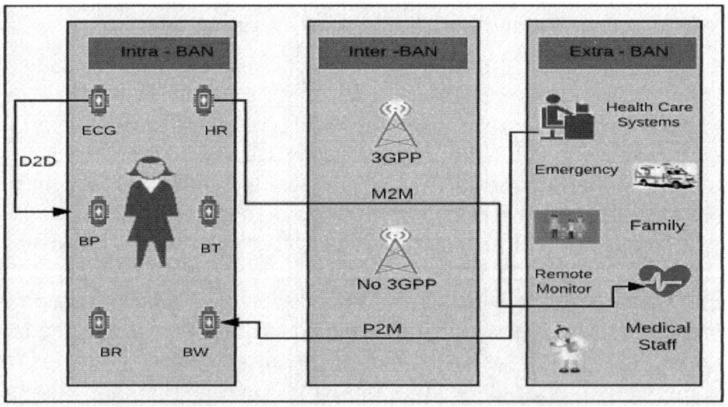

Figure 5. IoT networks for e-health services

The proposed of cognitive stimulation using IoT at present focuses on sending a set of photos, videos and phrases that allow the older adults to keep their memory active; one of the possible drawbacks that could be considered is the use of IoT is their available bandwidth, the delay times, and the privacy issues. We performed a simulation of an IoT environment composed of 16 sensors, with vital signs, video and voice data traffic. Based on the simulation, the bandwidth and delay consumption are not critical, which would allow a real-time interaction between the person with Alzheimer's and the IoT solution for cognitive stimulation. Figure 6, show simulated traffic for 16 sensors send voice, data and video traffic.

Application_metrics   Detailed View

Application Id	Application Name	Packet generated	Packet received	Throughput (Mbps)	Delay(microsec)
1	App1_SENSOR_APP	100	100	0.000400	4262.400000
1	App1_SENSOR_APP	100	100	0.000400	4262.400000
1	App1_SENSOR_APP	100	100	0.000400	4262.400000
1	App1_SENSOR_APP	100	100	0.000400	4262.400000
1	App1_SENSOR_APP	100	100	0.000400	4262.400000

Figure 6. Sensor Data in IoT simulation

The results in the Figure 7, show that the inclusion of encryption protocols to ensure the privacy of information of the older adult generate greater increase in delays, so this is a possible reason for which some IoT solutions are not used; we performed a simulation in the same environment indicated above applying AES 128 bits, the results are shown in figure 12, there are no significant increases in throughput or delay, that allow consider their exclusion in the design of IoT solutions.

Packet generated	Packet received	Throughput (Mbps)	Delay(microsec)
100	100	0.000400	4262.400000
100	100	0.000400	4262.400000
100	100	0.000400	4262.400000
100	100	0.000400	4262.400000

Figure 7. Data in IoT simulation with AES encryption

5. CONCLUSION

The IoT are technological resources that in the field of health allow respond to the diverse demands of patients, through the systems of ambient assisted living (AAL), facilitating the monitoring and control of diseases. We mentioned that Alzheimer's IoT solution could be focus on assisting patients and their families in problems related to memory, helping in the facial recognition of family members, in the geographical traceability of the person and in the detection of behavioral patterns and symptoms through of IoT sensors.

Cognitive stimulation is focus on keeping memory active, the periodic sending of images or sounds that allow establish a memory process could help the person with Alzheimer's reduces the impact of the disease. Through the use of IoT the objective is establish an acknowledgment of the activities of the older adult and based on this recognition send them visual and sound aids. In the scientific field there are few contributions to the moment on how IoT can contribute to the cognitive stimulation of people with Alzheimer's, it is important to emphasize future research in this context.

The third stage of Alzheimer's has symptoms that prevent anyone who performs routine activities with freedom, since that requires the support and care of a third party on a permanent basis, this is a major challenge for the investigation of technologies that may accompany this stage of the disease. IoT allows make important contributions to AAL systems and together with machine learning and BigData can improve the functionalities to improve the quality of life in people with Alzheimer's, so research not only should be focus on solutions for monitoring health, are necessary contributions in compensatory intervention processes.

The incorporation of security mechanisms is important to maintain the privacy of sensitive information, mechanisms must be evaluated so that their incorporation does not generate delays in sending information, algorithms such as AES are quite robust and may not strongly affect the performance of the IoT solution. In this context solutions like blockchain can be interesting for future works.

REFERENCES

Aitor Almeida, Rubén Mulero, Piercosimo Rametta, Vladimir Urošević, Marina Andrić, Luigi Patrono, 2019. A critical analysis of an IoT—aware AAL system for elderly monitoring, Future Generation Computer Systems, Volume 97, Pages 598-619.

Aljehani, S. Alhazmi, R. Aloufi, S. Aljehani and R. Abdulrahman, R *(2018)* "iCare: Applying IoT Technology for Monitoring Alzheimer's Patients," *1st International Conference on Computer Applications & Information Security (ICCAIS)*, Riyadh, Medina, Saudi Arabia, pp. 1-6.

Econsultancy.10 Examples of the Internet of Things in healthcare (2019). https://econsultancy.com/internet-of-things-healthcare.

European Union's Horizon 2020. City4age. www.city4ageproject.eu

Forbes (2019). 14 predictions about the impact of technology's growing presence in health care.

Forbes (2018). Internet of things market to reach $267b by 2020. https://www.forbes.com.

Health World Organization, HWO. Metal disorders (2019). https://www.who.int/es.

Kerchner, G. A. & Wyss-Coray, T. The Role of Aging in Alzheimer's Disease, 197–227 (Springer International Publishing, Cham, 2016).

Malavasi, M. et al. An innovative speech-based interface to control aal and iot solutions to help people with speech and motor disability. Ambient Assist. Living. ForltAAL 2016. 426, 269–278 (2017).

Medrano-Gil, A. et al. Definition of technological solutions based on the internet of things and smart cities paradigms for active and healthy ageing through cocreation. Wirel. Commun. Mob. Comput. 1–15, DOI: 10.1155/2018/1949835 (2018).

Mulero, R. et al. An iot-aware approach for elderly-friendly cities. IEEE Access 6, 7941–7957, DOI: 10.1109/ACCESS. 2018.2800161 (2018).

Muñoz E. Estimulación cognitiva y rehabilitación neuropsicológica. (2009). Editorial UOC.

Navarro, J., Doctor, F., Zamudio, V., Iqbal, R., Sangaiah, A. K., & Lino, C. (2018). Fuzzy adaptive cognitive stimulation therapy generation for Alzheimer's sufferers: Towards a pervasive dementia care monitoring platform. Future Generation Computer Systems, 88, 479–490. doi: 10.1016/j.future.2018.06.018

North Caroline State Center, NCSC. Health Statistics (2019). *https://schs.dph.ncdhhs.gov.*

Portellano, J, 2005, Introducción a la Neuropsicología, McGRAW-HILL/Interamericana de España, S. A. U, Madrid.

Roadmap. (2019). https://roadmap-alzheimer.org.

Roy, N, 2017, *Wearable and IoT for cognitive health assessment: Significance and challenges*, IEEE International Conference on Pervasive Computing and Communications Workshops (PerCom Workshops), Kona, HI, pp. 660-660. doi: 10.1109/PERCOMW.2017.7917641

Sinha, R., Yiqiao, W. & Hwang, S.-H. A survey on lpwa technology: Lora and nb-iot. ICT Express 3, DOI: 10.1016/j.icte. 2017.03.004 (2017).

Timeless (2019). https://www.timeless.care

Trendmicro. Patients and privacy: GDPR compliance for healthcare organizations. (2019). https://www.trendmicro.com/

Tsirmpas, C. et al. An internet of things platform architecture for supporting ambient assisted living environments. Technol. Heal. Care. 25, 391–401, DOI: 10.3233/thc-161270 (2017).

Tweri. (2019). http://www.tweri.com/home.aspx

United Nations, ONU. World popular ageing (2018). https://www.un.org.

United Nations, ONU. Global issues ageing (2018). https://www.un.org

Zhu, N. et al. Bridging e-health and the internet of things: The sphere project. IEEE Intell. Syst. 30, 39–46, DOI:10.1109/MIS.2015.57 (2015).

POTENTIALS EVALUATION FOR INSTANT USABILITY OF SMARTPHONE APPS IN HEALTH AND SPORTS ENDURANCE ACTIVITY TRACKING

Hans Weghorn
Baden-Wuerttemberg Cooperative State University Stuttgart
Kronenstrasse 53A, 70174 Stuttgart, Germany

ABSTRACT

Systematic control of physical workouts is required for achieving good curing or caring effects in health activities as well as for sustaining or even improving fitness level in non-elite and professional sports. For this purpose, respectively wearable medical monitors as also a wide range of personal sports computers were commonly in use since longer time. With the overwhelming evolvement of the smartphone market, health and sports monitoring was identified also as prospective working field for app developers. Accordingly, numerous smartphone apps are offered meanwhile for endurance training, many of them even as so-called freeware. An investigation is presented here about the required efforts in setup and in daily use as also about the reliability of such smartphone monitoring, taking especially into account the needs of possible elderly users. Furthermore, it is critically discussed in this work, whether smartphone apps really can replace the instant and seamless use mode of traditional sports computers.

KEYWORDS

Ubiquitous Health Monitoring, Sports Tracking Tools, Personalized Mobility, Elderly Care, Smartphone Training Apps

1. INTRODUCTION

Regular, weekly physical training preserves good health up to high living ages. This was repeatedly shown in medical studies. For instance, it has been particularly validated, that weekly brisk walking can reduce stroke risk for people at ages even above 70 (Jefferis et al., 2014). Numerous investigations report also positive effects on various other pathological conditions (Law et al., 1991), but especially diminishing stroke risks was repeatedly under research for men (Abbott et al., 1994) and women (Oguma, Shinoda Tagawa, 2004).

Endurance recreation sports like nordic walking or jogging can be performed easily without special facilities and without registering with any club or community. The particular benefit of both activities was investigated also early (Suter et al., 1994) (Church, 2010), but it was found later that too high training loads can be worse than avoiding physical exercises at all (Schnohr et al., 2015).

Accordingly, workout control appears mandatory, because the demand level has to be kept within certain limits for gaining positive effects, but also for preventing problems caused by too intense load (Suter et al., 1994) (Schnohr et al., 2015). Also for this reason, tiny sports computers were broadly commercialized since fairly long time. The modern version are powerful watch-like units with built-in GPS sensing and expandability by RF-linked body sensors, e.g. for registering also heart rate or food stride rate.

A converging commercial technology came up during the recent decades in the form of smartphones, which are meanwhile even equipped with various sensors, e.g., for detecting humidity and temperature of their environment, lighting conditions and others. Smartphones can trace their own movement by built-in acceleration sensors (Valentin, Howard, 2013), modern smartphones are typically equipped also with GPS positioning and network localization modes for improving the detected location. Furthermore, Bluetooth communication allows them to interlink with body sensors like chest straps for heart rate monitoring.

This set of technical features manifests a perfect base for a use of smartphones in fitness and sports tracking, which would help to overcome the need for other specialized devices like tiny sports computers. For the matter of completeness, it should be mentioned here, that other upcoming product segments like smart watches and fitness bracelets also aim to serve this application field (Alsulami, Atkins, 2016), but the severe limitations of such devices have been investigated and published recently before (Weghorn, 2019). The investigation reported here addresses the question, whether smartphones with appropriate application software tools are available yet, which can replace the established sports computers with the same degree of seamless and instant handling in daily endurance workouts.

2. EXPERIMENTAL EVALUATION OF RELIABILITY AND USABILITY

2.1 Field Testing Environment and Selected Devices

As experimental tracking devices a set of Android smartphones was selected from the middle price segment. Android phones were chosen, because this operating system has got the highest market share (statcounter, 2019) and stands therefore for the most typical device and handling surface, but the phones came from different vendors and had different major versions of Android for exploiting also the impact of the evolvement of the operating system. Table 1 lists these units, but for legal reasons neither brands nor particular device models are identified here.

Table 1. Devices used for tracking in experiments

Device	Vendor	OS level	Device description
A	X	Android 5	Multimedia consumer device
B	Y	Android 6	Robust outdoor smartphone
C	Z	Android 7	Universal standard device
T	G	proprietary	Triathlon sports watch computer

As established reference for instant use in tracking, the triathlon sports watch T (Figure 1b) manufactured by a reputed vendor, who has got long-term experience in this device market, has been used throughout the experimental series in parallel to A, B and C. This sports computer provides motion tracking through GPS and built-in acceleration sensors, as also additional optional monitoring through RF-coupled sensors for heart rate, pedal turn, foot strides and further more. After switching on, it can be used instantly by a single hard button press for initiating a workout tracking. If precise GPS trace is desired, start of workout better is delayed until sufficient signal receipt quality is reached and indicated.

All measurements were performed on a public course of 5 kms (Figure 1d), which was installed in 2018 as part of a health and sports supporting project in a joint venture between two neighbor cities and a sports club with long tradition, all located next to the Danube river in central Europe. This project is described on the web site www.donaurunning.de, where also recent results and long term of use statistics can be displayed.

Users of this parkour are automatically monitored during their running and walking activities through the four RF gates (Figure 1a), when they are wearing their particular ID chip (Figure 1c). All conducted laps are counted and can be inspected on the project web site. The official length of the track is 4.99 kilometers, but the project leader reminded in a private communication in December 2018, that the precise measure is more likely 4.98 kilometers. This is a quite typical distance for any walking workout, and a running round of roughly 30 minutes duration stands also for a basic training effort (Suter et al., 1994). The permanently available course represents a reproducible experimental field route for the investigation here, and it maps well the spirit of endurance work-outs in healthy living style and personal sports.

Figure 1. The running and walking parkour round, which was used for the tests, has got a length of 5 kilometers (d) and is equipped with four gates (a), through which sports activity is monitored automatically for people owning and wearing the corresponding RF ID chip (c)

2.2 Measurement Series on Precision of Movement Tracking

The original plan for the experiments was to first select a feasible smartphone fitness app and perform the following precision tests on base of this. It quickly turned out, that none of the available fitness apps provided such a seamless handling opportunity, so that this part of the investigation could be focused solely on the question of tracking precision. In the end, an app was found, which was intended more as general tracking tool, e.g. in car driving, for determining current velocity, elapsed time and distances and other values. For this, it uses a rather simple-styled display arrangement with big letters and minimal input buttons (Figure 2).

Figure 2. Screenshot of the simple app, which was used for validating the position tracking capabilities of the smartphones. The capture was taken after completing the Danube running round, it is clearly readable, the tracking screen consists of few UI elements only, and is primarily designed as assistive speed control during car driving

Furthermore, few GPS signal analysis apps were found for controlling the feasibility of the system platforms for detecting and tracing geographic location (GPS satellite positioning testing app is shown in Figure 4). In this configuration, the route length measurements in Tab. 2 have been recorded at varying jogging speeds.

Table 2. Measured distances in jogging around the Danube running loop

Device	# of laps	Moving speed	Average distance and statistical variation
A	5	12.6 km/h +/- 0.68 km/h	4,850 kms +/- 0,0363 kms
B	5	12.2 km/h +/- 0.42 km/h	4,883 kms +/- 0,0221 kms
T	20	12.4 km/h +/- 0.69 km/h	4,883 kms +/- 0,0206 kms

As seen here, the average errors in distance tracking of the 4.99 kms path range in the order of 100 meters absolutely or 2% relatively. Though the measures arise strictly too low, this all appears acceptable for tracing workouts in health or non-elite sports. There is almost no difference between the tiny sports watch and the bigger smart phone devices. All devices use GPS, which is assumed due to its low sampling speed to have intrinsic problems of reduced distance measures in, e.g., tight curves, which are present in the workout parkour also at several turning points.

Furthermore, the route path on the Danube running loop is not exactly or strictly defined, the walking paths have varying widths up to several meters. In the experiments above, the shortest connections between turning points were followed strictly, inner curve radiuses were used whenever possible, but the experimenters stayed always on the lanes.

For validating GPS errors, which are presumably induced by a low sampling rate especially in turns, the parkour was completed in the next tests at slower speed with even slower motion through all the tight curves. First, the shortest walking path was traced in a single shot experiment, and next, a path through the walking round was chosen with extended swings, but still remaining on the walking lanes. Table 3 list comparatively the corresponding results.

Table 3. Test rounds with slow motion especially in tight curves

Device	Course	Curve routing	Measured distance
B	Shortest	Innermost	4,924 kms
T	Shortest	Innermost	4,879 kms
B	Wide swing	Outer	4,948 kms
T	Wide swing	Outer	4,952 kms

These measurements do support the theory about the GPS sampling error problem, but show also the imprecision of the route itself. These distance values are closer to the official measure of the parkour, though Table 2 has shown already the capability of the smartphone devices for sufficient precision in route tracking.

Device A was working during the first four experiments well, but then suddenly it lacked completely of any localization capabilities. A lot of reboots and configuration checks couldn't fix the problem for many days, several experiments on workout tracking failed for this reason. In the end, the device started suddenly working again, when it was linked permanently to an Internet connection and was not moved physically until the GPS test tools aligned to the satellite signals sufficiently. This observation is interpreted as showing a fundamental problem with the operating system version and its localization method, especially with the internal improvement functionality of localization within the operating system that bases on additional information aside GPS, like e.g. surveillance of close-by WLAN networks. With device B and its newer OS version, such issues haven't been observed. Up to the project state now, experiments with device C couldn't be performed in a systematic way yet.

Another side observation concerns the handling of GPS localization in the devices in general. T activates this method immediately after being switched on. Localization reaches sufficient precision in the order of a minute, while with the smartpones it may take considerably longer. This is invoked by the powersaving approach of activating and de-activating the GPS module in smartphones on demand, which incurs a longer waiting period for GPS after a re-launch of the tracking app, despite the fact that it had been used just shortly before. Clearly, this kind of system behavior prevents that fitness apps on the smart phones can be used instantly, or at least quickly with only short preparation time.

2.3 Review of Handling Fitness Tracking Apps on Smartphones

The investigation here focuses on those smartphones, which are operated under the current market leader, the so-called Android operating system (statcounter, 2019). The official smartphone application software distribution source is operated by the Android project holder in a market place on the Web, which offers almost uncountable app tools for tracking in particular fitness activities of running and walking. For serving both categories, considerably more than 100 apps are available each, which can be sub-divided into those, where monetary costs apply for using the fully functional versions, while the others fall into the category of freeware in the sense, that no monetary costs are applied for all version levels.

Defining as handling reference wearable sports computers for endurance workouts like running, walking and cycling (like unit T in Figure 1b), the standard for instant use can be described as follows: After switching on - regardless of first or repeated use -, the system is available for immdiate use with pre-configured settings and display arrangements. The only delay, which is invoked by modern sports computers is, that the system needs some time for its alignment to the satellite navigation signal. After localization precision is reached, most computers can immediately be started with one single UI action (e.g. a hard or soft button press). Pausing or finalizing a workout appears as simple. In summary, instant use of the devices applies possible in every days use; some devices may require few setup information during first use, which is typically requested from the user by a very short guided process, where information has to be entered like display language and possibly some physiological parameters of the training person.

In clear contradiction to this kind of instant usability, handling of smartphone apps for endurance fitness workouts requires in general many stages more. Starting-up for the very first time with a particular app consists of the following handling steps (in the work here, this was practically experienced with a broader set of fitness apps like shown in Figure 3, and can also be logically deducted from the standard app programming concepts on smartphones):

- Analysis of available apps in the market place
- If not yet available, register account for market place
- Download and install particular app
- Revoke access rights of the app
- Place app launch on top UI level
- Start app
- Register new account for use of the app or link to a social network account
- Enter physiological information of the training person and possibly other personal data
- Select specific workout mode, in some cases from an extremely long selection list
- Await availability of localization system, since this is usually deactivated for energy saving reasons before this handling point

More configuration steps or other actions like de-selecting irrelevant presentation contents like commercial advertisements may be required additionally, before a first test of the tracking app renders possible.

In the case of daily use in workouts - when the first use has been successfully performed before - the following UI stages are required typically (also as practically validated in handling tests here):

- Unlock smartphone screen from sleeping mode
- Manually launch fitness app
- Select in a sub-menu the desired activity and display mode
- Start recording, but reasonable only after waiting time for positioning sensor precision
- Stop activity by simple UI actions
- Leaving the app usually requires to enter additional information for saving the trace

As further aspects it can be observed, that UI screens in such apps tend to be overloaded with information details, which leads to small text display and tiny input elements, which are especially not feasible for elder user with reduced haptic skills and difficulties in reading close-by surfaces. The arrangement of the UI elements also do not appear thoroughly elaborated in many app samples (Figure 3).

As intermediate result, it can be stated at this point, that the handling of the smartphones apps requires long series of tiny actions, therefore, the feasibility of such fitness apps for instant use has to be neglected. Furthermore, the readability, the clearness of the information presentation does not meet established standards of UI programming.

Few apps could be identified, which are offering a simple-styled UI with bigger text displays and less UI hierarchy. Unfortunately, such apps turned out being unstable, which leaded to a loss of workout trace because the app tool killed itself unexpectedly (middle app in Figure 3). Another observation is, that also popular and widely distributed apps for runners use to show sometimes spurious effects like displaying the just conducted running route through massive buildings.

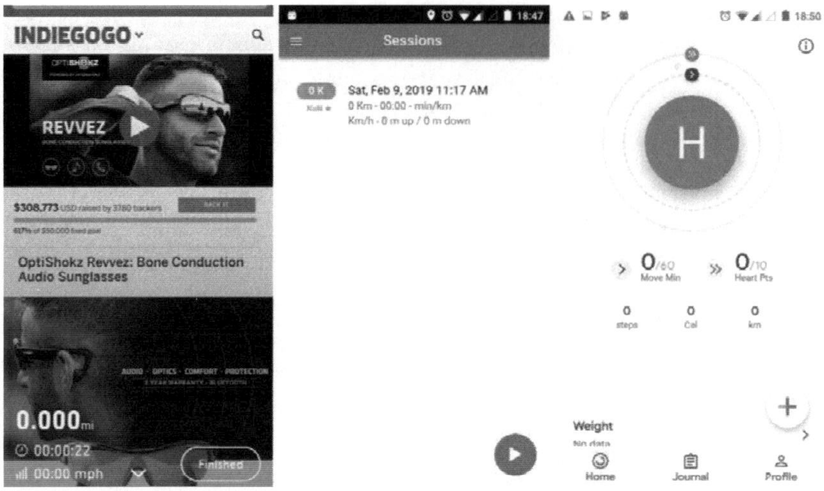

Figure 3. Screenshots of different fitness apps: on the left a "charity" app is shown, which spams the user not only by e-mails to the registered account, but also the activity display screen here is spoiled by commercial ads; the middle app is styled simplistic, but displays NaN as result of wrong internal processing and uses to kill itself, while the activity of the fitness app provided by the OS vendor is totally unclear in its controls, and uses - as all others - tiny text fonts and wastes most of the available screen surface

Concerning the needs of elderly users, it is also visible already in the few samples here (Figure 3) - and this was also the strict misfit in all other tested apps - that handling problems like inefficient or unintentional (multi)-touching in UI input as elements indicated early (Harada et al., 2013), is not served at all. Same finding applies for guiding through consecutive UI screen arrangements with missing clearly visible controls for canceling or back-navigation. The limited readability of the screen contents, which was discussed before, will render even more adverse for people with eyesight impairments (Adiseshiah, 2017).

3. CRITICAL DISCUSSION

The handling of specialized sports computers in endurance workouts like T, which are usually available in the housing of wrist watches, is well established for a long while already. Since such devices are intended for a specific purpose, they are properly pre-configured already directly out of the selling box for a first use; they are in general ready to use even more directly after power-on later in daily operation. Only waiting time applies for launching the satellite positioning, which is today used outdoor as mandatory standard. Though the displays of such devices have got small dimensions, they are optimized for readability in day light outdoor, and they are arranged clearly with big letters. The users can - for most devices - even assemble the desired screen components and by that influence also the readability according to best personal desires.

Within the many smartphones fitness tracking apps, the user is confronted with totally different demands. In most cases, at first a lengthy registration with common or proprietary social (sports) networks is required, and the user is forced to deliver lots of personal data. Other information may be collected by the app without giving clear notification about the extreme data thirst of the app, which in most cases has got no relation to the indented use scenario. Extreme samples are access to the entire phone book and contact lists of the sports app user or even the demand of issuing calls and messaging over the phone network.

In the end, the user is enabled only after a lengthy setup process to use the app for the desired purpose; using it again later for the same purpose usually also arises as complicate. Some apps - e.g. the fitness app distributed by the operating system vendor of the devices here - offer meanwhile an extremely long list of possible activities, which can be tracked and recorded. The selection among these requires another UI entering stage before the app can be used in the desired way. Furthermore, the screen arrangements do not really follow any reasonable UI concepts (Figure 3). It must be assumed that programmers in this market are developing tools, while they themselves don't have any experience or idea, what would be really required in using them efficiently.

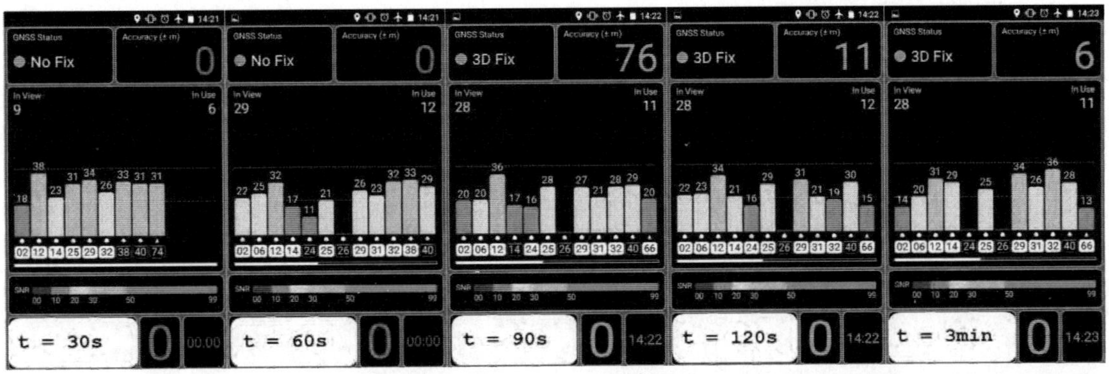

Figure 4. Set of consecutive display screens after launching the GPS analysis app and by that starting the localization sensing on unit B. The imprints at the bottom are listing the screen capturing times, and the corresponding positioning accuracy is displayed in the upper right corner as measure in meters. Accordingly, tracking is not possible in the first minute at all, and it requires three minutes for reaching a usable - but not good - precision of 6 meters

Some further fundamental restriction arises from the localization technologies used in smartphones. GPS usually is de-activated for saving energy consumption in this systems, but when it is launched only with the fitness app, it incurs an extended waiting time at this point of handling (Figure 4). Up to several minutes delay, which often applies, makes the entire tool worthless. Another effect disturbs this breaking effect even more: localization improvement in the underlying operating system mechanisms exploits additional information from other networks like the cellular phone system or from WLAN access points with the aim of improving and speeding up localization. This all uses to fail easily in outdoor environments, when no known networks are reachable or close by. It had to be observed, that in such situations some fitness apps refuse to work at all, if the training is started in a forest or a rural area, instead close to a city center with all its different digital RF networks.

In summary it has to be concluded, that actually available smartphone fitness apps are no prospective candidates for replacing the wearable endurance sport computers, since they do neither offer seamless, nor clear, nor instant usability.

4. CONCLUSION

As validated in the practical experiments here, recent generation smartphones are representing a feasible host base, which consists of a hardware platform with an operational software environment, for executing potentially reliable tracking tools for physical health and sports activities. Unfortunately, among a broad mass of software apps, no instant or easy-to-use solution is available for daily endurance workouts like health brisk walking or running.

Theoretically, such apps could be developed quite straight forwardly, but in the end there will always remain a last level of handling inconvenience, because smartphones usually hold simultaneously a mass of apps for quite different purposes; therefore, any tracking app has to be always launched manually from a broader choice before, and the device may not be ready or well prepared for the fitness tracking - especially sensors usually are not fully operational -, which then incurs further delay before efficient use.

On the other hand, the major issue appears today being the primary intent of the fitness app programmers, who are more focusing on collecting personal data and behavior patterns of the user, instead on aiming for an efficient, easy and clear handling of their app as tool for workout tracking. The latter would be even more important in supporting health activities of elderly with limited haptic capabilities and restricted eyesight.

Accordingly at the current market situation, monitoring units like specialized sports watches or personal medical monitors are representing the only solution, which yields seamless and instant use for daily physical workouts in endurance sports and health applications. As validated here, commercial smartphones could serve as sufficient hardware and software base as well, but appropriate apps are not available yet.

REFERENCES

Abbott, R.D et al., 1994. Physical activity in older middle-aged men and reduced risk of stroke: the Honolulu Heart Program. *Am. J. Epidemiol.*, Vol. 139(9), pp 881-893.

Adiseshiah, E.G., 2017. UX Design Thinking from A Senior Citizen's Perspective, UsabilityGeek, https://usabilitygeek.com/ux-design-thinking-senior-citizen-user/, last accessed 2019/03/20

Alsulami, M. H., Atkins, A. S., 2016. A model of a proactive ambient assisted living system to monitor elderly people in the Kingdom of Saudi Arabian. *IJRET*, vol. 4, no. 1 (March 2016), pp 35-41.

Church, T., 2010. Physical Activity and Heart Disease in Obesity, in: *Physical Activity and Obesity*, 2nd ed., Human Kinetics, Champaign (US), p282.

Colagiuri, S., et al., 2010. The cost of overweight and obesity in Australia. *Med J Aust*, 192, 5 (March 2010), pp 260-264.

Harada, S., et al., 2013 Characteristics of Elderly User Behavior on Mobile Multi-touch Devices, *INTERACT 2013*, Part IV, LNCS 8120, pp 323-341.

Jefferis, B.J. et al., 2014. Protective effect of time spent walking on risk of stroke in older men. *Stroke*, Vol. 45, pp 194-199.

Law, M.R. et al., 1991. Strategies for prevention of osteoporosis and hip fracture. *British Medical Journal*, Vol. 303, pp 453-459.

Oguma, Y., Shinoda-Tagawa, T., 2004. Physical activity decreases cardiovascular disease risk in women. *Am. J. Prev. Med.*, Vol. 26(5), pp 407-418.

Schnohr, P. et al., 2015, Dose of Jogging and Long-Term Mortality: The Copenhagen City Heart Study, *Journal of the American College of Cardiology*, Vol. 65(5), pp 411-419.

statcounter, 2019. Globalstats, http://gs.statcounter.com/os-market-share/mobile/worldwide, last accessed 2019/03/18.

Suter, E. et al., 1994. Jogging or walking - Comparison of health effects. *Annals of Epidemiology*, Vol. 4(5), pp 375-381.

Valentín, G., Howard, A. M., 2013. Dealing with Childhood Obesity: Passive versus Active Activity Monitoring Approaches for Engaging Individuals in Exercise. In *Proceedings BRC 2013* (Rio de Janeiro, Brazil, Feb. 2013).

Wannamethee, G, Shaper, A. G., 1992. Physical activity and stroke in British middle aged men. *British Medical Journal*, 304 (March 1992), pp597-601.

Weghorn, H., 2019, Unsubstantial Health and Sports Monitoring Reliability of Commercial Fitness Tracker Bracelets Induced by Their All-in-One Sensing Unit Approach - Experimental Evaluation of Measurement Accuracy in Dynamic and in Steady Physical Effort Scenarios, *icSPORTS 2016/2017: Sport Science Research and Technology Support*, CCIS, Vol. 975, Springer, Cham, pp 55-74.

ASTRONOMY FOR THE BLIND AND THE VISUALLY IMPAIRED

John Gialelis[1], Andreas Papalambrou[1] and Aristeidis Hlias[2]
[1]*Electrical & Computer Engineering Department, University of Patras, Patras, Greece*
[2]*Department of Informatics, University of Patras, Patras, Greece*

ABSTRACT

The need for growth and the social impact of assistive technology for the visually impaired and blind people is driven by the increasing ageing of the population and mainly of the blind population. Blind or visually impaired people cannot visually enjoy the magnificent astronomical objects and the universe and have great difficulty in engaging in the field of astronomy, which makes them feel even more socially excluded than they already are due to their condition, thus, enhancing their social introvertness. This work depicts a system which translates astronomical images into audible sounds. It involves recognition of the astronomical objects of interest inside a picture and separation from background sky and stars as well as detection of their shape, color and brightness. The detected images are then translated into audible sounds and vibration that correspond to the visual information detected. Our approach utilizes principles of state-of-the-art work, thus making sure the proposed system is embraced by the Visually Impaired and Blind People offering them the chance to significantly mitigate their social introvertness.

KEYWORDS

Visually Impaired People, Aural Interpretation of Astronomical Images, Digital Media

1. INTRODUCTION

Astronomy can promote and increase public awareness, understanding, and appreciation of science and technology, among people of all ages. Astronomy is an enjoyable, inexpensive hobby for millions of people with great impact on their social inclusion. Blind or visually impaired people cannot visually enjoy the magnificent astronomical objects and the universe and have great difficulty in engaging in the field of astronomy, which makes them feel even more socially excluded than they already are due to their condition, thus, enhancing their social introvertness. Therefore, diminishing this introvertness, by introducing inexpensive and unobtrusive means to blind and visually impaired people to access astronomy is imperative even though viewing an image of an astronomical object is a challenge for disabled people especially for people having vision problems. State of the art digital media can assist people with special needs towards this direction based on existing knowledge that visually impaired people can "see" through sound. Such voice signals exist in traffic lights, for instance, while audio books provide an opportunity to access written material.

The work described in this paper, adheres to well-grounded theory according to which an image of an astronomical object can be conceptualized through a mixture of sounds and leverages contemporary technology, so visually impaired persons can imagine what is being shown (Wagner, 1995, Levy-Tzedek, 2014), and following an innovative methodology of transforming images and photos into sounds, we introduce a system which transforms images into interactive sound files. More specifically, the proposed system can gradually scan an astronomical image and produce a designated sound when the scanning detects the background sky on the image, while when the scanning detects an astronomical object present on the image (e.g. a planet) then the sound characteristics (pitch, loudness, timbre) gradually change increasing or decreasing relatively to the shape, size and brightness of the object. This altering of the sound characteristics can provide an indirect impression of the object to a blind person. Certainly, frequencies and tones that meet the visually impaired preferences are chosen. Apart from the automated scanning option, which is in a way by default set, we provide the option to the user to scroll over with the mouse (or fingers on a touch enabled device) on the image, giving thus the ability to explore themselves each image. Thanks to this system, visually impaired people will be capable to "hear" an image but also interact with it as well as enjoy images together with their family and friends. Moreover, a vibrating function assists in easier identification of the borders of astronomical objects.

The rest of the paper is organized as follows. Section 2 briefly describes the background regarding assistive technology (AT) for the visually impaired (VI) and blind people (BP), Section 3 depicts our approach and decomposes our proposed system in its components while Section 4 reveals a demonstration of the proposed system. Section 5 summarizes the progress made so far along with the future steps to follow.

2. BACKGROUNG AND RELATED WORK

The need for growth and the social impact of assistive technology (AT) for the visually impaired (VI) and blind people (BP) is driven by the increasing ageing of the population and mainly of the blind population (Hakobyan et al. 2013). Information Technology (IT) was used to give innovative prospects of AT and recent developments on Mobile Devices provide the field of IT-based mobile AT in order to improve the opportunities in quality of life of the visually impaired (Billi et al. 2010, Bhowmick 2017).

By the emergence of the mobile phones, the situation for the VI and BP became more difficult, because the devices needed both visual reading options and handheld operations that were not ergonomic even for people without VI problems. The situation hasn't improved with the modern devices, taking into consideration the requirements of touch and vision being the main sense for their handling (Guerreiro, 2010). The evolution of mobile devices has either built-in the necessary facilities to assist VI and BP, or have been developed auxiliary methodologies and technologies that utilize the abilities of mobile devices with the lack of vision, such as speech recognition, non-speech auditory feedback, voice alerts, things and gesture recognition combining with voice, touch-based feedback, combinations of different sensory modalities, phone camera, etc. (Hakobyan et al. 2013). With ongoing advances in mobile technologies, it is becoming ever more feasible for the VI and BP to rely on mobile devices in order to communicate and understand their ecosystem (Bujacz, 2008).

The field of Human Computer Interaction (HCI) is developing eyes-free interaction methods that can apply to VI and BP, such as Foogue, which can be used to control a mobile device in two modes: menu mode for browsing and controlling a file system that is presented with spatial sound in front of the user, and listening mode, in which music, phone calls, and auditory notifications can be heard simultaneously and positioned around the head of the user (Dicke, 2010). The Slide Rule is a speech-based tool without visual representation. Users navigate through the screen and scan objects by brushing their fingers down the device surface interacting with on-screen objects. It recognizes a one finger scan for browsing information in a text or object list as the user slides a finger in a top-down direction, a second-finger tap for selecting items, a multi-directional flick for performing additional actions, and a L-select gesture for browsing hierarchical information. Its design methodology was user-centralized and came up of formative interviews with VI users (Kane, 2008).

AudioBrowser enables users to be navigate stored information and system commands combining both speech and audio feedback. Users are guided by speech and audio as they move around the screen that is split in two parts in order to allow the user to distinguish between information and commands. As users' fingers move across the screen, non-speech audio is used to inform them when they cross a boundary. Within a given segment of the screen, speech audio informs the user of the information contained therein (Chen, 2006). With the hierarchical structure that supports, the user is able to access different information, such as web pages, text files, audio files, etc. without the need to have visual view on the screen. This approach stemmed from the way VI and BP are looking for information on a touchscreen (Kulyukin, 2011). Another application for touchscreen mobile devices designed to route spatial paths in conjunction with finger-navigation sounds is the Timbremap. If the user wanders off the correct path, audio feedback is generated in order to set the user back to the path to continue tracing. In other mode, Timbremap informs the user about the number of paths around the edges of the screen, about gaps between path segments, and about the existence of any path intersections. While the user navigates the screen, they can hold one finger and by double taping anywhere on the screen with a second finger can listen to available information (Su, 2010). Another innovative approach is the Reconfigured Mobile Android Phone (R-MAP). R-MAP is a stand-alone tool for Android phones that is based on audio incentives. R-MAP uses the touchscreen of the device, without additional hardware requirements, and its fundamental advantage is that no internet connection is needed. The applications of R-MAP are associated with everyday activities of VI and BP, because it uses to pronounce labels of each packaging. In order for the blind to understand the orientation of the screen and launch the application, there are two buttons on the main activity screen; one in the top right corner of the screen that launch the application, and another in the bottom left corner that enter captured labels. The buttons are accompanied by audio feedback tones (Shaik, 2010).

Our approach utilizes principles of the state-of-the-art work described above, thus making sure the proposed system is embraced by the VI and BP offering them the chance to significantly mitigate their social introvertness.

3. SYSTEM DESCRIPTION

The system for aural interpretation of astronomical images translates astronomical images into audible sounds. It involves recognition of the astronomical objects of interest in the picture and separation from background sky and stars as well as detection of their shape, color and brightness. The detected images are then translated into audible sounds that correspond to the visual information detected. The audible sounds produced incorporate three basic elements of human sounds perception that is pitch (related to frequency), loudness (related to intensity) and timbre (related to the qualitative elements of the sounds). These elements are correlated to basic elements of visual perception such as light intensity, color and contrast. The system is provided as a service in the form of an Android application, which displays a full screen image that the users touch with their fingers and swipe across the screen while the system analyses location and speed of the fingers producing the appropriate sounds.

3.1 Vision and Hearing

Vision and hearing are the two richest senses of humans, providing a significant variety of sensory stimuli that can transfer a significant amount of information about surroundings including navigation, identification of people and objects, distance calculation, motion detection and much more. Hearing and vision share many similarities but also exhibit significant differences. Vision is based on reception of electromagnetic waves belonging to the visual window of the spectrum by the eye while hearing is based on the reception of mechanical waves travelling through air and belonging to the range of acoustic frequencies. Both vision and hearing can detect information by analyzing the amplitude, the frequency and the qualitative characteristics of waveforms.

A sound has three basic characteristics that are used from the brain to interpret its content. These are loudness, pitch and timbre. Loudness corresponds to the amplitude of the waveform, it's what is called volume in everyday language. Pitch corresponds to the frequency of the waveform, the closest common representation of it being a musical note. The timbre is the most difficult concept because it is defined in an opposite way: what makes two sounds of same volume and pitch apart is their timbre. It can be said that timbre depends on the qualitative characteristics of sound, details in the waveform that can, for instance, tell different instruments apart.

The goal of visual to aural translation is to transfer as much information as possible from the visual field to the aural including both the descriptive aspect as well as the "feeling" of the image.

3.2 Questionnaire

In order to better understand the expectations and preferences of the visually disabled, a questionnaire was prepared. The questionnaire contained questions that explored the associations the visually impaired make in their brain between aural and visual concepts, as well as practical issues regarding the format, time duration and complexity of an astronomical visual to aural translation tool. This questionnaire acted as a guide for the basic principles of the system.

A total of 132 questionnaires were filled in from visually impaired people in 5 countries (Greece, Latvia, France, Cyprus and Poland). 49% of the participants had a severe or total visual impairment while 51% had a low or moderate impairment. 42% of the participant were born with the impairment while 58% acquired it later on in life. The basic results of the questionnaires show that:

- The system should provide some kind of enjoyment and not be strictly educational.
- Each astronomical picture could be examined by the user for a period of 5-10 minutes.
- The visually impaired have a good understanding of principles such as size and shape but have difficulty understanding the concepts of color and brightness.

3.3 Conversion

The visual to aural conversion takes place by associating aural and visual properties as follows. Brightness is correlated to loudness (sound volume) while color is correlated to the sound frequency (pitch) using the correlations of the Sonochromatic Music Scale (Garcia, 2011, Modern Painters, 2008). Shape is perceived with the aid of vibration. The device vibrates briefly when the user transits from the background to the inside of the shape.

3.4 System Design

The proposed system comprises three (3) components as depicted below:

C1: User input detection and analysis component

This component detects and analyzes user touch input. Parameters of interest to detect are location, speed, acceleration, pressure, and single/multi touch. The analysis of the user input results in determining the proper visual/aural resolution to be used at any given time, e.g. slow finger movement will result in high aural resolution.

C2: Visual to aural translation component

This component makes a real-time visual to aural translation based on pre-calculated associations that are given in three (or more) different resolutions and stored in a table. As an input, this component takes the results of the C1 and associates user touch pattern into sound specifications based on pre-defined associations. The predefined analysis results for different visual/aural resolutions for each image are provided off-line and do not form part of this component.

C3: Sound generator component

This component generates the actual sounds based on the results of C2 in two ways:

-Generating tones/sounds based on software sounds generator.

-Generating sounds based on pre-recorded audio library.

4. IMPLEMENTATION

4.1 Astronomical Object Database and Image Preprocessing

In order to convert images to sound, the basic properties of the image that will be transformed to sound needed to be extracted in discrete way. These properties are shape, brightness and color. MATLAB software was used to extract these properties using algebraic calculations after converting the images to matrices.

To demonstrate the image preprocessing procedure, a sample image, as shown in Figure 1, is used of a globular cluster.

Figure 1. Sample Image

In the first step, a contour tracing algorithm analyzes the image and identifies the contours of the various components of the image as Figure 2 depicts. These can be used for marking the borders of each component to the user. In order to mark the color information of the image, the color resolution is reduced in order to avoid an overload of information in the acoustic field. An algorithm calculates the dominating color of each 20x20 area and replaces all pixels of the area with the dominant color.

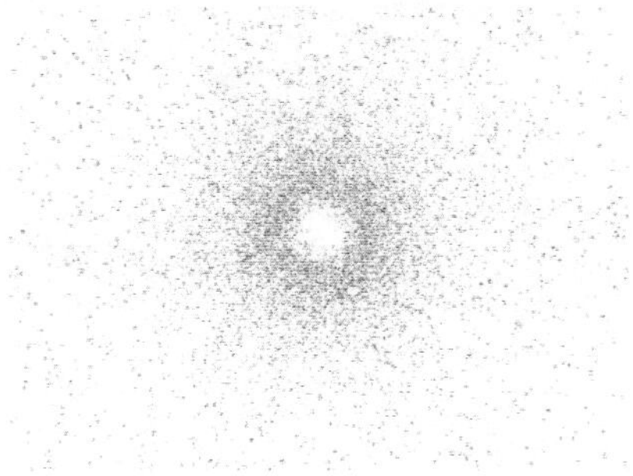

Figure 2. Image Contour Identification

A further color simplification, as shown in Figure 3, is performed by reducing the number of colors to the 7 colors of the rainbow.

Figure 3. Color Simplification

4.2 User Interface

The interface is user friendly and manageable and consists of simple touch commands and a vibration feedback. All screens are voice-assisted for the visually impaired whilst they also depict real images so that they can be used by users with no or non-total visual impairment.

When initially launched, the application presents a voice-guided image selection screen. The user needs only to touch the right image which will be translated to sound, as shown in Figure 4.

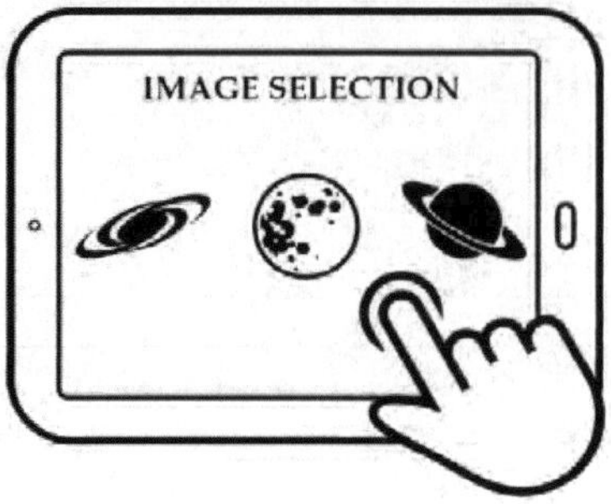

Figure 4. Voice-Guided Image Selection Screen

Once an image is selected, the user can use his finger to navigate the screen. A vibration function aids the user to detect the object edges as shown in Figure 5.

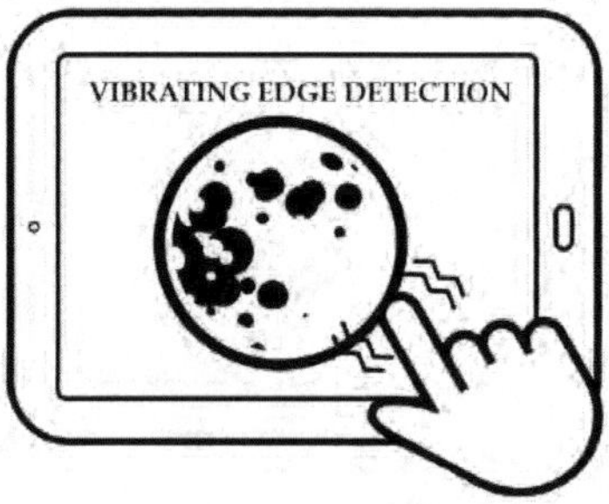

Figure 5. Vibrating Edge Detection Screen

Inside the object borders the app produces sounds as the finger is swiped across the image. The finder swiping speed detection algorithm in combination with the visual to aural translation algorithm, produce sounds that correspond to the visual information at the point of touch as presented in Figure 6.

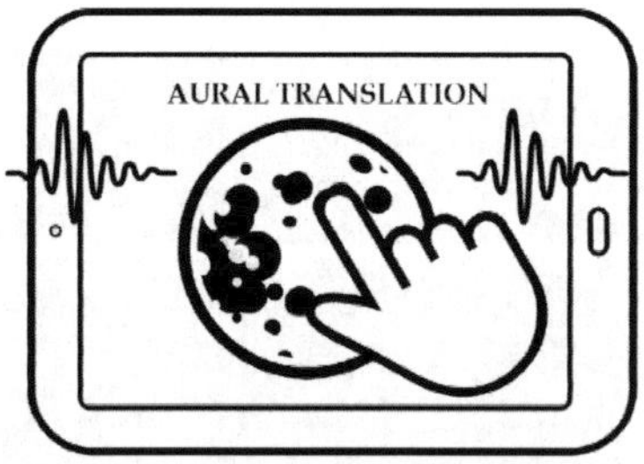

Figure 6. Aural Translation Screen

5. CONCLUSION

In this paper the main architectural and implementation aspects of a system, which transforms images into interactive sound files aiming at making visually impaired persons imagine what is being shown, are presented. System's main functions comprise image selection, finger speed detection, edge detection by vibration, finger center point assessment and sound generation. The system has been demonstrated to numerous visually impaired users and feedback has been selected mainly pointing the need for a new sound generation method. More demos are planned to follow so better fine tuning of the application is envisaged.

ACKNOWLEDGEMENT

This work is currently supported by the ERASMUS+ Program through the project 2017-1-EL01-KA201-036255- “ASTRONOMY FOR BLIND AND DISABLED (A4BD).

REFERENCES

Bhowmick, A., & Hazarika, S. M. (2017), "An insight into assistive technology for the visually impaired and blind people: state-of-the-art and future trends", Journal on Multimodal User Interfaces, Springer, June 2017, Volume 11, Issue 2, pp 149–172.

Billi M, Burzagli L, Catarci T, Burzagli, L., & Santucci, G. (2010), "A Unified Methodology for the Evaluation of Accessibility and Usability of Mobile Applications", Universal Access in the Information Society, 2010, 9:337-56.

Bujacz, M., Baranski, P., (2008), "Remote mobility and navigation aid for the visually disabled", Proceedings of the 7th International Conference on Disability, Virtual Reality and Assoc. Technologies with Art ArtAbilitation, Maia, Portugal, September 8-11, 2008, 263-70.

Chen, X., Tremaine, M., Lutz, R., Chung, J., Lacsina, P., (2006), "AudioBrowser: a mobile browsable information access for the visually impaired", Universal Access in the Information Society, 2006, 5:4-22.

Dicke, C., Wolf, K., & Y. Tal (2010), “Foogue: eyes-free interaction for smartphones”, Proceedings of MobileHCI ’10, pages 255–248, 2010.

Garcia, F.C "Nace una fundacion dedicada a convertir humanos en ciborgs", La Vanguardia, 1 March 2011.

Guerreiro, T. (2010), “Assessing mobile-wise individual differences in the blind”, Proceedings of the 12th International Conference on Human Computer Interaction with Mobile Devices and Services, Lisbon, Portugal, September 7-10, 2010; 485-6.

Hakobyan, L., Lumsden, J., O’Sullivan, D., & Bartlett, H., (2013) "Mobile assistive technologies for the visually impaired", Survey of Ophthalmology 58, Elsevier, 513-528.

Kane, S.K., Bigham, J.P., Wobbrock, J.O., (2008), "Slide rule: making mobile touch screens accessible to blind people using multi-touch interaction techniques. In Proceedings of the 10th International ACM SIGACCESS Conference on Computers and Accessibility (ASSETS), Halifax, Nova Scotia, Canada, October 13-15, 2008, 73-80.

Kulyukin, V., Crandall, W., Coster, D., (2011), "Efficiency or quality of experience: a laboratory study of three eyes-free touchscreen menu browsing user interfaces for mobile phones", The Open Rehabilitation Journal, 2011, 4:13-22.

Levy-Tzedek, S, Riemer, D, Amedi, A (2014), “Color improves “visual” acuity via sound”, Frontiers in Neuroscience, November 2014, Volume 8, Article 358, pp. 1-7.

Modern Painters, The International Contemporary Art Magazine pp 70–73 (New York City, June 2008).

Shaik, A.S., Hossain, G., Yeasin, M. (2010), "Design, development and performance evaluation of reconfigured Mobile Android Phone for people who are blind or visually impaired", Proceedings of the 28th ACM International Conference on Design of Communication (SIGDOC), Sao Carlos-Sao Paulo, Brazil, September 26-29, 2010, 159-66.

Su, J., Rosenzweig, A., Goel, A., De Laraet, E., & Truong, K.N., (2010) "Timbremap: enabling the visually-impaired to use maps on touch-enabled devices", Proceedings of the 12th International Conference on Human-Computer Interaction with Mobile Devices and Services (Mobile HCI), Lisbon, Portugal, September 7-10, 2010, 17-26.

Wagner, B. V. (1995), “Improving science instruction for students with disabilities”, Proceedings of a working conference on science for persons with disabilities. IA: University of Northern Iowa, pp. 80-83.

Short Papers

ERROR QUANTIFICATION TOOL OF HEALTH RECORDS FOR DEVELOPING COUNTRIES

Rose Nakasi[1], Ernest Mwebaze[2], Aminah Zawedde[1] and Gilbert Maiga[1]
[1]*Makerere University, Uganda*
[2]*Google AI*

ABSTRACT

In developing countries, the greatest information sources for disease prevalence are official Electronic Health Records (EHR). However, data is usually affected by human error which flaws decisions by stakeholders. The previous reviews considered data quality findings at only an aggregated level, however, this method is vulnerable to errors if wrong data is aggregated and therefore hard to trace the error. This exploratory study though limited, utilized a statistical approach for computing differences between paper and EHR records. A sample of selected health facilities was used as a base study and results indicated varying errors at each level of reporting tool.

KEYWORDS

Paper Health Records, Online Health Records, Error Quantification

1. INTRODUCTION

Electronic Health Records provide statistics about disease distribution cases which can adjust accuracies of disease modeling in a certain location at a given period of time. Therefore, for stakeholders to rely on such data, it must be reliable which isn't the case now (WHO, 2O11) and the fact that EHR is aggregated data, it is hard to track the errors from the source.

Taking a Case for Uganda, a developing country, the National eHMIS (electronic Health Management Information system) aggregates diagnosis cases monthly from health facilities that are in the system. The records are captured at health facilities daily and aggregated to weekly and then sent to the District Health officers who electronically capture these records into the system using data entrants. We believe this kind of system is prone to errors and must be investigated.

Error quantification in health records can contribute to a reliable source of health record data. It's even more useful if one wants to use the eHMIS data but has no idea of the error to account for in their analysis. The Uganda Ministry of Health (MOH) recognizes that improper data quality checks in the eHMIS system, unrealistic analysis using such data will affect true representation of the output from the analysis. Crucial in dealing with limitations in eHMIS, error quantification process must start from the health facility where the records are captured from (the laboratory) and then monitor captured and aggregated records on weekly and monthly basis both at facility and district level. In this research, we believe that proper tracking of daily records at health facility level provides a more realistic audit track and trail of the health records before and after they are aggregated in the weekly and monthly records.

Error quantification will undoubtedly increase the level of confidence to be placed on health data for effective disease distribution modeling. Here, we describe a simple statistical method based on the difference measure that makes a comparison of records at each reporting level implemented through a developed software tool that quantifies the error in reporting records. This method is utilized from the source of data at the health facility and spans through at the aggregated level at district level. We further incorporate the analysis tool into a web based platform that reliably helps to monitor and trace the error with a user interface.

2. RELATED WORK

With the increasing integration of electronic health records in health care, there is a growing interest in using such data sources to enhance the disease distribution modeling. Literature acknowledges the use of multiple data sources in modeling (Moore et al., 2016). The use of health records for disease distribution modeling is also gaining momentum in literature for example (Xia et al, 2013) showed how using Electronic Health Records (EHR) can improve modeling disease severity in multiple sclerosis, it's also demonstrated in another study using a Gaussian process framework to show how Health Management Information system data helps in mapping malaria across a changing territory delimitation in Uganda (Andrade-Pacheco, Mubangizi & Quinn., 2014). In this study however, the authors emphasize improving quality of data from the eHMIS if realistic response from decision makers is to be achieved.

2.1 Error Quantification

Recently, different studies (Stevenand et al., 2016) and (Jones et al., 2015) drew attention on citing biases associated with disease patterns found in EHR and associated complications arising if such data is used to calculate disease prevalence. In response, research efforts have gone into understanding the pattern of bias in the medical records. A 2017 study was conducted comparing paper based and electronic records without knowledge of the data in the electronic abstracts (Altrock et al., 2017). General coding rules were followed. Samples were picked from a government hospital and the study was conducted in 2001 for four months. Quantitative parameters were reported with absolute and relative frequencies, and distribution characteristics were reported using mean, median, and range.

World Health Organization has also carried out research on data quality assessment of health records in different countries including Uganda (WHO, 2011). For the 2010-2011 report, analyses were made on the district monthly reporting, missing and zero values and a computation of a percentage difference between monthly and end of year data at national level. A comparison among district reports was conducted and districts with poor data quality listed. A similar report by WHO has been conducted in Cambodia for 2012 to compare monthly paper records with the electronic results. It was found that Cambodia has a well-functioning system with a major recommendation of conducting facility data verification (WHO, 2012). While these studies provide impressive insights to error quantification, there is still lack for a systematic way to trace the reporting errors from source. Additionally, a software tool to quantify the errors is an added advantage to allow stakeholders easily and quickly account for the errors in eHMIS.

3. DATA AND METHODS

This research assessed the veracity of records by adopting statistical analysis of the raw records from daily, weekly and monthly records and the electronic records from weekly and monthly aggregated records to quantify the error rate. Comparison was made and presence of error reported as illustrated in Figure 1.

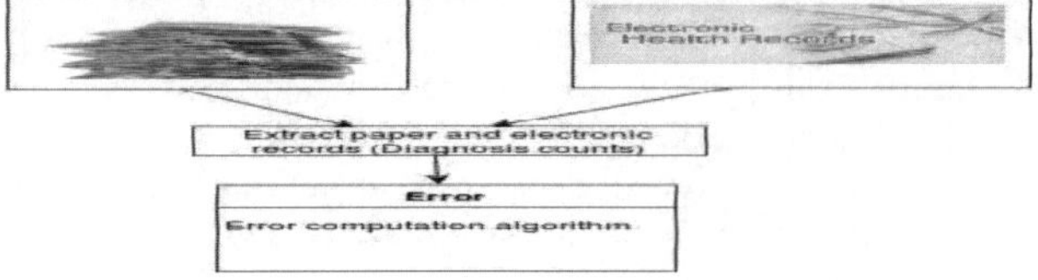

Figure 1. Methodology flow for error quantification

3.1 Data

1) Data variables to be extracted: Health records in a developing country like Uganda were captured at three levels, namely;

a) On a daily basis; these were captured in register books at the health facility. From these records, the study captured demographic data variables (date of diagnosis, name, age, gender) and clinical variables (disease, method of diagnosis, result of diagnosis). The patient level information like date of diagnosis, name, disease diagnosed, and diagnosis cases. Noteworthy is that this information is not captured in the eHMIS.

b) Weekly; these were aggregated records on a weekly basis from health facilities on paper and later taken to the district health centers for aggregation.

c) Monthly; these were aggregated from the weekly records and captured before recording at the district level. Here, we extracted only information on the disease under investigation.

2) Data extraction approach: From the facilities, scanned copies and records were manually extracted and tabulated in a csv file, uploaded into the health analysis tool and further analysis performed. Additionally, records in eHMIS, MOH provided data about the aggregated malaria cases from health facilities under study.

3.2 Analysis

Statistical methods; assessing the variability of the samples, a statistical quantitative parameter of difference was used to compare any two samples. Results were calculated using csv file.

$$difference = X_i\text{-}Y_J$$

where $X_i = X_1, X_2, \dots X_n$ and $Y_1, Y_2, \dots Y_n$.

The difference was computed over a given period of time and spatial environment and a comparison between the results in the raw paper records and the eHMIS computed and quantified.

The error difference rates isn't generalized across all districts but a similar approach can be adopted across similar tasks. Finally, we should weigh the magnitude of the error in different localities with in the country from the specified health facilities under study and indicate the rate of error in each.

3.3 System Development for Health Data Analysis Tool

The tool developed in the study (see interface in Figure 2) was designed to upload scanned paper and electronic health records at health facilities for all levels. The records were tabulated into excel or csv sheet and then analysis was performed on data with guidance of a user interface that was developed using PHP and JavaScript programming languages and can run on any web browser. It further provided visualization for the spatial magnitude of the error on the map using geo coordinates of the different health facilities on a google map.

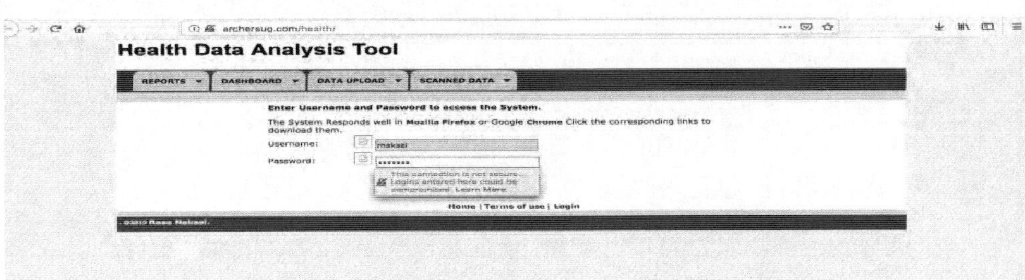

Figure 2. Health data analysis tool to visualise error magnitude between paper and electronic based records

4. RESULTS

A web based health data analysis tool was developed which quantifies error between diagnosis records of paper based and electronic health results. The tool facilitates paper based records at health facilities that have been scanned to be uploaded daily, weekly and monthly basis and equally electronic health records at district level and facility level on monthly basis. Using statistical measure of the difference, a comparison was done among the daily, weekly (paper based) and monthly records (from the eHMIS data).

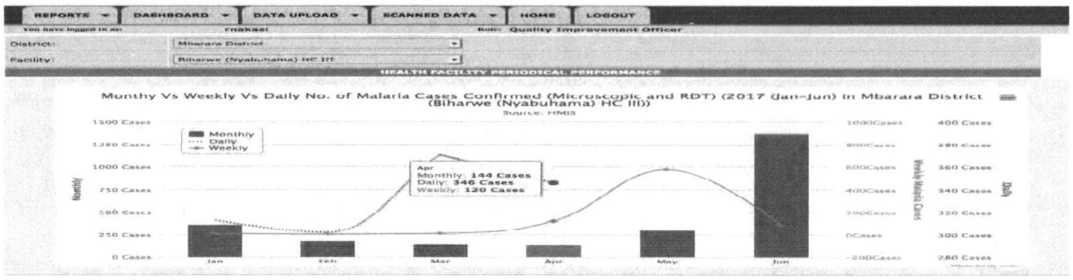

Figure 3. Monthly Vs Weekly Vs Daily number of malaria cases confirmed malaria positive in Biharwe HCIII

Sampling on Biharwe HCIII, a visualisation graph showed the records difference from the system between January and June 2017 (see Figure 3). From the graph, the results show a huge record difference from each of the datasets that is scanned paper records for aggregated daily per month, aggregated weekly per month and monthly (from the eHMIS data) from January to June 2017. Using the month of April, there were huge disparities in the number of malaria diagnosed cases for each dataset which indicate 346 daily paper records, 120 weekly paper records and 144 monthly eHMIS records.

The analysis tool also presented a difference between the health records for all the three health facilities in Mbarara district on a satellite map of Uganda using geo coordinates of facilities through Google maps. Sampling Biharwe HCIII indicates difference between records as visualised on the map in Figure 4 and the tabular representation of the same in Table 1.

Table 1. eHMIS monthly and scanned paper records of Biharwe per month from January to June 2017

Record	Jan	Feb	Mar	April	May	June
eHMIS	135	73	60	53	117	805
Paper	123	73	0	53	0	805
Difference	-12	0	-60	0	-117	0

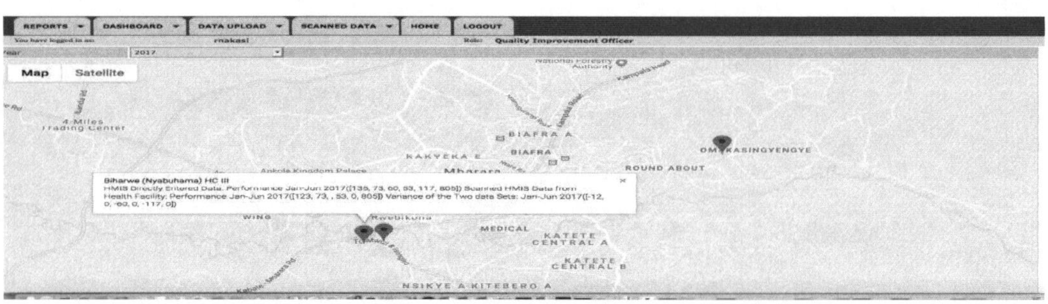

Figure 4. Spatial Visualisation of the difference at different health facilities

5. CONCLUSION AND FUTURE WORK

By analysing the results from this study, our approach retains the advantage of realistically analysing and quantifying the reporting error in health records which would provide accurate results to stakeholders. Additionally, the web based health data analysis tool acts a basic prototype that can be adopted in similar tasks with a lot of ease and act as a basis tool for designing new automated Electronic health systems. In the study the error does not only happen with aggregated data but out of a ripple effect from the source and through all levels of reporting as well. In future, semantic automation and analysis from a big spatial environment that covers a bigger part of the country should be conducted for a more accurate generalisation of the error margin. After collecting enough data for analysis, the web based reporting tool could be improved further to provide an error prediction mechanism. A point of care diagnostic system can also be developed which enables real time diagnosis and feed results directly into the electronic Health Management Information System. This way, manual entry of the health records is eliminated.

ACKNOWLEDGEMENT

The authors would like to thank Ministry of Health and Health facilities for providing data and authorization to use health records for this study and Swedish International Development Agency (SIDA) for funding this research.

REFERENCES

Altrock P.M, Ferlic J, Galla T, Tomasson M.H and Michor F, 2017. *A computational model of mgus progression to multiple myeloma identifies optimum screening strategies and their effects on mortality.*" doi: https://doi.org/10.1101/208645.

Andrade-Pacheco R, Mubangizi M and Quinn J, 2014. *Consistent mapping of government malaria records across a changing territory delimitation.* Malaria Journal.

Jones T.F, Dunn J, Jones S, Marder E and Garman K, 2015. *Assessment of administrative claims data for public health reporting of salmonella in Tennessee.* J. Am. Med. Inform. Assoc., 22 (e1) (, pp. e34-e38).

Moore A.W, Anderson B, Das K and Wong W.K, 2006. *Combining multiple signals for biosurveilance.* Ist ed., M. M Wanger, A. W. Moore and R. M Aryel eds. Elservier Inc, 2006, ch 15, pp. 321-331.

World Health Organisation, 2011. *Assessment of health facility data quality." Data quality report card Uganda.* WHO press, Geneva Tech Rep.

World Health Organisation, 2012. *Assessment of health facility data quality. data quality report card cambodia*, WHO press, Geneva Tech Rep.

Stevenand C, Bagley B, Russ and Altman B, 2016. *Computing disease incidence, prevalence and comorbidity from electronic medical records.* Journal of Biomedical Informatics; Volume 63, Pages 108-111.

Xia Z, Secor E, Lori B, Riley M, Bove M, Cheng S, Chitnis T, Cagan A and Vivian S, 2013. *Modeling disease severity in multiple sclerosis using electronic health records.* Doi: 10.1371.

TOTAL EXPOSURE HEALTH

Richard Hartman[1] and Mark Oxley[2]
Air Force Institute of Technology
WPAFB OH 45433, USA

ABSTRACT

This paper will introduce the origins and demonstrate how the concept and implementation of Total Exposure Health (TEH) is ushering in a bold solution to capture workplace, environmental, and lifestyle exposures to the individual using advances in science, technology, and informatics.
It will also introduce and describe the power behind Total Exposure Health, which is a mathematical process to quantitatively evaluate individual health risks based on genetics, occupational, lifestyle, and environmental exposures, medical disposition, protective factors, etc., forming the Individual Exposure Health Risk Profile (IEHRP).

KEYWORDS

Genomics, Informatics, Noise, Precision Health, Risk Assessment, Sensors

1. INTRODUCTION

After the military's experience of health consequences relating to Agent Orange exposure during the Vietnam War and the significant adverse health claims by Service members deployed to the Persian Gulf War, the necessity to monitor and document exposures was codified into Public Law in 1997 whereupon the Department of Defense (DoD) was mandated to develop a deployment health surveillance system to detect, document, prevent, or minimize health problems arising as a result of occupational and environmental exposures during deployments and operations.

To respond to the aforementioned mandates and improve the long-term health and welfare of the warfighter, the Department of Defense (DoD) created the Total Exposure Health (TEH) Initiative, an innovative approach to protect its military members and beneficiaries, promote readiness, and deliver personalized risk information in support of enhanced primary care (Figure 1).

---

[1] Richard Hartman, PhD, is an independent consultant for the Air Force Institute of Technology.
[2] Mark Oxley, PhD, is a full professor of Mathematics in the Department of Mathematics and Statistics, in the Graduate School of Engineering and Management at the Air Force Institute of Technology, Wright-Patterson Air Force Base, Ohio.

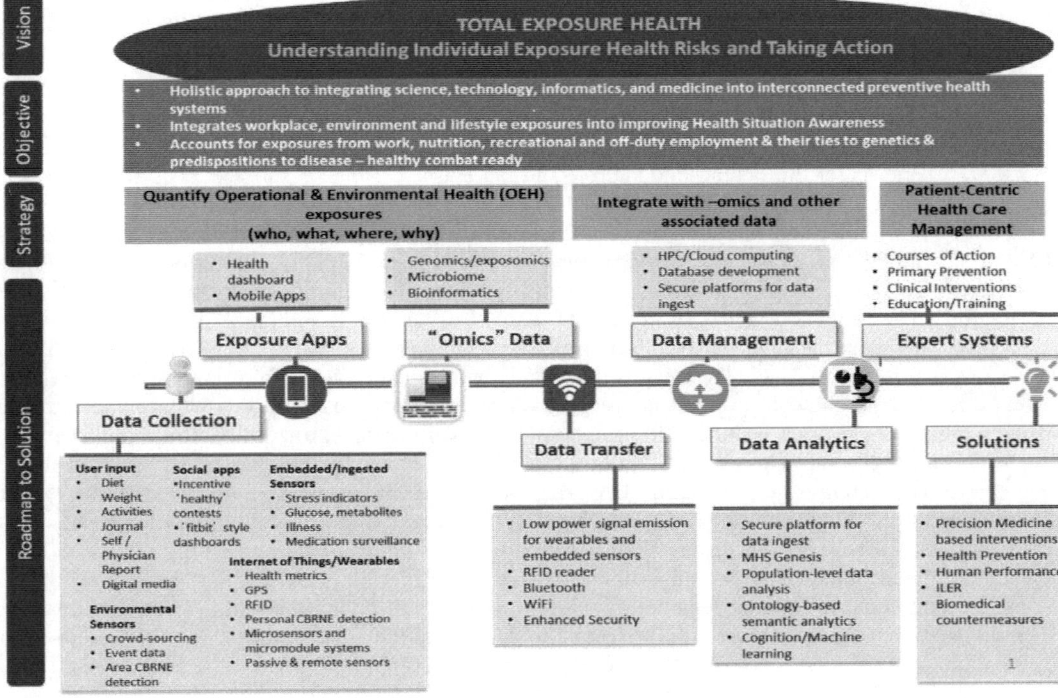

Figure 1. TEH end-to-end system development

2. TOTAL EXPOSURE HEALTH (TEH)

While the DoD made advances in health surveillance, the issues to collect, document, and act upon occupational and environmental health exposures while deployed remains a challenge. This is due to the nature of deployments, which are 24/7 workplaces, so the need to develop a more comprehensive view of what exposure meant beyond the typical 40-hr workweek and workplace was necessary. This meant expanding the practice of monitoring occupational and environmental exposures to include lifestyle choices.

This new way of characterizing worker well-being originated at the National Institute for Occupational Safety and Health (NIOSH) through Total Worker Health (TWH) as "a holistic approach to worker well-being that acknowledges risk factors related to work that contribute to health problems previously considered unrelated to work." (NIOSH, 2015)

The DoD recognized the validity of this approach but sought to realize a greater level of personalization by utilizing recent advances in the exposure sciences, genomics, sensor and data technologies, and health informatics which would aid in a more complete understanding of an individual's health risks, root causes of disease and injury, and innovative but accessible methods for primary prevention. The result is what is now referred to as Total Exposure Health (TEH).

Operationalizing TEH required partnering across multiple program areas, using traditional and emerging exposure assessment technologies (including sensors and –omics-based molecular biology), as well as leveraging a Big Data infrastructure and advanced analytics.

While the near-term goal of TEH is to improve exposure characterization and understand individual variability, susceptibility, and vulnerability to cumulative exposures and risks factors, the first achievement was to demonstrate the feasibility of TEH and the ability to collect exposure data 24/7 to better understand the effects of total exposure. In the long-term, TEH adds genomics and other data sets to translate these findings into clinically actionable recommendations for health prevention to improve both the health and well-being of both the workforce and also their beneficiaries.

Though the ultimate implementation of TEH will monitor a wide range of physical risks such as radiation and chemical exposure from the workplace and environment as well as lifestyle risks from recreational activities and diet, the first focus was on noise exposure.

2.1 Noise Exposure Demonstration Project (NEDP)

Noise was the focus for this demonstration project because it is: (1) an exposure that does not discriminate by age, race, gender, and/or socioeconomic status; (2) costly (currently hearing health expenditures range from $1.5 billion to $2 billion annually for the U.S. Department of Veterans Affairs in benefits and medical costs) (Alamgir, et al, 2016) and; (3) affects approximately 10% of the U.S. population according to a 2017 report by the Centers for Disease Control and Prevention, where 40 million American adults show signs of noise-induced hearing loss (NIHL). (Carroll, et al, 2017).

NEDP used sensor technology to collect total noise exposure 24 hr a day, seven days a week. Unique to our noise monitoring devices, the wearable sensor collects sound events from both ambient noise in the workplace and environment as well as in-ear sounds from ear buds used with smartphones and other devices playing digital media (such as music) with a smartphone app recording those decibel (dB) readings.

Nineteen subjects participated in this study at Moody Air Force Base, Georgia over 10 days in June 2018. We captured 12,680 noise events of 70 dBA or above, with 2,968 events, or 23%, were 95 dBA or above. (A-weighted decibels are adjusted for human hearing.) The average daily (24-hr) decibel dose was about 75 dBA. About 10%of the subjects had total daily noise exposures under 70 dB, and 10 percent over 80 dB, with the majority in the middle.

We found high noise exposure both at the workplace (46%) and off-duty (52%), and significant cumulative high noise individual exposure (three to 27 hr over the 10-day study period). We also identified geospatial "hotspot" locations of exposure across the population.

Overall, the NDEP met its objectives as a successful TEH demonstration project. We developed a low-cost noise dosimeter/sensor that monitors external and media smartphone device noise sound levels around the clock; used advanced analytics to collate multiple sensor devices, with geospatial layering; and managed participant compliance.

Recognizing that development of NIHL varies among populations who have been exposed to the same levels of noise, we wondered if there were genetic markers associated with a predisposition. We identified ten published studies, with small to modest sample sizes, which have indeed established a link between multiple genetic variants associated with NIHL. The effect is surprisingly large, with odds ratios of 5.2 to 22.36 indicating an elevated risk. (PLOS One, 2015).

Furthermore, within the broader USAF population, of which the NDEP participants are a subset, 17% of the 2,000 who have had their genes fully sequenced showed a particular gene variant, rs7598759, indicating they may be at a substantially increased risk for a hearing threshold shift—that is, of a hearing loss susceptibility due to noise.

This finding led to a new question: What if we could take a person's cumulative, daily exposure to noise and combine this data with their genetic predisposition for NIHL to optimally protect against it? More broadly speaking, what if we could predict, by aggregating big data and genomics, a person's susceptibility to external health risk factors—and take action to alleviate it using individualized health protocols?

We determined that by using an individual's medical records we could identify pre-existing conditions that would also increase the risk of NIHL, thus bringing together exposure data, genomics data and medical record data together to describe a more holistic montage of the persons risk to a particular exposure.

This estimate/calculation of individual health risk factors involved the need for a mathematical/informatics process that could merge and analyze various data from sensors, medical records, and unstructured information as well as genomics data to identify relationships between them at the individual level called the Individual Exposure Health Risk Profile (IEHRP).

2.2 Individual Exposure Health Risk Profile (IEHRP)

The IEHRP integrates exposure data from both traditional and cutting edge emerging exposure assessment technologies (including sensors and "–omics" based molecular biology) with clinical and genomic data. Combining the three: 1) the exposure measurement, 2) the genetic proclivity associated with the exposure and 3) current clinical history associated with the exposure, provides a better description of the individual's risk than would a focus on any one variable. This combination originally resulted in the Individual Exposure Health Risk Index (IEHRI) defined by

$$IEHRI_1 = v_{1,1} + v_{1,2} + v_{1,3} \quad (1)$$

where $v_{1,1}$ is the exposure measure, the $v_{1,2}$ is genetic proclivity to the exposure and $v_{1,3}$ is the clinical effect per the evidence in the individual's medical record (medical history). For example, when evaluating noise, the IEHRI becomes:

$$IEHRI_{(Noise)} = v_{(Noise,\ measure)} + v_{(Noise,\ genes)} + v_{(Noise,\ history)} \quad (2)$$

Accounting for multiple exposures lead to the development of the IEHRP, which combines multiple IEHRI's for an individual, represented by the equation (3) for (i) exposures.

$$IEHRP = (IEHRI_1, IEHRI_2, \ldots, IEHRI_i) = ([v_{1,1} + v_{1,2} + v_{1,3}], [v_{2,1} + v_{2,2} + v_{2,3}], \ldots, [v_{i,1} + v_{i,2} + v_{i,3}]) \quad (3)$$

With the IEHRP equipped to address multiple exposures, the limitations to the IEHRI became evident, which only accounted for three variables to describe an individual exposure. This was observed early in the NEDP, which revealed other variables that would affect the $IEHRI_{(Noise)}$, such as, ototoxins or family history. This required the IEHRP to be modified so it could account for multiple confounding factors, and is represented by the equation (4) with (j) confounding factors.

$$IEHRP = (IEHRI_1, IEHRI_2, \ldots, IEHRI_j) = ([v_{1,1} + v_{1,2} + v_{1,3} + \ldots + v_{1,j}], [v_{2,1} + v_{2,2} + v_{2,3} + \ldots + v_{2,j}], \ldots, [v_{i,1} + v_{i,2} + v_{i,3} + \ldots + v_{i,j}]) \quad (4)$$

With the development of the IEHRP for multiple exposures and multiple variables two key questions still needed to be addressed: (1) Which variable ($v_{i,j}$) was the most important? That is, should some weigh more than others? and; (2) How do we account for the variability of each variable?

To answer the first question (e.g. are genes more important than exposure measure?) the equation was enhanced with a "Weighting Factor" ($WF_{i,j}$), a numerical value that would account for importance of each variable ($v_{i,j}$). To account for the variability (confidence) the equation included a "Correction Factor" ($CF_{i,j}$), a numerical value. Combining the CF with the WF resulted in the most recent iteration of the IEHRP, equation (5):

$$\begin{aligned} IEHRP = {} & (IEHRI_1, IEHRI_2, \ldots, IEHRI_j) \\ = {} & ([(v_{1,1})(CF_{1,1})(WF_{1,1}) + (v_{1,2})(CF_{1,2})(WF_{1,2}) + \ldots + (v_{1,j})(CF_{1,j})(WF_{1,j})], \\ & [(v_{2,1})(CF_{2,1})(WF_{2,1}) + (v_{2,2})(CF_{2,2})(WF_{2,2}) + \ldots + (v_{2,j})(CF_{2,j})(WF_{2,j})], \\ & \vdots \\ & [(v_{i,1})(CF_{i,1})(WF_{i,1}) + (v_{i,2})(CF_{i,2})(WF_{i,2}) + \ldots + (v_{i,j})(CF_{i,j})(WF_{i,j})]) \end{aligned} \quad (5)$$

This IEHRP is for an individual, so if we imagine N individuals with similar or different exposures we would expect to see N distinct IEHRP's. Therefore, observing the collection of IEHRPs

$$\{IEHRP_1, IEHRP_2, \ldots, IEHRP_N\} \quad (6)$$

would be of interest.

For example, if we visualize the IHERPs for two individuals N=2 (Figure 2), individual A has a high risk for noise whereas individual B has a high risk for radon. This visualization allows the provider or IH/OH professional to target and prioritize interventions based on the individual's highest risk exposure(s).

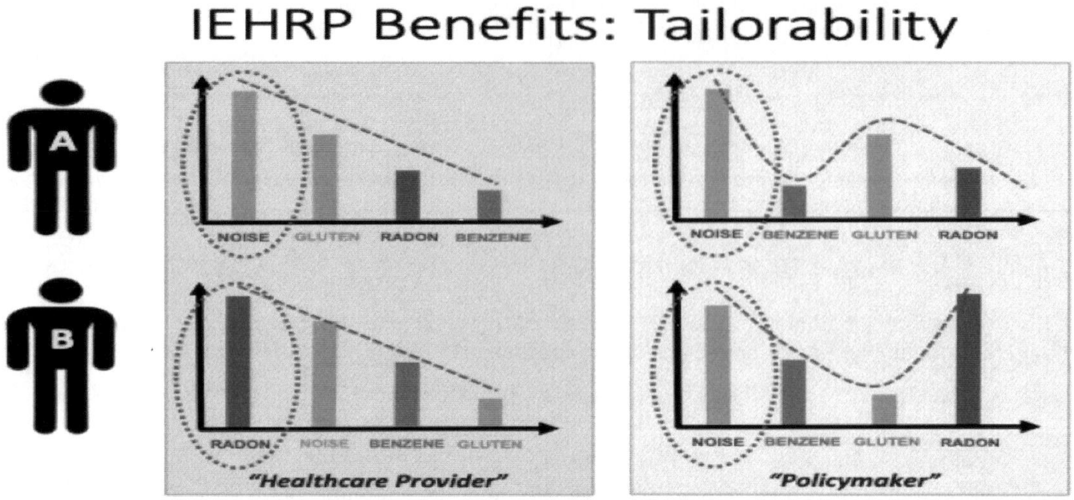

Figure 2. IHERP data visualization for personalized health and population health

If we modify the visualization with the same data to accommodate a policy maker, we notice noise is prevalent in both A and B, allowing the policy maker to identity the high priority exposure for either policy development/modification or proper resourcing versus focusing on a lower priority exposure.

When fully developed, the IEHRP will provide an enhanced capability to describe individual health risks based on genetic factors and occupational, lifestyle, and environmental exposure factors, medical disposition, protective factors, and other variables that affect exposure health risk.

3. CONCLUSION

Through initiatives like TEH we will uncover a new understanding of the relationships between an individual's genetic predispositions, epigenetic factors, and exposures from lifestyle, occupation, and the environment to support the development of diagnostic approaches, treatment methods, and intervention strategies to truly institute primary prevention to improve worker health, performance and productivity.

Through TEH, the healthcare profession will leverage and advance genomics, sensor and data technologies, data analytics, and health informatics into a future-focused and progressive approach that represents a disruptive but necessary paradigm to shift from a healthcare delivery system dependent heavily on infrastructure and staffing to one that allows individuals to take ownership of their health and drive their behaviors and health-related choices.

REFERENCES

Alamgir, H., D. L. Tucker, S.Y. Kim, et al.: Economic Burden of Hearing Loss for the U.S. Military: A Proposed Framework for Estimation. *Military Medicine, Volume 181, Issue 4, 1 April 2016, Pages 301–306.*

Carroll Y.I., E. J., Scinicariello, et al.: Vital Signs: Noise-Induced Hearing Loss Among Adults – United States 2011-2012. *MMWR Morb Mortal Wkly Rep 2017;66:139-144.*

National Institute for Occupational Safety and Health (NIOSH): "Total Worker Health." Available at https://www.cdc.gov/niosh/twh/default.html

PLOS ONE: "Genetic Polymorphisms Associated with Hearing Threshold Shift in Subjects during First Encounter with Occupational Impulse Noise," http://bit.ly/plosonenoise (June 2015).

USING ASSOCIATIVE LINKS FOR STORING PERSONALIZED MEDICAL INFORMATION

Aleksandra Vatian, Anna Ratnikova, Aleksey Gruntov, Svyatoslav Osipov, Anatoly Shalyto and Natalia Gusarova
ITMO University, 49 Kronverksky pr., Saint-Petersburg, Russia

ABSTRACT

The existing medical information systems do not provide the possibility of operational fixation and storage of such associative information. The paper proposes a system for storing the creative associations of the doctor, which are of interest both from the point of view of personalizing the approach to treating a patient, and from the point of view of forming a personalized physician knowledge base. We have developed an association storage system consisting of three levels: timestamps, semantic labels from external plug-in dictionaries, and a dictionary of doctor's own associations, being dynamically filled in by the system.The ensembling of search results in all connected dictionaries allows the user to expand the coverage and enrich the results of search results (enrich search results).
The proposed structure of storing associations is universal and allows the user to refer to any objects. It can be easily embedded in the database of a medical institution or a specific physician, as well as in other structures for storing medical information - for example, in the form of links to external files or the doctor's own files.paper must have an abstract.

KEYWORDS

Personalized Medicine, Healthcare, Associative Links, Nosql, Medical Database

1. INTRODUCTION

Modern medicine is experiencing an informational boom. The flow of new information necessary for practical use by a specific practitioner is steadily increasing. The traditional methods of mastering information assumed that every new incoming information source - whether it is an article, a monograph, etc. - is systematically and thoroughly analyzed in all aspects. However, the resources for such an approach, especially for a clinician with a lot of practice, have now been exhausted. Therefore, as practice shows, most of the potentially useful articles either pass by the doctor, or are aimlessly recessed in his personal informational vaults.

In this situation, the primary viewing of the newly incoming literature is of great value for the doctor. As practice shows, at this time associations arise in his brain highlighting aspects that are potentially important for his practical and / or research activities. Such associations being manifestations of the doctor's creativity are of great value. They allow, if necessary, to return to the article found before and study it in a more detail way, which is important for personalized decision making.

Consider an example. Dr. Green is the attending physician of Mr. Brown. Mr. Brown suffers from pneumonia, for the treatment of which Dr. Green uses traditional tactics. Dr. Green finds in the article a reference to the fact that this tactic is in some cases not effective enough, and suggests that Mr. Brown may be just such a case (due to other diseases in his anamnesis), but it will become clear only during the treatment. Dr. Green's desires to fixate the information found, in order to be able to subsequently, depending on the course of Mr. Brown's treatment, return to it, study in a more detail way and make the appropriate personalized decision. Obviously, it is important here to fixate not only the information objects themselves being linked (for example, Mr Brown's EHR and the article found), but also the content of the association in a form convenient and understandable for the doctor - in the form of meta tags, which he can choose or even come up with personally. Note that these tags can be both semantic ("anamnesis", "pneumonia") and non-semantic ("see after a week", "page 3 of the article", "good scheme" etc.).

Associations are one of the main mechanisms of human creative activity. The development and use of the creative component of the medical professionals is one of the trends of modern healthcare [Zuber, Leber].

For their development in physicians, a wide range of organizational and psychological procedures are proposed [Creativity in Healthcare]. But, as the analysis shows, there are currently no tools for supporting the manifestations of creativity of a particular doctor in his current work.

Various means of associative search are considered in literature for a long time [Kawamoto, Berger]. They are based on a pre-compiled Learning Matrix containing a list of all potentially possible interrelations (usually – correlations) between associated objects, and a specific association is formed as a sample from the matrix for the desired set of parameters of these objects. They already have an instrumental implementation [QlikView] including for medicine [Aldape-Pérez]. In addition, tools have been proposed for evaluating semantic proximity not only by the correlation matrix, but also by external sources [Mikolov, Wang], as well as by the internal vector of associations (for example, by mutual information [Damani]). They allow, among other things, solving the tasks of accounting for synonymy, as well as multilingualism (it is easier here than the Germans). However, in relation to the problem under discussion, these systems have a number of limitations: a large amount of statistical tests is required for their formation, the origin of the association is not detected, its type is not fixed. In other words, these systems only fix associations previously formed by other users, and single out the most frequent ones.

Thus, a review shows that existing information systems that are available to the practitioner do not provide the possibility of operational fixation and storage of such associative information. The article proposes a structure for for this purpose that is implemented as an add-on over various storage structures and allows storing the mentioned associations, which are of interest both from the point of view of personalizing the approach to treating a patient, and from the point of view of forming a personalized physician knowledge base. We propose an association storage system consisting of three levels: timestamps, semantic labels from external plug-in dictionaries (SNOMED, WordNet), and a dictionary of doctor's own associations, which can be dynamically filled in while working with the system. The ensembling of search results in all connected dictionaries allows to expand the coverage and enrich search results.

2. PROPOSED SYSTEM

2.1 System Structure and Data Model

The data model used in the system is shown in Figure 1, where the structure of the system itself and its connected external environment are shown. Associations storing is implemented in the structure of No-SQL type, which gives an opportunity to dynamically add storage objects without specifying their type in advance. A DBMS MongoDB was chosen as a means of implementation, since in this case high performance is not required, but orientation towards the storage of heterogeneous data is required.

The central object is *Association*, the pseudo code of which is presented in figure 2. This object stores a list of tags, a list of medical objects related to personal doctor's association, meta-information (doctor's notes about this connection), as well as a history of changes. *Validation* allows to separate private associations of a particular doctor from associations that can be visible to all doctors working with the system.

The *Change* object contains the time of the change and the description of the change. Thus, it is possible to trace the entire history of changes in associations, since its inception. *Tag* object may have different semantic connotations. For example, it may be a term from external ontology (like SNOMED), from a lexico-semantic dictionary (like WordNet), as well as a term or other entry reflecting the doctor's own associations. *MedicineObject* allows associations to refer to a real-life medical facility. The *Name* attribute indicates the name of the object, *Link* is a link to an object, which can be a file system object, an Internet resource, an object of a hospital medical system, etc. The *Time* field indicates the time the object was created. The *Extract* field can contain information about the specific fragment that caused the association. This will allow the doctor later not only to view the entire article, but immediately float the desired fragment, which was the source of the association.

The process of entering associations that have arisen at the doctor into the system looks as follows. Through the interface provided by the system, the doctor enters into it information about the objects between which the association needs to be fixed. At the same time, for each of them, a *MedicineObject* is created, for which the doctor enters the name and adds, if necessary, meta information. The creation time is filled automatically.

The *Link* field is filled in according to the type of source. For example, if the source is a database object of a relational type, then the *Link* field contains the names of the table, in which the object is stored, and the ID of the object itself (for example, Figure 1 shows the relation between *MedicineObject* and

HospitalObject). If the referenced object is stored in the structure of archetypes (as, for example, in OpenEHR), then a reference to a node in an EHR in the form of a URI or other path identifier is used. In the case of an Internet resource, its URL is indicated; for the file system object, the path to the file is specified (in Figure 1 – relation between system *MedicineObject* and *ThirdPartyObject*). Note that applicational implementation of the link will certainly depend on the interface provided by the structure in which the object is stored. Besides, specific steps may vary depending on the content of the association and the technological implementation of the structures being connected. The created object is added to the *ObjectList*.

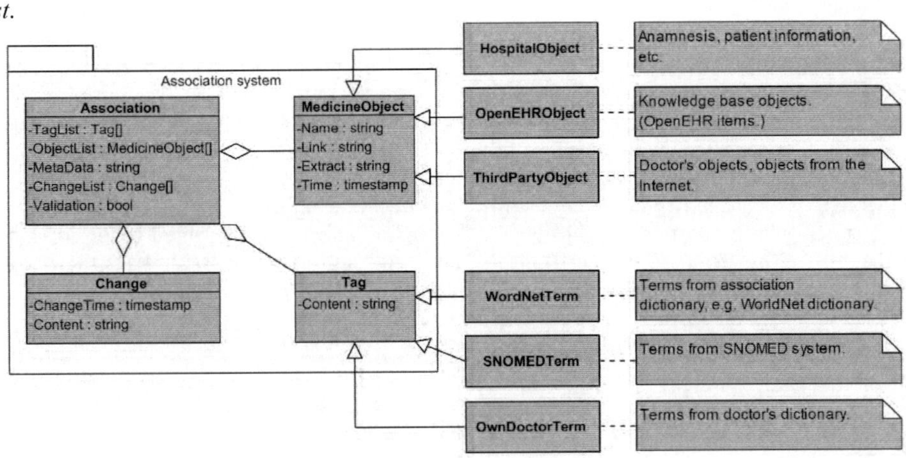

Figure 1. Data model

```
public class Association: MongoEntity
        {
                public string MetaData { get; set; }
                public List<Tag> TagList { get; set; }
                private List<MedicineObject> ObjectList;
                public List<Change> ChangeList { get; private set; }
                public bool Validation { get; set; }
        }
```

Figure 2. Structure of Association object

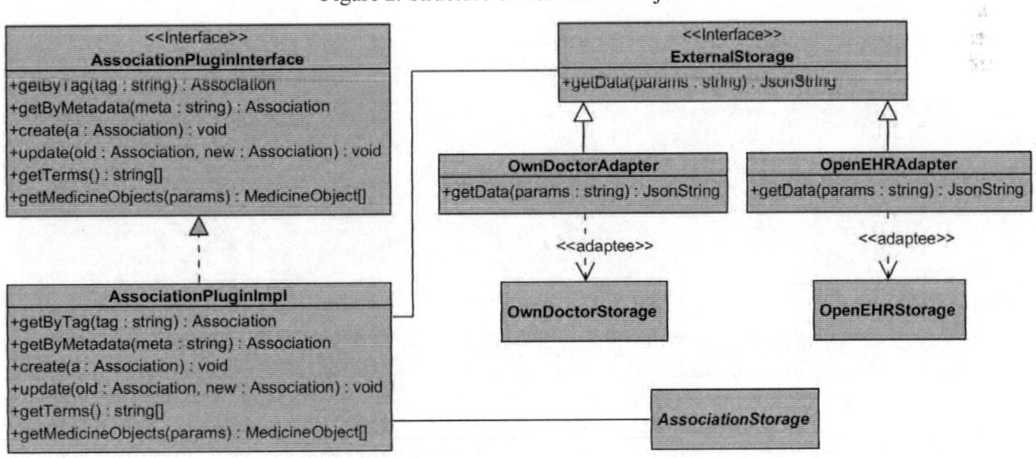

Figure 3. Class model

Next, the doctor forms an association between objects by filling in the *Content* field of the *Tag* object. The tag value can be listed in plug-in dictionaries (Figure 1 shows *SnomedTerm* and *WordNetTerm*) or be added by a doctor to indicate his individual interpretation of association (*OwnVectorTerm*). The created object is appended in the *TagList* association. If desired (optionally), the doctor adds Metadata, i.e. a brief description of the association, as well as the value of *Validation*, to the *Association* object. In order to avoid duplicated tags, it is desirable for the tags taken from different lists not to match textually. To do this, the doctor may specify his own links with the indicator, for example, "pneumonia_my".

The system creates a *Change* object with the current date and time, and appends it to the *ChangeList*. Further changes to the association will also be recorded using *Change* objects. When the association is saved, it is recorded in *MongoDB* as a JSON object (see Figure 4, 5).

The 'Adapter' pattern is used to provide access to a range of external data sources. The diagram of Class model is depicted in Figure 3.

2.2 Association Search

The specificity of the proposed system is that the search is being carried out by ensembling, i.e.the doctor is able to search all connected dictionaries of terms. Consider an example of the implementation of such a search using the connected resources presented in figure 1.

Let the system recorded two associations, presented in the form of JSON in Figure 4 and 5, respectively. *ObjectList* stores an array of medical objects. The following objects are presented in Figure 4: article, doctor, patient, diagnosis. The article in the *Link* field contains an example of a URL link, in the *Link* field of other objects the names of the tables of the local hospital database and ID are indicated.


```
{
  "MetaData": "Mr Brown's symptoms of pneumonia resemble with
              those, which are indicated in this article.",
  "TagList": [
    { "Content": "Mr. Brown" },
    { "Content": "pneumonia" }
  ],
  "ObjectList": [
    {
      "Name": "Article about pneumonia",
      "Link": "https://medicine/pneumonia/symptoms/",
      "Extract": "",
      "Time": "2019-01-18T16:00:00Z"
    },
    {
      "Name": "Doctor",
      "Link": [ "Doctor", "51" ],
      "Extract": "",
      "Time": "2019-01-18T16:00:00Z"
    },
    {
      "Name": "Patient",
      "Link": [ "Patient", "108" ],
      "Extract": "",
      "Time": "2019-01-18T16:00:00Z"
    },
    {
      "Name": "Pneumonia diagnosis",
      "Link": [ "Diagnosis", "108" ],
      "Extract": "",
      "Time": "2019-01-18T16:00:00Z"
    }
  ],
  "Validation": false,
  "ChangeList": [
    {
      "Time": "2019-01-18T16:03:00Z",
      "Changes": "Association was created"
    }
  ]
   }
```


Figure 4. Example of Association object


```
{
  "MetaData": "The Ambrobene medicine did not help the patient Mr. Red
in the treatment of pneumonia.",
  "TagList": [
    { "Content": "Mr. Red" },
    { "Content": "pneumonia" },
    { "Content": "Ambrobene" }
  ],
  "ObjectList": [
    {
      "Name": "Ambrobene medicine",
      "Link": "ehr:1234567/87284370-2D4B-4e3d-A3F3-
303D2F4F34B::rmh.nhs.net::2",
      "Extract": "",
      "Time": "2019-01-18T16:00:00Z"
    },
    {
      "Name": "Doctor",
      "Link": [ "Doctor", "51" ],
      "Extract": "",
      "Time": "2019-01-18T16:00:00Z"
    },
    {
      "Name": "Patient",
      "Link": [ "Patient", "191" ],
      "Extract": "",
      "Time": "2019-01-18T16:00:00Z"
    },
    {
      "Name": "Pneumonia diagnosis",
      "Link": [ "Diagnosis", "108" ],
      "Extract": "",
      "Time": "2019-01-18T16:00:00Z"
    }
  ],
  "Validation": false,
  "ChangeList": [
    {
      "Time": "2019-01-18T16:03:00Z",
      "Changes": "Association was created"
    }
  ]
}
```


Figure 5. Example of Association object

Figure 5 presents the objects: medicine, doctor, patient, diagnosis. The medication in the *Link* field contains an example of a link to a node in the EHR in the form of a URI; in the *Link* field of other objects, links to objects from a local database are also indicated. The *ChangeList* field stores an array of changes. In these cases, associations have just been created and have not yet changed, so we see only one the element in which the time of association creation as well as the information that the association has been created are stored. The *Validation* field in both Figure 4 and Figure 5 stores the value false.

To search by existing associations, Dr. Green performs a query on the tag "pneumonia" and the time interval: 04/01/18 - 08/01/18. In response to a query, the system searches through the entire ensemble of connected dictionaries. As an example, here are the connected dictionaries listed in Figure 1. SNOMED dictionary search allows you to return not only associations containing the tag "pneumonia", but also associations whose tags are SNOMED terms, hierarchically associated with the term "pneumonia". WordNet dictionary search in turn allows you to find associations whose tags are semantically associated with a query. Along with the third-party dictionaries, the system searches through the doctor's own dictionary of tags, which contains its individual interpretation of association. All search results, the date of creation (last modified) which does not fall within the requested time interval, are filtered by the system.

3. CONCLUSION

The paper proposes a system for storing the creative associations of the doctor, which are of interest both from the point of view of personalizing the approach to treating a patient, and from the point of view of forming a personalized physician knowledge base. We have developed an association storage system consisting of three levels: timestamps, semantic labels from external plug-in dictionaries, and a dictionary of doctor's own associations, being dynamically filled in by the system.

The ensembling of search results in all connected dictionaries allows the user to expand the coverage and enrich the results of search results (enrich search results).

The proposed structure of storing associations is universal and allows the user to refer to any objects. It can be easily embedded in the database of a medical institution or a specific physician, as well as in other structures for storing medical information - for example, in the form of links to external files or the doctor's own files. This supports the currently sought-for personalized medicine approach.

The system is developed as plugin, so it can be easely integrated into any other system. Currently, it is included in the medical decision support system being developed by the authors, which is currently undergoing clinical trials.

The considered system supports the approach of personalized medicine and is included in the system developed by the authors to support individual medical decision-making, which is currently undergoing clinical trials.

ACKNOWLEDGEMENT

This work was financially supported by the Government of Russian Federation, Grant 08-08.

REFERENCES

Zuber C. D., Moody L., 2018. Creativity and Innovation in Health Care. Tapping Into Organizational Enablers Through Human-Centered Design // *Nursing Administration Quarterly*, Vol. 42, Is. 1, pp. 62–75.

Leber J., 2014. Why Health Care Needs Creative Thinking More Than Ever // *Most Creative People*, 20.05.14 [Internet]. Available from: https://www.fastcompany.com/3027327/why-health-care-needs-creative-thinking-more-than-ever

Creativity in Healthcare. Canterbury Christ Church University, 2018. [Internet]. Available from:

https://www.canterbury.ac.uk/health-and-wellbeing/england-centre-for-practice-development/creativity-in-healthcare/creativity-in-healthcare.aspx

Kawamoto M. et al., 2004. An Implementation of a Semantic Associative Search Space for Medical Document Databases // Proceedings of the 2004 Symposium on Applications and the Internet-Workshops (SAINT 2004 Workshops). P. 488–494.

Berger H., Dittenbach M., Merkl D., 2004. An Adaptive Information Retrieval System based on Associative Networks // First Asia-Pacific Conference on Conceptual Modelling (APCCM2004), New Zealand, pp. 27–36.

Aldape-Pérez M. et al., 2018. An Associative Memory Approach to HealthcareMonitoring and Decision Making // *Sensors*, V. 18, pp. 2690.

QlikView: Using Associative Search as List Box Search Mode. 2015. [Internet]. Available from:

https://www.qgate.co.uk/knowledge/qlikview-using-associative-search-as-list-box-search-mode/

Mikolov T. et al. 2013. Efficient Estimation of Word Representations in Vector Space // *Proceedings of Workshop at ICLR*, 2013.

Wang Z. et al. 2015. An Inference Approach to Basic Level of Categorization // *ACM International Conference on Information and Knowledge Management (CIKM), ACM – Association for Computing Machinery*, October 2015.

Damani O.P. 2013. Improving Pointwise Mutual Information by IncorporatingSignificant Co-occurrence // *Proceedings of the Seventeenth Conference on Computational Natural Language Learning*, Bulgaria, Aug. 2013. pp. 20–28.

CERTIFICATION OF COMPONENT-BASED PARTICLE THERAPY SOFTWARE

Carlos F. Afonso[a b], Luigi Casalegno[b] and Cristiana Larizza[a]
[a] *Department of Electrical, Computer and Biomedical Engineering, University of Pavia, Pavia, Lombardy, Italy*
[b] *Centro Nazionale di Adroterapia Oncologica, Pavia, Lombardy, Italy*

ABSTRACT

The operation of particle therapy facilities requires complex control systems, often spanning several software development environments, each containing large number of applications. While the component-based development approach brings many benefits, institutions often struggle integrating it during the certification process.
As part of the technological update of one of its control system environments, the Centro Nazionale di Adroterapia Oncologica (CNAO) treatment center chose to develop the software applications of this environment in a component-based approach, while focusing on strategies to also optimize the certification process.
In this document, we examine the challenges of compositional development in the medical software certification process. Additionally, the approach taken by CNAO during the update of one of its control system's environments is presented, along with the evaluation of the expected impact on its certification as medical software.

KEYWORDS

Software Engineering, Medical Software, Certification

1. INTRODUCTION

Under the regulatory environment for medical software in member states of European Union, certification is accredited on the basis of complying with the development process defined in the standards IEC 62304+Amendment1(IEC, 2006, 2015), ISO 14971(ISO, 2007), and ISO 13485(ISO, 2016). Since certification is based around complying with the processes defined in the certification standards, applications that share software components are still required perform the certification tasks independently.

Still, several authors propose architecture strategies for efficient validation and certification of critical software that reuses components (Nadkarni and Miller, 2007; Hatcliff *et al.*, 2012). This document focusses on the component-based approach, which proposes an architecture that emphasizes separation of concerns and well defined interfaces (Liu *et al.*, 2005).

In a particle therapy facility for cancer treatment, such as CNAO, control system software performs a wide range of functions such as the control of the accelerator subsystems, dissemination of treatment parameters, and facility configuration. These software packages are often separated into groups, depending on which equipment they run on top of, or their role in the control system. The slow application environment is one of the software environments in CNAO, it is comprised of several applications that share the same development environment and possess similar constraints. Such applications do not generally possess real time constraints and typically perform tasks such as accelerator configuration, data management, or specialized monitoring. At CNAO, applications in the configuration and support environment are developed by the control systems group, are in general expected to receive a low safety class, but are numerous. The work here presented is a part of the OMA project (Cockcroft Institute, 2016) and aims at the design and development of a software infrastructure that the control systems group will exploit for the development of large amounts of such applications.

In this paper we present in detail the compositional approach used in the development of CNAO's configuration and support environment software, which, in addition to improving the reliability and maintainability of these applications, is aimed at reducing the amount of time required for validation and certification steps of the development process.

2. INTEGRATING REUSABLE COMPONENTS INTO MEDICAL APPLICATIONS

In an ideal world, reusable components would be certifiable and then reused without requiring extra certification steps. However, this is not the case in safety critical software (Rushby, 2002). Certification process standards for medical software dictates that risk analysis has to be performed for the software system as a whole, and that risk originated by a component failure can only be determined once the component has had its role defined in regards to the software architecture (IEC, 2006).

Because of this, separation of concerns between software components is not sufficient to allow for reuse of validation and certification artefacts without taking the entire system architecture into consideration. However, well applied compositional development strategies should be able to facilitate the integration of reusable components into new systems (Åkerholm and Land, 2009), and there are several currently adopted industry practices to facilitate the integration of reusable components in safety-critical systems (Land, Åkerholm and Carlson, 2012).

After reviewing the guidelines provided in certification standards and those presented by several authors, we have developed a strategy for designing the new configuration and support environment, outlined in this document. This approach focuses on enabling software reuse, while minimizing the total effort for certification. Later, we evaluate this approach based on the certification standards.

2.1 CNAO's Component-based Development Approach

In order to minimize the effort and tasks necessary for development and certification of future configuration and support environment applications, two main strategies were applied while designing the infrastructure necessary for the software development. The first is the separation of reusable software elements into well-defined *components*, implemented as *libraries*. The second is the definition of a *reference architecture* that is to be followed by all future configuration and support environment applications.

2.1.1 Libraries as Reusable Components

After careful analysis, the following operations have been designed into libraries, becoming reusable components, referred in this document as *standard service libraries*:

- *External communication*, such as database access, local network filesystem access, and communication with other control system equipment, via the OPC-UA (Mahnke *et al.* 2009) protocol.
- *User authentication and authorization* using the OpenID Connect standard (Sakimura *et al.*, 2014).
- Other *standardized operations*, such as system logging and configuration.

2.1.2 Standardization of Software Architecture

In order to provide a set of software patterns that should be followed by developers, we have defined a three-tiered *reference architecture* for configuration and support environment applications. The reference architecture describes each software layer (Presentation, Model, and View Layers), the relationships between each layer, and what software libraries should be used for each part of the final application. By understanding the reference architecture, the facility's developers are able to understand the architectural patterns enforced in the configuration and support environment, and have shared terminology to use when discussing the application's architecture.

A diagram illustrating a class diagram from the reference architecture is shown in Figure 1(a). In order to describe the proposed architecture, we start from the Presentation Layer. The *Presentation Layer* follows the MVVM (Smith, 2009) software pattern. This pattern decouples presentation layer code from domain classes, and allows model and view model classes to be fully unit tested (Smith, 2009), thus aiding the verification process.

We have developed the Framework2020 set of libraries to support the usage of the reference architecture. It provides dedicated base class definitions for MVVM, as well as access control and initialization of the standard services. This is done to assist developer in complying with the reference architecture, as well as segregate the application domain classes from the services used by them.

In the *Domain Layer*, developers should define domain specific classes to implement the final application's features. Very few constraints are placed upon these classes by the architecture, allowing developers to decide how to organize their application. The Specific Domain Classes address the domain requirements of final application. These classes have access to data via the standard services, present results to users via the presentation layer, or communicate/persist the results using the standard service libraries.

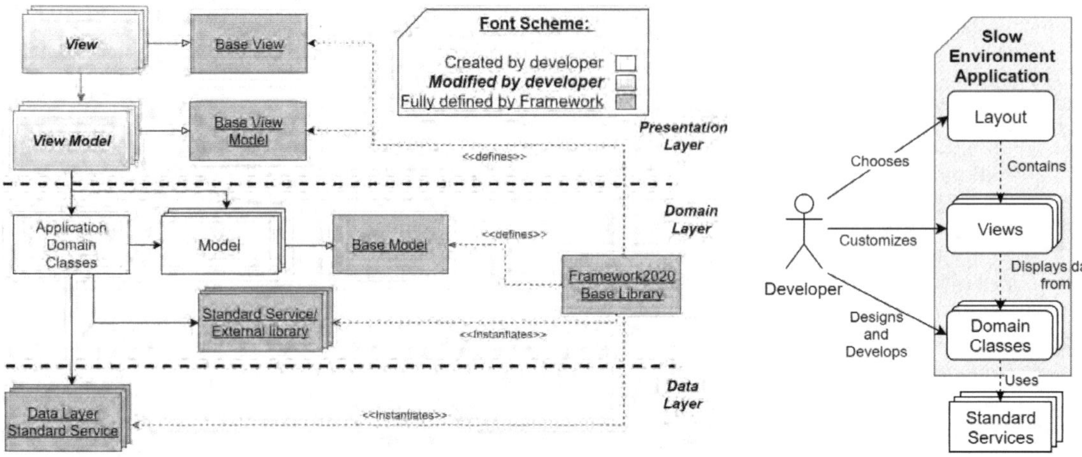

Figure 1. (a) Architecture of a configuration and support environment application (left). (b) Application development lifecycle (right)

2.1.3 Application Skeleton Generator

In order to make the applications development easier and more reliable, developers will be provided with an out-of-the-box configured implementation of the reference architecture, by a configurable skeleton generator, currently in development. It should allow developers to specify the general requirements of their configuration and support environment application, such as the application layout, navigation method and standard services required following the lifecycle depicted in Figure 1 (b). By enforcing the development lifecycle developers have to design their application around a custom-made skeleton that is compliant with the reference architecture.

2.1.4 Evaluation Plan

After reviewing the proposed approach and verifying the usability of the application generator, we have decided to proceed with the development approach, using the previously mentioned reference architecture as basis of the configuration and support environment's new applications. In the following months, after the development of the first applications is concluded, a testing and validation phase begins, in which the medical certification documentation is to be prepared. During this phase, the approach presented in this paper is to be evaluated in comparison to previous medical certification for software projects of the CNAO facility. In particular, the evaluation shall focus on two main aspects: First, instances of reuse of certification artefacts shall be documented, as well as the effort necessary for adapting them. Secondly, the overall certification process shall be compared by querying the certification team about the length and complexity of the certification process.

3. DISCUSSION

The IEC 62304/Amd1 certification standard emphasizes that, when defining the software architecture, strategies for component segregation should be used to avoid unsafe interactions between them. Additionally, when credible arguments can be given to justify that the architecture separates components into separate software items in a correct manner, these software items may receive a different software safety class from the system (IEC, 2006). The certification process has different levels of granularity depending on safety class, and some certification activities may even not be required for low safety levels (Hrgarek, 2012).

By implementing commonly used software operations into reusable components, these become well defined software units, often of the lowest safety class (class A). Thus, several certification tasks are not mandatory for these units, such as the task 5.4.2, entitled "Develop detailed design for each software unit" (IEC, 2006). Additionally, we argue that certification artefacts from developing and certifying reusable components can be the basis for their counterparts in the final application, such as unit-test verification (activity 5.5), and establishment of a software maintenance plan (activity 6.1).

Limitations of the component-based approach were also identified. Namely, the scope of most certification tasks is the software system as a whole, therefore a compositional approach is not able to affect the granularity or necessity of system wide tasks.

Regarding the usage of a reference architecture, we argue that its adoption and enforcement reduces effort required for certification. The importance of knowledge of the system's architecture in the certification process is highlighted multiple times. Such as in the annex B of the IEC 62304 standard, where it is stated that, concerning software safety classification, risks associated with each software item can only be determined once the software architecture "defines the role of the software item in terms of its purpose and its interfaces with other software and hardware items" (IEC, 2006). Furthermore, in the context of the activity 7.1, which is the analysis of software contributing to hazard situations process activity, the guidelines argue that hazard situations can only be fully identified once the software architecture has been designed, and, only then, proposed risk control measures can be completely evaluated (IEC, 2006).

By defining the reference architecture and enforcing it during the design of applications, developers have access to a 'used-and-tested' architecture, accompanied by several internal documents explaining its usage, interaction between components and validation procedures. As developers design configuration and support environment applications using the enforced general architecture, the arguments provided to justify the safe interaction between developer modules and reusable components should be written using the reference architecture's documentation as basis.

Altogether, the authors argue that a composition-based approach provides clear benefits to the development process and that, when managed correctly, reduce the effort related to the medical software certification process. In the future, we plan to empirically evaluate the impact of the presented development approach during the development and certification of the new Accelerator Cycle Designer application.

4. CONCLUSION

We have analyzed challenges of compositional development in the certification process and presented the approach taken by the CNAO facility during the ongoing technological update of its control system.

While composition-based development provides several benefits, its application in medical software creates a few challenges that must be addressed. With the approach presented in this document, we expect to reduce the effort necessary for complying with the certification process of software applications in the new control system environment at the CNAO facility.

We conclude that, while the great majority of certification tasks must still be performed for component-based applications, we have identified several areas where the certification artefacts of components are able to support the activities required for medical certification of these applications. This will be verified during the validation we have planned in the coming months.

ACKNOWLEDGEMENT

This project has received funding from the European Union's Horizon 2020 research and innovation programme under the Marie Skłodowska-Curie grant agreement No 675265.

REFERENCES

Åkerholm, M. and Land, R., 2009, Towards systematic software reuse in certifiable safety-critical systems, *RESAFE-International Workshop on Software Reuse and Safety*, Falls Church, VA, United States.

Cockcroft Institute, 2016, *OMA Project*, Available at: https://www.liverpool.ac.uk/oma-project/about-us/ , Accessed: 24 May 2019.

Hatcliff, J. *et al.,* 2012, Rationale and architecture principles for medical application platforms, *Proceedings of the 2012 IEEE/ACM Third International Conference on Cyber-Physical Systems*, pp. 3–12.

Hrgarek, N., 2012, Certification and regulatory challenges in medical device software development, *Proceedings of the 4th International Workshop on Software Engineering in Health Care*, pp. 40–43.

IEC, 2006, *IEC 62304:2006 - Medical device software -- Software life cycle processes.*

IEC, 2015 *IEC 62304:2006/Amd 1:2015 - Medical device software - Software life-cycle processes / Amd 1.*

ISO, 2007 *ISO 14971:2007 - Medical devices -- Application of risk management to medical devices.*

ISO, 2016 *ISO 13485:2016 - Medical devices -- Quality management systems -- Requirements for regulatory purposes.*

Land, R., Åkerholm, M. and Carlson, J., 2012, Efficient software component reuse in safety-critical systems--an empirical study, *International Conference on Computer Safety, Reliability, and Security*, pp. 388–399.

Liu, J. *et al.,* 2005, Component-based Medical and Assistive Devices and Systems, *Proceedings of High Confidence Medical and Systems (HCMDSS) Workshop*. Philadelphia, PA, United States.

Mahnke, W., Leitner, S.-H. and Damm, M., 2009, *OPC unified architecture,* Springer Science & Business Media.

Nadkarni, P. M. and Miller, R. A., 2007, Service-oriented architecture in medical software: promises and perils, *Journal of the American Medical Informatics Association*, Vol. 14, No 2, pp. 244–246.

Rushby, J., 2002, *Modular certification.* Menlo Park, CA, United States. Available at: https://ntrs.nasa.gov/search.jsp?R=20030001129.

Sakimura, N. *et al.,* 2014, *OpenID Connect Core 1.0 incorporating errata set 1, The OpenID Foundation, specification.* Available at: https://openid.net/specs/openid-connect-core-1_0.html (Accessed: 27 March 2019).

Smith, J., 2009, Patterns-wpf apps with the model-view-viewmodel design pattern, *MSDN magazine*, 72.

SECURE AND EFFECTIVE CONSENT SYSTEM

Hyun-A Park
Department of Medical Science, KyungDong University
WonJu-City, Kangwon-Do, South Korea

ABSTRACT

The data recorded in a medical information system has a distinctive feature from other kinds of sensitive information. For example in EMR(electronic medical record) of HIS(hospital information system), the data subject is the patient and the persons who generated the medical information about the subject are his doctor and other medical practitioners who perform the treatment prescription of the doctor. Hence, before disclosing the data to outside, doctor's decision and the patient's consent should be made first. However, some researchers have developed secure and cryptographic schemes without considering this feature of medical information. In this paper, we propose a secure and effective consent protocol which is applicable to a real world under the current medical laws, where the server manager cannot guess and disclose the data without the patient's consent and the doctor's decision.

KEYWORDS

Patient's Consent, Doctor's Decision, Data Subject, Data Producer, Data Disclosing

1. INTRODUCTION

Generally known as medical information of HIS (hospital information system) or EHR (electronic health records) includes very sensitive and personal things. Nevertheless, it should be devoted as medical data for researches, other statistical surveys or developments of medical systems. The problem at this point is the ownership of data because the data subject is different from data producer in medical information. The data subject is the patient and the producer is his doctor or other medical staffs, so that the methods and processes for managing information and disclosing it to others are different from general sensitive data. When disclosing medical information to the outside, the decision of the data producer and the consent of the patient should be made first, and then the information should be published to the outside after the privacy protection technology is applied.

However, some researchers have proposed security technologies without considering these specificities; the research area of 'patient-controlled (or centered) EHR system' assumes that data owner is the patient and the management/decision on disclosing the data to others is also up to the patient (Szolovits 1994, Eom 2016, Omotosho 2017, Vimalachandran 2017, Seol 2018, Ming 2018, AlThqafi 2016). In (Eom 2016), even a doctor or service provider cannot access the patient's information without patient's consent even if the information was written or produced by the doctor or medical staffs. The authors made it possible by encrypting the information with the patient's private key. That is because the researchers might consider their database model as 'DAS (database as a service), where server manager never know the contents because all cells are encrypted with users' private key, and the data subject and the producer are same as the user. DAS model assumes that server manager is untrustworthy and an inside attacker, so all data are encrypted with users' private key. However, HIS or EHR system have different types of entities; a patient (data subject), a doctor or medical staffs(data producer), a server manager.

In this paper, we propose the consent system for disclosing data to outside, which considers the properties of medical information. Therefore, it can be applicable to current medical laws and systems.

2. PRELIMINARIES

2.1 Security Definitions

Definition 1. PRF(Pseudo Random Function) We say that '$F : Kf \times X \rightarrow Y$ is (t,q,e) -secure pseudorandom function' if every oracle algorithm A making at most q oracle queries and with running time at most t has advantage $Adv_A < e$. The advantage is defined as $Adv_A = | Pr[A^{Fk} = 1] - Pr[A^R = 1] |$where R represents a random function selected uniformly from he set of all maps from X to Y, in which the probabilities are taken over the choice of k and R (Song, 2000).

Definition 2. ***PRG*** G_r **(Pseudo Random Generator)**: We say that *'*G_r*:* $K_{Gr} \rightarrow S$ *is a (t,e)-secure pseudorandom generator'* if every algorithm A with running time at most t has advantage $Adv_A < e$. The advantage is defined as $Adv_A = |Pr[A(G_r(U_{KGr})) = 1] - Pr[A(U_S) = 1]|$. Where U_{KGr}, U_S are random variables distributed uniformly on K_{Gr} and S (Song, 2000).

2.2 Application Scenario

Our application environment will be applicable to both EHR and HIS, if the shared key is set appropriately. This paper takes one example as an EMR system of HIS. The participants are a doctor, a patient, a server manager, and other medical staffs such as a nurse and an examiner.

Each patient has a private key and the shared keys with other medical staffs. A doctor has his private key and the shared keys with each patient, other medical staffs, and a server manager respectively. A server manager has the institute's secret key and the shared keys with each medical staff. Normally, EMR system is encrypted with the doctors' secret key or the shared key between a doctor and other medical staffs. All private keys and shared keys are registered in advance.

When a request is made for information to be disclosed, the doctor decides whether or not to disclose it and then asks his patients the consent for disclosing data to outside. After decrypting the agreeing data, privacy protection technologies are applied to them.

2.3 Notations

Table 1.

Dr.	doctor
Pt.	patient
Sr.	server manager
$d_i p_j$	shared key between a doctor i and a patient j
$d_i s$	shared key between a doctor i and a server manager
a_s	*s-th* attribute
r	random number
α_{pj}	temporal secret key for patient j
α_{cj}	temporal secret key for patient j's consent
f	pseudo random function
S.A	selected attributes
A.A	all attributes

3. CONSTRUCTION

3.1 Whole Protocol

1. Notice : disclosing information and requesting consent
2. Dr. → Pt. : $y = (x - d_i p_j) + r_j$
3. Pt. → Dr. : $f_{r_j}(d_i p_j || S.A(a_1,..a_s,..a_r)||\alpha_{pj})$
4. Dr. → Pt. : $f_{\alpha_{pj}}(S.A(a_1,..a_s,..a_r))$
5. Pt. → Dr. : $f_{\alpha_{pj}}(\alpha_{cj})$
6. System process : $f_{\alpha_{pj}+\alpha_{cj}}(S.A(a_1,..a_s,..a_r))$
7. Dr. → Sr. : $y = (x - d_i s) + \alpha_{cj} || j$
8. Sr. → Dr. & Pt. : $f_{\alpha_{cj}}(A.A(a_1,..a_r,..a_m))$
9. Dr. & Pt. → Sr. : $y = (x - \alpha_{cj}) + \alpha_{pj},\ f_{\alpha_{pj}}(S.A(a_1,..a_s,..a_r))$
10. Sr. : Decrypting S.A and Applying privacy protection technology

3.2 Protocol Description

1. The request of consent for disclosing information is noticed to all members of HIS
2. A doctor i challenges each of his patients with an equation; $y = (x - d_i p_j) + r_j$. The patient j inserts the shared key $d_i p_j$ to the equation, then he gets the random value r_j. This process is a kind of mutual authentication between doctor i and the patient j.
3. A patient j sends the encrypted message with r_j to the doctor i: $f_{r_j}(d_i p_j || S.A(a_1,..a_s,..a_r)||\alpha_{pj})$. This message contains the selected attributes $S.A(a_1,..a_s,..a_r)$ for the patient j to allow the disclosure, and his temporal secret key α_{pj} which is generated randomly.
4. After receiving the encrypted message, the doctor decrypts it with r_j which was generated randomly by the doctor himself in the step 2. He got to know the patient's temporal secret key α_{pj} and encrypted the patient j's S.A with α_{pj}, and sends it to the patient; $f_{\alpha_{pj}}(S.A(a_1,..a_s,..a_r))$. This is also a mutual authentication between the doctor and the patient, and the reconfirming process for the allowed S.A by patient j.
5. The patient j verifies the received message and randomly generates another temporal secret key α_{cj} for consent, which is encrypted with α_{pj} and is delivered to the doctor.
6. A doctor i decrypt a patient j's S.A, then the system automatically encrypts the patient's S.A with a new encryption key $(\alpha_{pj} + \alpha_{cj})$
7. A doctor i challenge a server manager with $y = (x - d_i s) + \alpha_{cj} || j$. This is also a mutual authentication between doctor i and a server manager. Through this process, the server manager gets the temporal secret key for patient j's consent. Here, j means the patient j's ID.
8. The server manager asks doctor i and patient j about the S.A of patient j's consent through encrypted message $f_{\alpha_{cj}}(A.A(a_1,..a_r,..a_m))$ with α_{cj}.
9. The doctor i and the patient j send the equation $y = (x - \alpha_{cj}) + \alpha_{pj}$, which is for mutual authentication through α_{cj} and the provision of another temporal secret key α_{pj} for patient j. And they send the encrypted S.A of patient j with α_{pj} respectively.
10. The server manager authenticates mutually with the doctor i and the patient j and finds new value α_{pj}. He verifies whether the received messages from the doctor i and the patient j are the same or not. All of the above tests are successful, the server manager decrypts S.A with the secret key $(\alpha_{pj} + \alpha_{cj})$ and applies privacy protection technologies to the selected attributes S.A.

3. DISCUSSSION AND CONCLUSION

This protocol is applicable to the current medical laws and medical information systems. At normal time, all data are encrypted with doctor's private key or the shared key between a doctor and other medical staffs, because the doctor or medical staffs are the producer and the manager for the data. However, the data subjects are patients, so that the permission is needed from the data subject when disclosing medical data to the outside. Our proposed protocol shows the doctor's decision and requesting the patient to permit disclosing the data using with their shared key and temporal secret keys. This process enables the server manager not to guess and not to disclose the data without the patient's consent.

As to security respects, the challenging equation makes the mutual authentication and sharing a new key possible. When the shared key in advance is inserted to the equation, the remaining value is a new temporal secret key. The temporal secret keys are generated randomly. With these keys and pseudorandom function, sending messages and data are encrypted. Hence, the security of our protocol depends on it of pseudorandom generator and pseudorandom function.

The proposed protocol is also efficient because all of the used functions are not heavy but light; simple equation and pseudorandom function which falls under the symmetric key encryption method.

Conclusively, our consent protocol can be said available in a real world, secure and efficient. In addition to the consensus system, secure medical information management technologies for normal time and personal information protection technologies for disclosing should be studied to be applicable to current medical laws and medical systems.

ACKNOWLEDGEMENT

This work was supported by the National Research Foundation of Korea (NRF) grant funded by the Korea Government (Ministry of Education, NRF-2017R1D1A1B03029488).

REFERENCES

AlThqafi, N. et al, 2016. *Achieving Patient-Centered Fine-Grained Access Control in Hospital Information Systems - Using Business Process Management Systems*, 9th International Joint Conference on Biomedical Engineering Systems and Technologies, Rome, Italy.

Eom, J. et al, 2016. *Patient-Controlled Attribute-Based Encryption for Secure Electronic Health Records System*, Journal of Medical Systems, Vol. 40, No.12, pp.1-16

Ming, Y. et al, 2018. *Efficient Privacy-Preserving Access Control Scheme in Electronic Health Records System*, Sensors, Vol. 18, No.10, doi:10.3390/s18103520

Omotosho, A. et al, 2017. *Ensuring patients privacy in a cryptographic-based-electronic health records using bio-cryptography*, International Journal of Electronic Healthcare (IJEH), Vol. 9, No. 4, pp. 227 – 254.

Seol, K. et al, 2018. *Privacy-Preserving Attribute-Based Access Control Model for XML-Based Electronic Health Record System*, IEEE Access, Vol. 6, pp. 9114 – 9128.

Song, D. et al, 2000. *Practical techniques for searches on encrypted data*, IEEE Symposium on Security and Privacy, pp.44-55.

Szolovits, P. et al, 1994. *Guardian angel: Patient-centered health information systems*, Technical report, Cambridge, MA, USA.

Vimalachandran, P. et al, 2017. *Preserving Patient-centred Controls in Electronic Health Record Systems: A Reliance-based Model Implication*, ICOT2017, Singapore.

HOW INSPIRING IS YOUR APP? A USABILITY TAKE ON AN APP FOR ASTHMA MEDICATION ADHERENCE

Ana Ferreira[1,2], Rute Almeida[1,2], Cristina Jácome[1,2], José Fernandes[4], João A. Fonseca[1,2,3] and Pedro Vieira-Marques[1,2]
[1]CINTESIS - Center for Health Technology and Services Research
[2]Departamento Medicina da Comunidade, Informação e Decisão em Saúde (MEDCIDS)
Faculty of Medicine, University of Porto, Portugal
[3]Medicina, Educação, I&D e Avaliação, Lda (MEDIDA), Porto, Portugal
[4]Bloomidea, Braga, Portugal

ABSTRACT

The low level of awareness and usability of existing mobile health (mHealth) applications still hinders people, especially patients, to take advantage of such applications to improve their daily lives and health outcomes. This is also true for patients with asthma however, the use of smartphones and gamification, can be a solution to change behaviours regarding inhaled medication adherence and to improve asthma monitoring closer to the patients' needs. This paper presents a usability study of *InspirerMundi*, an mHealth gamified app developed to include various monitoring functionalities to improve inhaled medication adherence in adolescents and adults with asthma. A well-defined protocol with specific tasks was applied to 8 participants, who never used the app before nor had asthma (4 female and 4 iOS users), and results show that it was possible to find the most pressing issues regarding usability, right from the beginning. This work helps to realize that mHealth applications need to be tested for usability and functionalities adequacy with a varied sample closely related to technology and personal characteristics (e.g., type of smartphone, gender, age, mobile game-oriented), for which those applications need to be prepared.

KEYWORDS

Usability, Mobile Health Application, Asthma Medication Adherence

1. INTRODUCTION

The rapid growth of mobile health (mHealth) applications (apps) has allowed new ways to improve medication control, monitorization and health promotion. However, there is still a lot to be done regarding the level of awareness and usability of such apps (Kayyali et al., 2017) to make sure that: a) patients know more/better ways to treat their condition; and b) apps do not go into the market without being tried for their majority of problems (Boulos et al., 2014). Especially in the healthcare domain, effects of using mobile apps can be crucial as the lack of standardization and adequate scientific validation can cause serious problems to the user's physical and mental health (Parati et al., 2017). In terms of respiratory diseases (e.g., asthma), patients' low adherence to inhaled medication is one of the main factors that can hinder the clinical improvement and maintenance of patients' quality of life (Morton et al., 2014) (van Dulmen et al., 2007). Applications that can help this to be concretized are critical. The use of smartphones and gamification approaches can be relevant options to consider for changing behaviours to improve asthma medication adherence closer to the patients' needs (Haase et a., 2017) (Marcano et al., 2013) (Morton et al., 2017).

With all this in mind, the mHealth app *InspirerMundi* was developed with various monitoring and gamification functionalities aiming to improve inhaled medication adherence in adolescents and adults with asthma (Jácome et al., 2017). This paper presents a first round of usability tests that were performed on the *InspirerMundi* app in order to identify: functionalities that are hard to execute, lack of clarity in language and design as well as the average time tasks take to complete, to identify areas that need improvement.

2. METHODS

2.1 Sampling

This was a cross-sectional study. The participants were recruited using a convenience sampling procedure and were included if they were over 18 years old, frequently used a smartphone, did not have asthma and had never used the *InspirerMundi* app before. The focus was to have a balanced sample regarding gender and the use of different operating systems users (e.g., iOS and Android). The sample integrated non-asthmatic people to avoid previous users experience both with such apps and different type of inhaled medication as well as any other type of emotional and behavioural involvement.

2.2 Protocol and Materials

The study protocol comprised 4 steps: 1) researchers presented the study and its objectives to the participants by providing the same information they would find when downloading the app from the online stores[1]; 2) participants read and signed an information consent describing the main goal of the study and how collected data would be processed and analysed; 3) participants performed nine pre-defined tasks (Section 2.3) with smartphones provided by the research team, which were video recorded with a web cam and *Mr. Tappy* support (Figure 1); and, at the end, 4) participants filled a usability and satisfaction questionnaire.

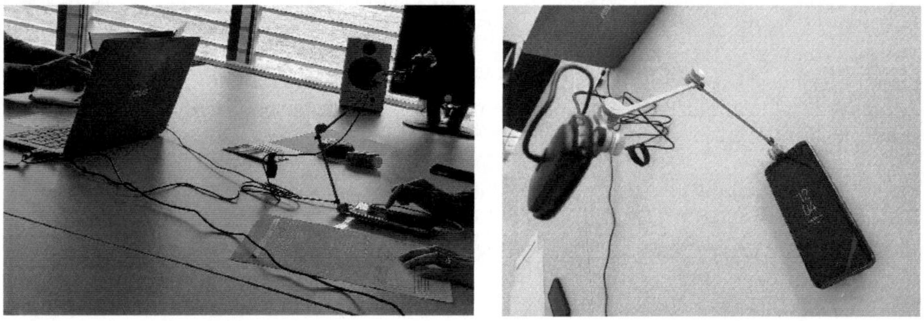

Figure 1. Setup of the video recording of the participants' performance and smartphone display

2.3 Performed Tasks and Survey

The nine tasks focused on activities to be daily executed by patients who want to improve their adherence to asthma medication. These were: 1) login with anonymous credentials; 2) answer the Control of Allergic Rhinitis and Asthma Test (CARAT) (Fonseca et al., 2010); 3) answer a weekly symptoms survey; 4) insert two inhaled medications and one pill medication; 5) register the taking of the three previously inserted medications; 6) modify the posology of one of the inserted medications; 7) register an SOS medication; 8) make an Inspirer request; and 9) check the statistics for the medication adherence. Every task was completed by two experts (one using Android, the other iOS), who knew the application, to create a baseline for each task duration. During the study, no help was provided to the participants when performing the tasks. There were two observers in the room and the participants could change the sequence of the tasks' execution and give up or stop whenever they thought fit, without any intervention or suggestion from the observers.

An anonymous post-study survey was given to the participants and consisted of four parts: i. demographic data; ii. use of smartphones; iii. satisfaction and usability of the app; and iv. security and privacy issues.

[1]https://play.google.com/store/apps/details?id=com.bloomidea.inspirers&hl=en or https://itunes.apple.com/pt/app/inspirermundi/id1441138269?mt=8

3. RESULTS

3.1 Participants

The sample comprised eight participants (a number that helps to initially detect between 50 to 80 percent of the problems the app may have (Ritch, 2009)) with an average of 36 (28-44) years old, 50% were female, 50% were Android users and all of them had a high education degree and frequently use a smartphone.

3.2 Tasks Scores

Most tasks were successfully performed by all the participants (7 out of 9), however, one of the most relevant ones (Task 5) had two failures and 4 partial successes. This task implied the registration of a video image of the taken medication and only when that medication was in the format of a pill case did the participants understood that they were supposed to use the smartphone camera to make a short video/photograph. Another issue concerned Task 6, where it was not obvious to the participants how this could be achieved since the functionality that allows to do it was "hidden" and not close to the available list of medications. Figure 2 shows the variability in time needed to complete a task when comparing the users and the experts' performance. Tasks 5, 6 and 7 are the highest duration tasks involving more interactions with the app from the participants, probably for the reasons already presented.

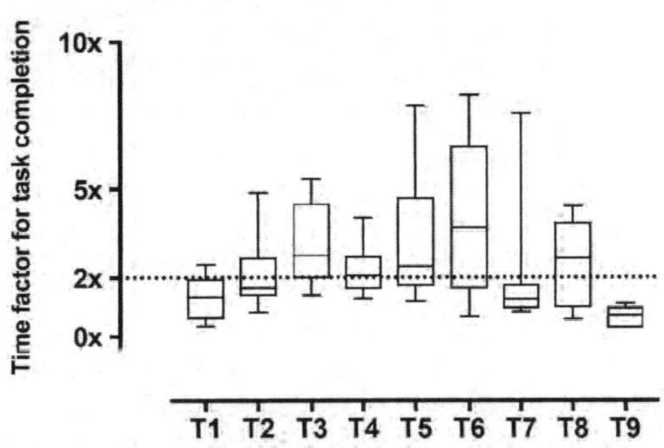

Figure 2. Time factor for task completion (experts vs participants) (T5 was finished by 6 participants and T6 by 5)

3.3 Post-Study Survey

Most participants use smartphones mainly to: a) access the Internet (more Female and Android users); b) use mobile apps (more Male and Android users); c) check social networks (more Female and IOS users as well as non-gamers); and d) to take pictures (more Female and IOS users as well as non-gamers).

Since the evaluated app focus on gamification elements to promote adherence to medication, the main usability aspects that need improvement are detailed for gamers and non-gamers in Table 1. Despite some differences, both types of users agreed that learning and application usage and app interactions were critical to improve. Regarding the security issues, some participants avoid giving similar apps access to the stored contacts and photos (more Female, IOS and non-gamers users) while all the participants had no problems in allowing those apps accessing their location as well as sending them notifications.

Table 1. Critics of usability characteristics depending on the type of participant (mobile gamer or not)

	Never plays games in smartphone		**Regularly plays mobile games**	
Usability characteristics	More critical	Less critical	More critical	Less critical
Size and disposition of buttons/icon/menus/content are appropriate		X	X	
Quality of buttons/icon/menus/content				X
Functionalities, menu and button components are efficient			X	
Learning and application usage is fast	X		X	
Language used in instructions and menus/icons is clear				X
Display changing is correct/logical/appropriate and without interruptions	X			X
Interactions are consistent and intuitive	X		X	
App is fun to use and includes interesting gaming components		X	X	

4. DISCUSSION AND LESSONS LEARNED

In terms of usability of the mHealth *InspirerMundi* app, this small study already gave important clues on what were the main issues to solve regarding two of the main features. One of the core features, the recognition of the inhaled medication, was not that straightforward to grasp and so needed attention. Besides being a common gesture in mobile app usage (taking a photo or making a video) the subject being captured is not. The inclusion of an explanatory description with images/video before the actual process takes place will allow for the first-time users to better understand what is expected. This shows how important these types of studies are before releasing the apps and how they can help to identify a lot more issues than anticipated. Another important aspect is the fact that the study was applied to a sample population completely alienated from the theme at hand which, besides avoiding some cognitive or behavioural bias, made us have more confidence in the obtained results as there was no pressure for those participants to do well on that specific app. To also notice that an average expert baseline time for the performed tasks helps comparing times and efficiency on a normal daily usage. In some tasks the users took on average between 1,5 to 2 times more to complete when compared with experts, this situation is natural as the interface is new to the users and is expected that after the initial steps the difference will decrease. However, this can also be a good indicator regarding what aspects of the app need to be improved to prevent frustration and dropout on the initial stages of its use. Finally, although there is the necessity to focus on the basic functionalities first, allowing the removal of obstacles that may hinder the correct assessment of the app's effective purpose, there is the necessity of assessing gamification components, that will need, nonetheless, a more extensive period of testing.

For the satisfaction, perception and attitudes of study participants towards the app, it was clear that the majority of users were satisfied with most functionalities however, different attitudes arose for mobile gamers. Non-gamers focus more on the consistency and easiness of actions, while gamers also focus on those characteristics, plus on the aesthetics aspect such as size and disposition of graphical components and how useful and fun they are. This was expected, so, if the app is a health serious game, recommendations are to tackle usability issues preferably with a sample integrating both views.

In terms of security issues, although the number of security questions was very small, results show that gender, together with the type of smartphone (e.g., iOS or Android) or mobile game habits, can influence the perception of providing identifiable personal data. This must be further analysed in future research.

Limitations comprise the use of an equipment (e.g., Mr. Tappy) that was not helpful to perform the task of video/photographing an object. Nevertheless, this may only partly explain the problems executing this task, as the equipment was used only with three participants. Also, only healthy adults were included and it was not possible to compare their performance with patients with asthma, who need to use such app on a daily basis in the interest to improve their symptoms and quality of life. Future studies should include participants both healthy and with asthma, from distinct age groups.

5. CONCLUSION

This work helped to understand how mHealth apps, integrating gamification, need to be assessed not only for the basic functionalities, but also to the adequacy of each individual user. In spite that participants were not patients with the condition addressed by the app, they can have other relevant characteristics to be taken in consideration (e.g., gender, age, type of smartphone, game-oriented) and for which this type of apps need to be prepared. Future work includes performing a longer study with patients with asthma, from adolescents to older adults, in order to better adequate and personalize the app according to their lifestyles and needs.

ACKNOWLEDGEMENT

This work was financed by ERDF (European Regional Development Fund) funds through the COMPETE 2020 - Operacional Programme for Competitiveness and Internationalisation (POCI), and by Portuguese funds through FCT - Fundação para a Ciência e a Tecnologia in the framework of the project POCI-01-0145-FEDER-029130 ("mINSPIRERS—mHealth para medição e melhoria da adesão à medicação nas doenças respiratórias obstrutivas crónicas—generalização e avaliação de tecnologias de gamificação, suporte por pares e processamento avançado de imagem").
It was also supported by FCT through the Project TagUBig - Taming Your Big Data (IF/00693/2015) from Researcher FCT Program funded by National Funds through FCT - Fundação para a Ciência e a Tecnologia.
In addition, Cristina Jácome has a post-doctoral grant (SFRH/BPD/115169/2016) funded by FCT, co-financed by the European Social Fund (POCH) and Portuguese national funds from MCTES (Ministério da Ciência, Tecnologia e Ensino Superior).

REFERENCES

Boulos, M. N., Brewer, A. C., Karimkhani, C., Buller, D. B., & Dellavalle, R. P. (2014). Mobile medical and health apps: state of the art, concerns, regulatory control and certification. Online journal of public health informatics, 5(3), 229. doi:10.5210/ojphi.v5i3.4814.

Fonseca JA, Nogueira-Silva L, Morais-Almeida M, Azevedo L, Sa-Sousa A, Branco-Ferreira M, et al. (2010). Validation of a questionnaire (CARAT10) to assess rhinitis and asthma in patients with asthma. Allergy. 65:1042-8.

Haase J, Farris KB, Dorsch MP. (2017). Mobile Applications to Improve Medication Adherence. Telemed J E Health. 23:75-9.

Jácome C, Almeida R, Teixeira J, Vieira-Marques P, Vilaça R, Fernandes J, et al. (2017). Inspirers: An app to measure and improve adherence to inhaled treatment. Proceedings of the International Conference e-Health 2017; Lisboa.135-9.

Kayyali, R., Peletidi, A., Ismail, M., Hashim, Z., Bandeira, P., & Bonnah, J. (2017). Awareness and Use of mHealth Apps: A Study from England. Pharmacy (Basel, Switzerland), 5(2), 33. doi:10.3390/pharmacy5020033.

Marcano Belisario JS, Huckvale K, Greenfield G, Car J, Gunn LH. (2013). Smartphone and tablet self-management apps for asthma. The Cochrane database of systematic reviews. CD010013.

Morton RW, Elphick HE, Rigby AS, Daw WJ, King DA, Smith LJ, et al. (2017). STAAR: a randomised controlled trial of electronic adherence monitoring with reminder alarms and feedback to improve clinical outcomes for children with asthma. Thorax. 72:347-54.

Morton RW, Everard ML, Elphick HE. (2014). Adherence in childhood asthma: the elephant in the room. Arch Dis Child. 99:949-53.

Parati, G., Torlasco, C., Omboni, S. et al. (2017). Smartphone Applications for Hypertension Management: a Potential Game-Changer That Needs More Control. Curr Hypertens Rep. 19: 48. https://doi.org/10.1007/s11906-017-0743-0.

Ritch Macefield. 2009. How to specify the participant group size for usability studies: a practitioner's guide. *J. Usability Studies* 5, 1 (November 2009), 34-45.

van Dulmen S, Sluijs E, van Dijk L, de Ridder D, Heerdink R, Bensing J. (2007). Patient adherence to medical treatment: a review of reviews. BMC Health Serv Res. 7:55.

METHODOLOGY FOR DESIGNING AAL-IOT SOLUTIONS FOR OLDER ADULTS

Roberto Omar Andrade, Sang Guun Yoo and Jhonattan J. Barriga A.
Facultad de Ingeniería de Sistemas, Escuela Politécnica Nacional, Ecuador

ABSTRACT

Population growth worldwide is considered a health issue in this moment and it will have greater impact in the following years due to the limited number of specialists in the medical field. Health organizations present a special concern in older adults over 65 years because most of them live alone and their growth rate is the highest one in the world. In this circumstance, AAL solutions will become an important contribution to the society since those solutions will provide support to older adults. In this sense, the use of IoT is allowing to provide better alternatives and new features to AAL solutions. However, IoT still presents unsolved security risks that must be managed, especially for AAL solutions, since they manage sensible information related to the privacy and health of users. In this research project, our interest is to propose a methodology for designing AAL-IoT solutions according to physical, logical, procedural and legal aspects of security that could guarantee the privacy and life improvement of the older adults.

KEYWORDS

IoT, AAL, e-Health, Methodology

1. INTRODUCTION

According to United Nations Organization, by the year 2050, world population will grow approximately 2.1 billion of people in relation to 2017. Figure 1 shows the worldwide population growth. We can see that the group of people over 65 years of age grow faster than all younger age group (Zhu, 2015). People over 65 years old are the 13 % of the world's population and this age-group has a growth rate of 3 % per year. Another important aspect is the "Expectancy Life", which has increased from 77.8 years old in 1940 to 79.4 in 1960, 81.5 in 1980 and 84.3 in 2017; this means that 65 years old people are expected to live another 19 years more (Sinha, 2017; Malavasi,2017). The concern generated in health organizations with respect to older adults is not just focused only on the type of diseases that they could suffer such as diabetes, hypertension and heart disease but in the high growth rate of such diseases in the older adults' group (Mulero, 2018).

The implementation of Ambient Assisted Living (AAL) solutions using Internet of Things (IoT) technology has been considered to improve the quality of life of older adults and increase the length of time that this group of people can live independently (Medrano-Gil, 2018). For example, the SPHERE project introduce an AAL solution that combines e-Health with Internet of Things (Tsirmpas, 2017). SPHERE includes sensor data acquisition, ontologies, interoperability, and privacy. Medical devices based on IoT solutions uses communications infrastructures such as LORA, Sigfox and NarrownBand IoT. Summarized technical characteristics of IoT technologies are shown in Table 1. Risk factors are associated to different aspects such as smoking, lack of exercise, high blood pressure, high cholesterol, and obesity that can affect the health of older adults; those aspects are important to be considered for continuous monitoring (Mathisa, 2015; Kerchner, 2016; Cooper, 2015; Sanai, 2016; Sohrabji, 2016; Clark, 2018).

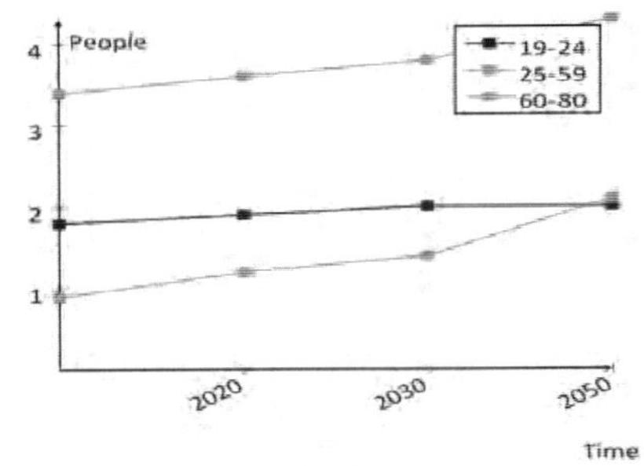

Figure 1. Population by Age-Groups (Thousands of Millions)

Table 1. Communications Technologies for IoT

	LORA	SIGFOX	NB–IOT
Spectrum (MHz)	863-870	868.1-868.3	700-900
Bandwidth (KHz)	125	0.1	180
Max Payload	51	12	128
License	Unlicense	Unlicense	License

2. IC-IV METHODOLOGY PROPOSAL FOR AAL-IOT

AAL solutions uses physiological, biochemical and environmental (temperature, lighting, motion, and pressure) sensors for recognizing activities and behavior of older adults with the goal of executing actions for maintaining or improving their quality of life (Panwar, 2017). AAL can directly or indirectly infer the physical or cognitive conditions of a person in real-time (streaming) based on the analysis of information obtained from the sensors. This kind of solutions can alert family members or medical personnel to reduce the risk of dangers of the people. AAL-IOT solutions includes three aspects that are shown in figure 2. The design of an AAL-IoT solution should ensure compliance with technical, security, privacy and legal issues (Paneni, 2017; Pierce, 2017; Chatterjee, 2018; Incalzi, 2004; Jiang, 2016; Raji, 2016; Von Chong, 2019).

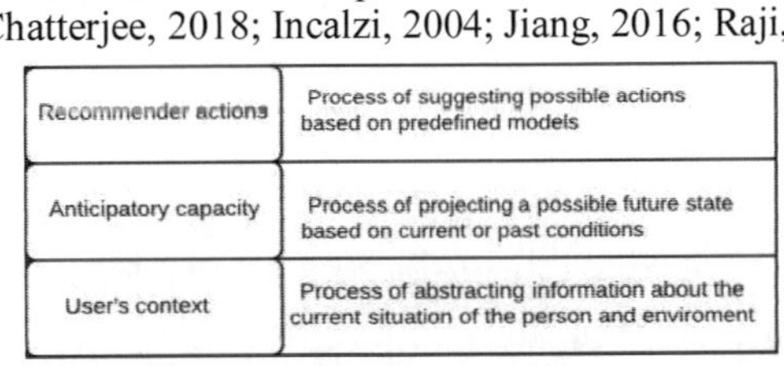

Figure 2. Functional Aspects of e-Health Solutions based on IoT

In this work, based on the review of different AAL solutions presented in the academic field, we propose a design methodology that considers a top-down approach that include technical, security, functionality and compliance aspects related to variables of the health conditions. The proposed methodology has been called IC-IV (Identify-R, Characterize-Identify-T, Validate) and consists of four phases. The first phase "Identify-R" contemplates a process of identification of technical, security, and legal aspects related to medical information handling. The second phase "Characterize" considers the characterization process that allows to abstract the information related to the medical condition and behavior pattern of the person. The third phase "Identify-T" focuses on identifying the IoT components that can be used to implement AAL, including technical aspects such as the location and type of the IoT devices, as well as the communication protocols. The fourth phase "Validate" considers the establishment of parameters that allow to validate the technological, security and operation compliance of the proposed solution for generating a functional prototype. The figure 3 presents the six processes defined for the "Identify-R" phase. The first process is focused on determining the environmental variables such as temperature, humidity, light and smoke that are found in the place where the older adult lives. The second process is related to determining parameters of behavior or activity of the older adult, such as time they spent in a room, physical areas for walks, and number of times they leaves the house. From these two processes, the most relevant vital signs and risk factors are identified. The following three lower processes are focused on identifying local, national and international regulations.

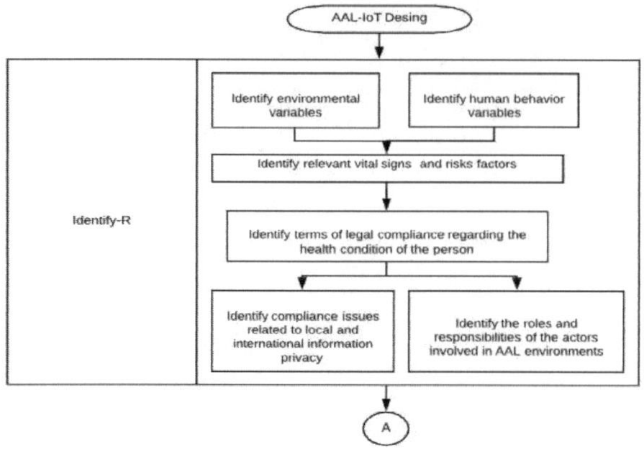

Figure 3. Identify-R Phase for AAL-IoT design

The second phase of the methodology "Characterization" is show in the figure 4. The scope of this phase is the abstraction of information from the characteristics of the activities, risk factors and vital signs of the older adults that allow determine an affectation in the health or a physical or cognitive problem.

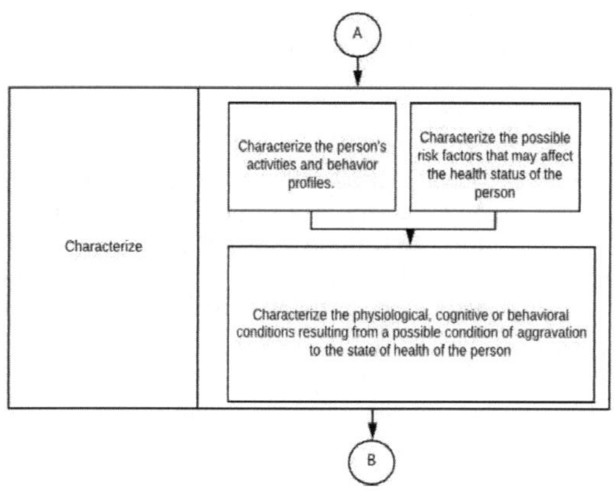

Figure 4. Characterize Phase for AAL-IoT design

The third phase "Identify-T" is shown in figure 5. Its scope includes the identification of the communication devices and protocols that can be used to implement the AAL-IoT solution. For this phase, three types of Body Area Networks (BAN) are considered i.e. Intra-BAN, Inter-BAN and Extra-BAN (Beyond) networks. The AAL solutions have different actors and components. Machine to Machine (M2M), Person to Machine (P2M), and Device to Device (D2D) communications are present; so these interactions imply that sensitive information of older adults should be protected in the involved devices and communications protocols (Dawaliby, 2016; Lippuner, 2018; Ratasuk, 2016; Khurpade, 2018).

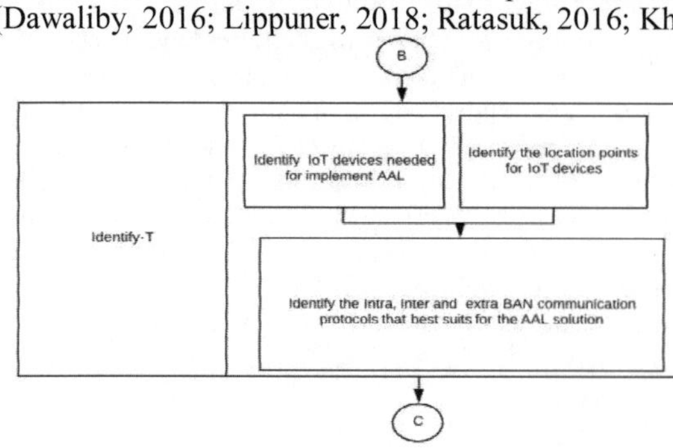

Figure 5. Identify-T Phase for AAL-IoT design

The fourth phase "Validate" is show in figure 6. It considers the validation processes for AAL-IoT. The result of this phase is the creation of a function that accomplishes the technical, security and functionality aspects.

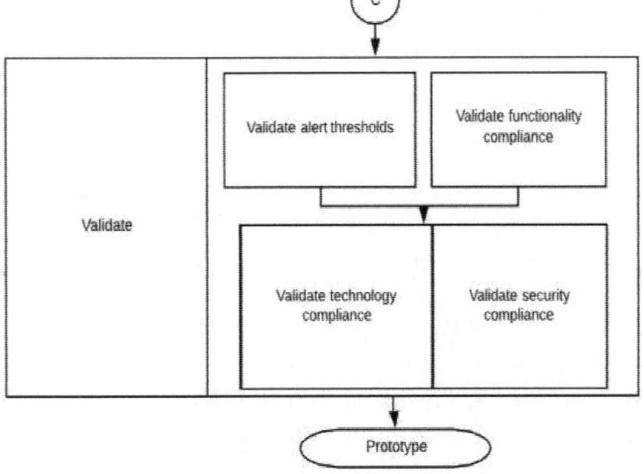

Figure 6. Validate Phase for AAL-IoT design

3. CONCLUSION

The impact of technology in the healthcare field is becoming more important each day, especially for older adults since many of them are living alone in their homes far away from friends and family. The AAL solutions based on IoT is proving a part of the solution of such problem making possible to send alerts in case of emergencies or automate devices for helping older adults in case of injury or health issue. In this sense, it have been created different IoT solutions that can be used in the implementation of AAL. This situation generates the need for architectures with heterogeneous components, and therefore the need for having a methodology that allow taking into account the medical, technical, security and legal aspects that must be considered in the design phase of an AAL solution based on IoT.

The proposed methodology follows a top-down approach, that start from the understanding of the needs of the older adult, their daily activities, and environmental factors and current medical conditions. Depending on the categorization of aforementioned characteristics the abstraction is executed to determine the IoT devices, communication protocols and security aspects that make possible the designing of a personalized AAL-IoT solution for each person. We have proposed a methodology adapted from the IoT reference model created by ITU in their Y2060 recommendation. We have considered the importance of making decisions based on the analysis of the data provided by IoT sensors. We have added an "Analytical Layer" that covers the processes of acquisition, maintenance, testing and management of data, and "IoT Governance" feature to cover the governance issues that must be managed in each layer of the IoT architecture such as policies and definition of roles and responsibilities.

REFERENCES

Chatterjee, S., Mudher, A. Alzheimer's disease and type 2 diabetes: A critical assessment of the shared pathological traits. *Front. Neurosci.* 12, 383, DOI: 10.3389/fnins.2018.00383 (2018).

Clark, B. C. Neuromuscular changes with aging and sarcopenia. *The J. Frailty & Aging* 8, 7–9, DOI: 10.14283 /jfa. 2018.35 (2019).

Cooper, Jason F., Dues Dylan J., Spielbauer, Katie K., Machiela, E., Senchuk, Megan M. Delaying aging is neuroprotective in parkinson's disease: a genetic analysis in c. elegans model. Nat. Res. 1–15, DOI: 10.1038/npjparkd.2015.22 (2015).

Dawaliby, S., Bradai, A., Pousset, Y. In depth performance evaluation of lte-m for m2m communications. In *2016 IEEE 12th International Conference on Wireless and Mobile Computing, Networking and Communications (WiMob)*, 1–8, DOI:10.1109/WiMOB.2016.7763264 (2016).

Incalzi, R. A. Management of chronic obstructive pulmonary disease in the elderly. *Aging Clin. Exp. Res.* 16, 13–21, DOI:10.1007/BF03324526 (2004).

Jiang, M. *et al.* Iot-based remote facial expression monitoring system with semg signal. In *2016 IEEE Sensors Applications Symposium (SAS)*, 1–6, DOI: 10.1109/SAS.2016.7479847 (2016).

Kerchner, G. A., Wyss-Coray, T. *The Role of Aging in Alzheimer's Disease*, 197–227 (Springer International Publishing, Cham, 2016).

Khurpade, J. M., Rao, D., Sanghavi, P. D. A survey on iot and 5g network. In *2018 International Conference on Smart City and Emerging Technology (ICSCET)*, 1–3, DOI: 10.1109/ICSCET.2018.8537340 (2018).

Lippuner, S., Weber, B., Salomon, M., Korb, M., Huang, Q. Ec-gsm-iot network synchronization with support for large frequency offsets. In *2018 IEEE Wireless Communications and Networking Conference (WCNC)*, 1–6, DOI: 10.1109/WCNC.2018.8377168 (2018).

Malavasi, M. *et al. An innovative speech-based interface to control aal and iot solutions to help people with speech and motor disability. Ambient Assist. Living. ForItAAL 2016.* 426, 269–278 (2017).

Mathis, S., Vallat, J.M, Ingrand, P., Neau, J.P, Bouche, P. Causes of neuropathy in the elderly: A retrospective study with 785 patients. *Eur. Geriatr. Medicine* 1–15, DOI: 10.1016/j.eurger.2015.02.006 (2015).

Medrano-Gil, A. *et al. Definition of technological solutions based on the internet of things and smart cities paradigms for active and healthy ageing through cocreation. Wirel. Commun. Mob. Comput.* 1–15, DOI: 10.1155/2018/1949835 (2018).

Mulero, R. *et al.* An iot-aware approach for elderly-friendly cities. *IEEE Access* 6, 7941–7957, DOI: 10.1109 /ACCESS. 2018.2800161 (2018).

Paneni, F., Diaz Cañestro, C., Libby, P., Lüscher, T. F. & Camici, G. G. The aging cardiovascular system. *J. Am. Coll. Cardiol.* 69, 1952–1967, DOI: 10.1016/j.jacc.2017.01.064 (2017). http://www.onlinejacc.org/content/69/15/1952.full.pdf.

Panwar, L. S., Kala, S., Panwar, V., Panwar, S. S. & Sharma, S. Design of mems piezoelectric blood pressure sensor. In *2017 3rd International Conference on Advances in Computing,Communication Automation (ICACCA) (Fall)*, 1–7, DOI:10.1109/ICACCAF.2017.8344698 (2017).

Pierce, G. L. Aortic stiffness in aging and hypertension: Prevention and treatment with habitual aerobic exercise. *Curr. Hypertens. Reports* 19, 90, DOI: 10.1007/s11906-017-0788-0 (2017).

Raji, A., Kanchana Devi, P., Golda Jeyaseeli, P. & Balaganesh, N. Respiratory monitoring system for asthma patients based on iot. In *2016 Online International Conference on Green Engineering and Technologies (IC-GET)*, 1–6, DOI:10.1109/GET.2016.7916737 (2016).

Ratasuk, R., Vejlgaard, B., Mangalvedhe, N. & Ghosh, A. Nb-iot system for m2m communication. In *2016 IEEE Wireless Communications and Networking Conference*, 1–5, DOI: 10.1109/WCNC.2016.7564708 (2016).

Sanai, S. A. *et al.* Aging and multiple sclerosis. *Multiple Scler. J.* 22, 717–725, DOI: 10.1177/1352458516634871 (2016). PMID: 26895718, https://doi.org/10.1177/1352458516634871.

Sinha, R., Yiqiao, W. & Hwang, S.-H. *A survey on lpwa technology: Lora and nb-iot. ICT Express* 3, DOI: 10.1016 /j.icte. 2017.03.004 (2017).

Sohrabji, F. *The Impact of Aging on Ischemic Stroke*, 161–196 (Springer International Publishing, Cham, 2016).

Tsirmpas, C. *et al. An internet of things platform architecture for supporting ambient assisted living environments. Technol. Heal. Care.* 25, 391–401, DOI: 10.3233/thc-161270 (2017).

Von Chong, A., Terosiet, M., Histace, A., Romain, O. Towards a novel single-led pulse oximeter based on a multispectral sensor for iot applications. *Microelectron. J.* 1–15, DOI: 10.1016/j.mejo.2018.03.005 (2018).

Zhu, N. *et al. Bridging e-health and the internet of things: The sphere project. IEEE Intell. Syst.* 30, 39–46, DOI: 10.1109/MIS.2015.57 (2015).

Reflection Papers

A NOVEL 4-SENSOR SYSTEM FOR FALL DETECTION AND ABNORMAL MEDICAL SITUATIONS IDENTIFICATION

Gairat Kuttumuradov, Arslan Mukhamatnurov, Nurmukhamed Tursumbek and Laurent Billonnet
University of Limoges, Limoges, France

ABSTRACT

Many organizations and industries already provide turnkey solutions for the detection of falls, but many of these systems are too expensive. Most of people like living independently at home. However, many people are not able to afford expensive systems to detect or prevent falls. Many people are not even thinking about using it. Some activities in our daily life can lead to accidents, such as falls. The fall are defined as unintentionally coming to rest on the ground, floor or other lower levels. Falls can lead people to fatal conditions, even death. We propose here a four-sensor concept to improve the efficiency of the system and reduce its global cost. A basic concept of fall detection system using three inclinometers and a heartbeat sensor based on Smartphone is presented. Inclinometers and heartbeat sensors are worn in different part of human body to detect fall more accurately. Automatic alert call will be sent to family members or caregivers when a fatal condition is identified. Our system is quite innovative since it can distinguish several situations critical or not by combining data coming from the four sensors. Thus, the system can also monitor daily activities to improve daily life of the users.

KEYWORDS

Inclinometer, Heartbeat Sensor, Falls, Abnormal Situation

1. INTRODUCTION

The fall is an accident that threatens health, especially for older people. Fall detection has thus become an important research field. Every year around the world, about 646,000 people die from falls, of which more than 80% are in low and middle-income countries. Although not fatal, an estimated 37.3 million falls become serious enough to demand medical care every year. Very often, the high mortality rate among older people is due to a fall and falls are the second leading cause of death from accidental or unintentional injuries worldwide. This problem affects older people aged 65 and above. The reason is the decrease in strength and stability of the human body. Preventing a fall is also a difficult problem. At least, a reliable detection of falls can prevent the worst consequences (severe physical or psychological impacts and in some cases death).

Nowadays, different types of fall detection system based on cameras exist. But most of people worry about the privacy of information of themselves so that no one wants to stay under camera in his daily living. However, this is not the main reason to doubt to this fall detection system. An omni-camera is a camera with 360-degree field of view. Successful rate of this system is approximately 80%, which means that in 1 case out of 5, the fall is not detected and can lead to user's death. The causes of falls are often the result of complex interactions between many factors. World Health Organization has identified more than 400 factors that contribute to a fall. In addition, falls of elderly result from the effect of additive factors.

Thus, screening the factors may help reduce the risk of falling. In fact, the risk of falling and the number of factors are linearly correlated. Risk factors are often classified in three dimensions. There are factors related to the behaviour, other factors linked to the health status of the elderly and finally factors linked to the environment (extrinsic factors). The first two dimensions can be merged together. They score the factors associated to the person (intrinsic factors). Falls happen because of the interaction among these dimensions as illustrated in figure 1.

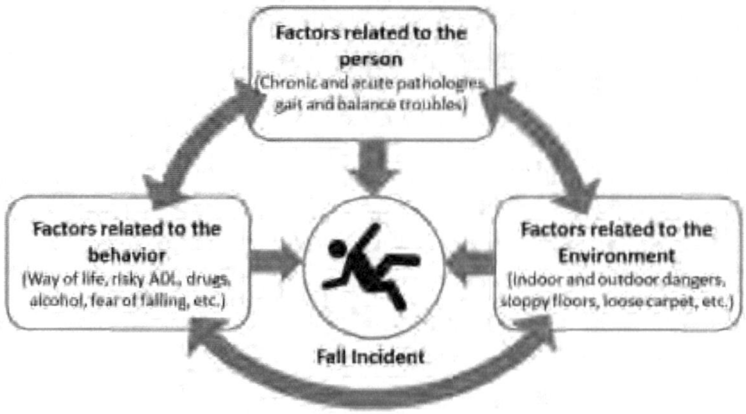

Figure 1. Three dimensions of fall risk factors

Because people usually want to feel safe in their house, indoor system can be more relevant. However, probability of the fall increases outside. In this case, wearable devices have to be integrated in the global solution. It means that wearable device has to be useful and comfortable to wear. Luckily, the development of wearable devices and mobile Internet gives chance to deal with the problem. Wearable devices based on health care have seeped in our daily life and changed our lifestyle a lot with great benefit.

2. DATA TRANSMISSION

Our system is based upon a 4-step process: data collection, data transmission, data processing and fall alert.

Modern Smartphones have excellent information processing via to their advanced processors. By collecting data from the inclinometers and sending it to the Smartphone, the Smartphone's application is able to analyse the data from the inclinometers and extract decision parameters. To accurately determine the fall, the system reads data coming from the inclinometers every second. Otherwise, we should expect a delay of several seconds, which will adversely affect the quality of this system. After processing of information, the application recognizes the user's position and situation and in case of alert criterion identification (i.e., the parameters recognizing a True positive situation of fall) the system sends a fall alert in real-time.

In addition, the heartbeat sensor gives more accuracy and relevance. Its focus is the determination of abnormal medical situations in the pre fall phase. This is very important since it proves that our system can also prevent from falls, or, at least, identify the high probability of fall before it occurs.

The connection between the sensors will be established by Bluetooth low-energy, which allows the sensors to transmit information at a frequency of 2.4 GHz with low power consumption. The packet transfer rate in this case is about 1 Mbit/s. Consuming less power; Bluetooth low-energy technology will offer long-lasting connectivity and connect small devices such as sensors. It should also be noted that the standard Bluetooth network supports one master and maximum of seven slaves. In our case, we use four sensors as slaves driven by the smartphone that is the main device (i.e. the master).

3. MONITORING ACTIVITIES AND HEALTH

As mentioned above, our system is, strictly speaking, unable to prevent a fall, but thanks to information collection in real time, it is able to almost predict a fall (through the use of the heartbeat sensor) and at minimum to determine a person's position in real time.

To do this, three inclinometers are placed on different parts of the human body and a heartbeat sensor is worn through a connected smartwatch, on a soft skin area (with a cluster of blood vessels) for greater sensitivity of the sensor (Figure 2).

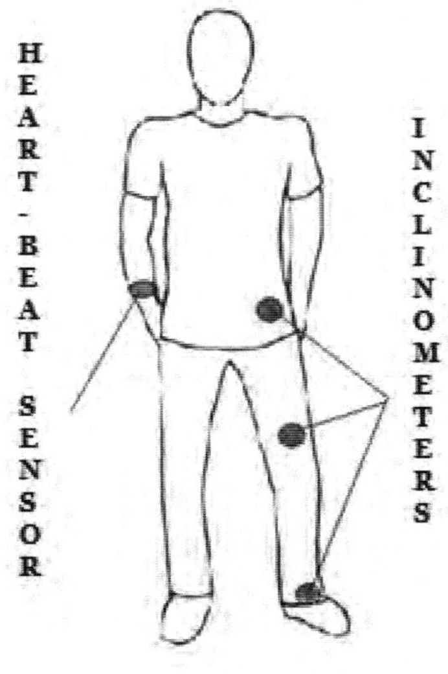

Figure 2. Four-sensor system

The location of the sensors plays a significant role in this system. Moreover, in order to the sensors to be mounted firmly on the human body, the attachment to patches is implied. We cannot consider attaching the sensors to clothes, since in this case the sensors will not be on a good mount and will start to hang out, which will lead to a disruption of the system and to a possible wrong analysis of the data.

Even despite the cheapness of this system, the convenience of the size of the sensors and its effectiveness, it is not suitable for all people. The main problem is people with cognitive disorders, leading to impaired mental activity and memory loss. It means that people entering this category will periodically forget to wear these sensors to ensure their safety.

An inclinometer is a device designed to measure the angle of inclination of various objects relative to the gravitational field of the Earth. They are used in cameras, aircraft flight controls, automobile security systems, switches and are used for platform levelling, boom angle indication, and in other applications requiring measurement of tilt. One of the inclinometers is installed on the thigh to calibrate the user's torso in order to understand at what angle the upper part of the human body is located. The second is on the upper leg. The third inclinometer is placed on the lower leg. Then, three inclinometers are enough to understand in what position the user's body is.

Table 1 shows the user's basic positions such as standing, sitting and lying positions. When a user moves the angle of inclinometers changes with time. The variety speed of angle inclinations depends on the movements of the user. By receiving data from the inclinometers in real-time, the system recognizes the position and variety speed of angle inclination, thereby understanding in which of the positions the user is. In addition, using time as a parameter, the system is able to monitor in real-time the user's activity. This analysis thus enables not only to evaluate the activity itself (number of steps, walking pace, calories burned, etc.) but also to show a drift in the wandering of the person and then to predict the probability of a fall.

The three inclinometers enables to efficiently identify main positions in almost all classical daily activities. In particular, stand up positions, sitting and lying position can effectively tracks position of the body. In correlation with the heartbeat sensor, the system is able to link the position and the health status of the user. For example, a user can be sitting with a normal heartbeat pulse. This means that he is normally sitting.

But if the inclinometers identify a sitting position while the heartbeat pulse is too high or too low, this corresponds to an critical situation. The efficiency is the same in the lying position. If the user is lying with a normal heartbeat pulse, then he is just resting or sleeping. In the same position, if the heartbeat is too high or too low, then a critical situation is also identified.

Table 1. Usual positions of human body for daily activities

Position	Sensor 1 angle	Sensor 2 angle	Sensor 3 angle
Standing up	90°	90°	90°
Sitting	90°	0°	90°
Lying	0°	0°	0°

Pulse measurement using optical plethysmography is the most common way to measure the pulse and is now implemented in sports watches, trackers, mobile phones. Light entering the bloodstream will be quite predictably scattered when the blood flow rate changes (for example, an increase in cardiac output). The optical technology measures the pulse using LEDs that evaluate blood flow to the wrist. This means that you can measure your pulse without using a chest strap. In practice, the optical sensor on the back of the watch emits light on the wrist using LEDs, and measures the amount of light scattered by the bloodstream.

At final step, user's data is automatically read and transmitted to the smartphone which application processes the data for output parameters. In the case of high probability of an abnormal situation, the application can also send a request to a heartbeat pulse reading to the watch. At this point, the user can ask for help or reject the request if everything is all right. If the user does not validate in any manner after 15 seconds, the system automatically calls for help.

Using time as a parameter, our system is also able to monitor the user's activity. In particular, for fragile people, it is able to measure a drift in the rhythm of daily life processes such as walk. But it is also able to determine the duration of the activities and to quantify the level of activities. That can also be useful for people doing sports to measure the performance itself but also to ensure the activity is done safely.

Using the same approach, the system can be used with benefit for people in reeducation or rehabilitation (for example after an accident when they are back home). The system is then used as a coach to validate the efficiency of the physical exercises and activities of the user in terms of intensity and rhythm.

That shows that our system is then much more than a fall detection system but can also be used with great flexibility for a lot of health and well-being (even well-ageing) applications at low cost.

4. CONCLUSION

Nowadays, the gerontechnology sphere is actively developing because we are facing the problem of active aging of the population. This paper focuses on the problem of falling. The fall is one of the main causes of unintentional injuries for the elderly. After analyzing the disadvantages of the existing solutions, we propose a system capable of compensating for this range of deficiencies.

This paper considers problems of falls and presents the fall detection system based on three inclinometers, pulse meter and smart watch, designed for the elderly. The system provides automatic alerts in case of fall. In comparison of other existing solutions, our concept provides several advantages, such as possibility of monitoring outdoor, ability to determine the abnormal medical records of the user through his heartbeat, determination of user location (in critical situations), comfort of use, ethical design.

The future possibility of this technology relies in the introduction of artificial intelligence in the system. Artificial intelligence will significantly improve the performance of our system and will allow not only to identify the fall, but also to help in improving the user's physical activity based on statistics compiled.

REFERENCES

Chen, K. et al., 2016, A System of Fall Detection Using a Wearable Device Based on Bluetooth Communication. *Proceedings of the 13th IEEE International Conference on Solid-State and Integrated Circuit Technology (ICSICT).* Hangzou, China, p. 1.

Florent Lachal, https://prezi.com/7rk3a2uehhel/fall-sensors/

Li, C. et al., 2018, Novel sensor-fusion system concepts for indoor and outdoor fall detection. *Proceedings of the 1st International Conference on Human System Engineering and Design (IHSED 2018).* Reims, France.

Matti Siekkinen, Markus Hiienkari, Jukka K. Nurminen and Johanna Nieminen Dept. of Computer Science and Engineering and Nokia Research Center, Finland, How low energy is Bluetooth low energy? Comparative measurements with ZigBee/802.15.4

Rakhman, A. Z., 2014. Fall Detection System Using Accelerometer and Gyroscope Based on Smartphone. Proceedings of the 1st International Conference on Infonnation Technology, Computer and Electrical Engineering, Semarang, Indonesia, pp. 99-100.

World Health Organization, http://www.who.int/news-room/fact-sheets/detail/falls, 2018.

https://www.elprocus.com/heartbeat-sensor-working-application/

ENERGY MANAGEMENT FOR DEVICES SUPPORTING DEMENTIA PATIENTS

Diogo Silva, Fernando Luís-Ferreira, João Sarraipa and Ricardo Gonçalves
CTS, UNINOVA, Dep.º de Eng.ª Electrotécnica Faculdade de Ciências e Tecnologia, Universidade Nova de Lisboa 2825-529 Caparica, Portugal

ABSTRACT

The advances in medicine and the improvements in healthcare have provided treatment and cure for many pathologies and awarded a larger expectancy of life. It represents a considerable improvement in ensuring treatment for diseases otherwise mortal and thus becoming, in some cases, chronic disorders that a person can live with for many years. This evolution has a tremendous impact in the overall healthcare environment and thus promoting longer life expectancy especially in developed countries. It is necessary, however, to consider that by increasing the width of the top level of the aging pyramid, it is therefore expected an increased incidence of the ageing related disorders. In that matter, the raising prevalence of Dementia cases is noticeable and is among the most prominent worries for the health community, as expressed in the World Health Organization reports. It is therefore of major relevance to provide assistance to citizens with dementia, especially in what concerns with losing cognitive functions such as the notion of self, awareness of current location and identifying the need to return home or to known places. The present research work is bases on the usage of smartphones to track people and has a specific target, in this matter, in extending battery life and for that matter the time a person can be located and assisted. The strategy, applicable to smartphones or other devices, relies on ensuring that energy is wisely managed so that batteries have an extended duration, increasing the chances that devices are still connected and accessible when most needed, such as while a person is experiencing a wandering episode. The aim of the present work is to focus on energy consumption to increase the chances to locate a person that is lost, with limited cognitive functions and most likely at risk for health or when even survival could be endangered.

KEYWORDS

Dementia, Wandering, Smartphone App, Energy Optimization

1. INTRODUCTION

Today the expectancy of life has reached levels that where unthinkable centuries ago. The major part of the population in developed countries have access to healthcare that allows prevention and treatment for most of the pathologies once considered deadly, sooner or later. It is thus expectable that with the extended lifetime the prevalence of aging associated pathologies becomes more problematic for people and the society. Dementia, in its various forms is a major problem recognized by the World Health Organization as a public health priority. Dementia is becoming a problem for healthcare systems that will become worse with time as people's lives will become longer. The benefit of better healthcare is putting a burden on the quality of life for those living longer and a challenge to support healthy longer life. Dementia is a syndrome that affects memory, thinking, behavior and ability to perform everyday activities. The number of people living with dementia worldwide is currently estimated at 35.6 million. This number will double by 2030 and more than triple by 2050 (Alzheimer's Disease International, 2012) . There are several types of dementia, depending on the causes and symptoms but the consequences are rather similar and the problems they rise are almost the same. Alzheimer is the most prevalent with about 60 to 70% of the cases. Then it comes Vascular Dementia as the second most frequent cause for dementia, about 15% of the cases but, unlike Alzheimer, without any licensed treatment (O'Brien and Thomas, 2015), The other types are front temporal dementia, Dementia of Lewy corps which is a neurodegenerative disorder resulting in slowly progressive and unrelenting dementia until death (Weisman and McKeith, 2007) and the well-known Parkinson's disease. Those types of dementia share common characteristics that emerge from cognitive impairments. It is likely that patients in different stages present different symptoms and that, as time passes, the patterns and activity level of the disease are changing, in most of the cases getting worse over time. The common behaviors that worry carers and families are the tendency to forget basic information, such as name and address and, for that reason, the tendency to wander without known destination as the person forgets what was doing or where is home. Those are risky situations that can be addressed by devices that a person can carry and help in getting their location.

2. ANALYSIS OF ENERGY CONSUMPTION

Currently, energy consumption is a hotly contested topic all over the world. The objective is to find effective methods to extend the life of the devices in each load. Patients with dementia are no exception and, due to the problems in remembering paths and places, it is important that they can be located. For that it is necessary that devices have tracking information and, more important than that, devices must be working. For these patients that need daily follow-up due to lack of orientation it is important that unless they are accompanied, it is possible to reach them by any available device. In order to keep their independence and continue to become integrated in society, it is better to use electronic devices that most people use in daily routine and, at the same time, keep them trackable and connected to the caregivers. The easiest devices to find by the end consumer that meet these requirements are smartphones. For this reason, it's decisive to optimize resources for a better battery management. That is the aim of the current work.

In previous works (Luis-Ferreira *et al.*, 2018) research has been focused in the devices and applications that can provide location and help the patients with the adequate features to detect risks and provide timely communication of detected risks and location, to carers and family. The problems then identified were related with the extension of the absence of notice and the time a person can be lost by wandering in unknown places. That significantly arises the question that, when the person most need it is possible that the devices are shut down by lack of energy.

A reasonable approach to address this problem is to manage, optimized through machine learning, the usage of sensors and antennas in smartphones. In our research study, as observed in the following graphs, energy consumption of the sensors and antennas may vary in different situations.

The consumption patterns will take in account not only the devices but also the applications being used and the result, in terms of consumption of both software applications and hardware devices being used for the same objective. It was noticed a dramatic increase in energy consumption once the smartphone actually starts transmitting data over the Wi-Fi antenna. The value climbs almost three times if we run an application such as Google Maps, which uses GPS and Wi-Fi simultaneously, while rendering images dynamically (Corral *et al.*, 2013). In Figure 1 it is possible to observe our early basis of work by observing the different consumption patterns for devices in a smartphone (Nokia N95) that gives clues about where energy is spent and what benefits can be obtained with a proper management of those devices (Perrucci, Fitzek and Widmer, 2011).

Technology		Action	Power [mW]	Energy [J]
Wireless data	Bluetooth	BT off	12	
		BT on	15	
		BT connected and idle	67	
		BT discovery	223	
		BT receiving	425	
		BT sending	432	
	WiFi IEEE802.11 (infrastruct ure mode)	In connection	868	8.2
		In disconnection	135	0.4
		Idle	58	
		Idle in power save mode	26	
		Downloading@4.5Mbps	1450	
	WiFi IEEE802.11 (ad hoc)	Sending @ 700 kB/s	1629	
		Receiving	1375	
		Idle	979	
	2G	Downloading@44Kbps	500	
		Handover 2G->3G	1389	2.4
	3G	Downloading@1Mbps	1400	
		Handover 3G->2G	591	2.5

Figure 1. Energy Consumption in smartphone communication devices

It is observed that the use of GPS causes major losses of battery when compared with other sensors like accelerometers that are known to provide guidance for short periods or time or small to medium displacements by evaluating relative displacements. So, it is deductible that its usage should be kept to strictly necessary. It is still interesting to note that GPS, although more precise on the outside, consumes more energy in this situation. There are still other antennas that help locating the device such as the Wi-Fi antenna. Figure 2 presents the different energy consumption patterns for a user moving and standing either inside or outside a building.

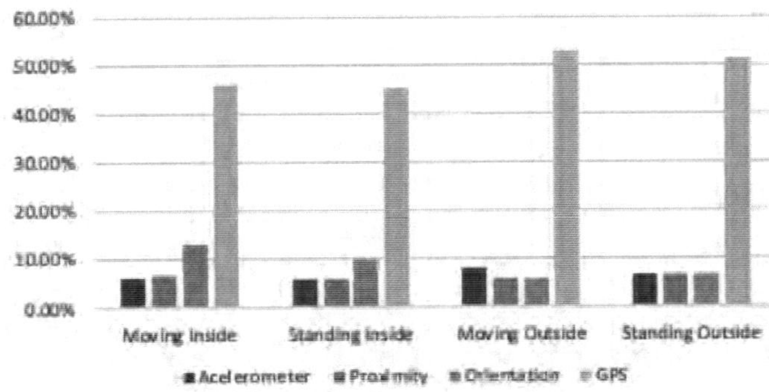

Figure 2. Energy Consumption with data transmission

In the previous chart, it becomes evident that the Wi-Fi, GSM and Bluetooth antennas consume less energy than GPS. For this reason, they can be used, when near to other recognizable devices, to locate the patient's device it can thus be applied to smartwatches that can use the smartphone location devices. Another important strategy to be applied in these cases is the intermittent activation of the GPS that can minimize the impact of that device in energy consumption patterns and still ensure that location is stored and provided from time to time. Table 1 presents the usage of different devices according to the present situation, making possible to extend the battery life by switching on and off the sensors that are not needed in the present moment. The different devices in a smartphone have different consumption levels meaning different weight in the overall consumption of a device's battery. The consumption patterns also vary according to the activity a person is executing while indoors or outside and such results can be observed next in Table 1 (Inayat Khan et al, 2016).

Table 1. Sensors relative Energy consumption

Sensor	Interior Movement	Standing Inside	External Movement	Standing Outside
Accelerometer	6,35%	5,95%	8%	6,6%
Proximity	6,7%	5,9%	5,9%	6,9%
Orientation	13,23%	9,7%	5,9%	6,9%
GPS	46%	45,22%	53%	51%

Beyond comparison, there are different profiles of consumption according to behavioural status. This is the case, for the same person, that moves, goes outside or returns inside buildings. Those patterns as observed next in Figure 3.

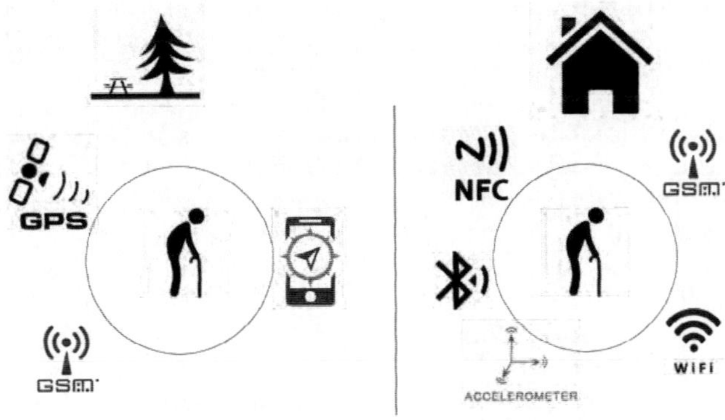

Figure 3. Devices selected according to location

The energy management information becomes of major importance by allowing an optimization of resources in smartphones for people who are in situations that need constant monitoring. The present work comes as a continuation of the early work on the development of personalized technological architectures to assist dementia patients based on energy consumption profiles which become essential for the successful implementation of the current work where those profiles are in fact brought to implementation (Luis-Ferreira *et al.*, 2019).

3. CONCLUSION

The presented research is being carried in the scope of an ongoing MSc in Electronics and Computer Science and is applied in the research work for the CARELINK AAL project. Work carried so far passed early stages as it was verified the high potential of energy management strategies to extend the autonomy of mobile devices. This smart management of energy is of most importance for patients with Dementia, as sometimes, cognitive problems appear without notice in episodes where the person manifests lack of orientation, lost sense of time and place and in general inability to return home or to a known place. In these cases, people also tend to distance themselves from a non-controllable situation and refuse to admit their situation and thus are not likely to ask for help even while carrying a communications device. In those cases, it becomes important to have an active device to make possible location and contact, to make possible reaching this people while in a situation that can be of risk or even life threatening. It is known that the first 72 hours that a person is lost are critical, with time passing, without any option to locate them time runs against the possibilities of a good outcome. The most common and used solutions for communication and location services are in fact the smartphones but other customized dispositive can also be used. The study hereby presented addresses the need to extend, as much as possible, the autonomy of the devices which, when dry out of energy, become completely useless for tracking and rescue. From the current tests it Is possible to envisage an extension of battery above 20% making a contributing factor to promote the possibility that lost people can be found and become safe before it is too late.

ACKNOWLEDGEMENT

The authors acknowledge the European Commission for its support and partial funding and the partners of the research project CARELINK, AAL-CALL-2016- 049 funded by AAL JP, and co-funded by the European Commission and National Funding Authorities of Ireland, Belgium, Portugal and Switzerland.

REFERENCES

Alzheimer's Disease International, W. H. O. (2012) *Dementia : A Public Health Priority*. World Health Organization.

Corral, L. *et al.* (2013) 'A method for characterizing energy consumption in Android smartphones', in *2013 2nd International Workshop on Green and Sustainable Software (GREENS)*. IEEE, pp. 38–45. doi: 10.1109/GREENS.2013.6606420.

Inayat Khan et al (2016) 'Sensors are Power Hungry: An Investigation of Smartphone Sensors Impact on Battery Power from Lifelogging Perspective', *Bahria University Journal of Information & Communication Technologies*, 9(2), pp. 8–19. Available at: https://www.bahria.edu.pk/ojs/index.php/bujict/article/view/52 (Accessed: 28 February 2019).

Luis-Ferreira, F. *et al.* (2018) 'Assessment of Risk for Alzheimer Patients by Monitoring Heartbeat and Identifying Falls', in *Proceedings of 10th International Conference on e-Health 2018*. Madrid, Spain.

Luis-Ferreira, F. *et al.* (2019) 'Personalised Technological Architectures to Assist Dementia Patients based on Energy Efficiency', in *ICIST 2019 - 9th International Conference on Information Society and Technology*. Kopaonik, Serbia.

O'Brien, J. T. and Thomas, A. (2015) 'Vascular dementia', *The Lancet*. Elsevier, 386(10004), pp. 1698–1706. doi: 10.1016/S0140-6736(15)00463-8.

Perrucci, G. P., Fitzek, F. H. P. and Widmer, J. (2011) 'Survey on Energy Consumption Entities on the Smartphone Platform', in *2011 IEEE 73rd Vehicular Technology Conference (VTC Spring)*. IEEE, pp. 1–6. doi: 10.1109/VETECS.2011.5956528.

Weisman, D. and McKeith, I. (2007) 'Dementia with Lewy Bodies', *Seminars in Neurology*. Copyright © 2007 by Thieme Medical Publishers, Inc., 333 Seventh Avenue, New York, NY 10001, USA., 27(1), pp. 042–047. doi: 10.1055/s-2006-956754.

DESIGN OF NEW GENERATION OF SERIOUS GAMES FOR PATIENTS WITH MODERATE COGNITIVE IMPAIRMENT

Barbara Carillon[1], Brune Dirian-Angeli[1], Gabrièle Breda[2], Samuel Benveniste[3] and François Deparis[1]
[1]*ARPAVIE, Issy-Les-Moulineaux, France*
[2]*ALTRAN Resarch Department, Vélizy-Villacoublay, France*
[3]*CEN Stimco – Broca Living Lab, Hôpital Broca, Paris, France*

ABSTRACT

Among the new challenges created by the aging population, which continues to increase significantly, many researchers have taken an interest in designing new services for people suffering from Alzheimer's disease such as serious games. They have proposed a diverse selection of guidelines for designing cognitive serious games, considering cognitive complexity and motivation. Generally, existing work rely on virtual reality or video interfaces. However, most of these games are not simple and effective enough to suit the needs of people suffering from Alzheimer's disease.
In this paper, a set of specific guidelines for designing and implementing effective serious games targeting patients with Alzheimer's disease and different disabilities are proposed. These guidelines aim to design serious games offering natural interaction helping people to communicate with the ability to capture their attention and to improve their self-esteem by encouraging them and highlighting their preserved abilities.

KEYWORDS

Alzheimer's Disease, Robot-Assisted Therapy, Humanoid Robot, Cognitive Stimulation, Serious Games

1. INTRODUCTION

Among the new challenges created by the aging population, which continues to increase significantly, many researchers have taken an interest in designing new services for people suffering from Alzheimer's disease. Alzheimer's disease and related diseases are part of the dementia category. They are characterized by a gradual deterioration of cognitive abilities caused by brain damage, with repercussions on the emotional states and behavior of patients. Thus, the symptomatology is diverse and it evolves slowly and gradually. The services designed for this population aim to lessen the impact of their disease in order to increase their autonomy and to improve their safety and well-being (Chibani A. et al, 2015).

In this context, many research projects focus on designing serious games for people with Alzheimer's (Castilla D. et al., 2013, Tong T. et al. 2017, Kort H., and Huisman C. 2017). These studies show clearly how such serious games can be beneficial for persons with Alzheimer's and their caregivers. Eichhorn et al demonstrate, in (Eichhorn C. et al, 2018), that exposure to these games improved people's cognition even weeks after they played them, partly countering the typical cognitive decline associated with Alzheimer's disease. Three main types of serious games are considered in the literature: cognitive, physical and social/emotional (McCallum S. and Boletsis C. 2013).

For people with Alzheimer's, cognitive exercises are very important and serious games play a key role to generate a positive atmosphere. Many researchers have proposed a diverse selection of guidelines for designing cognitive serious games, considering cognitive complexity and motivation. Tong et al. address different types of cognitive decline with potential game concepts such as object recognition, shifting and inhibition (Tong T. et al. 2017).

Generally, existing work rely on virtual reality or video interfaces. However, most of these games are not simple and effective enough to suit the needs of people suffering of Alzheimer's disease. Indeed, according to Pyae et al., the game context should be familiar and game interactions should be easy (Pyae A. et al. 2016). In this context, some studies introduced social robots as a game, such as the Zora robot. Although it is successful in promoting physical exercise (Kort H., and Huisman C. 2017), a first trial with this robot for use in cognitive games showed its limitations due to ineffective verbal human-robot interaction.

Designing robotic system-based cognitive games for patients with Alzheimer's disease requires sophisticated and novel approaches. In order to capture the attention of patients and foster positive interactions, these systems need to be endowed with advanced, robust natural language processing, contextual non-verbal interaction (body language) and better decision-making capabilities.

To address these challenges, this paper presents a set of specific guidelines for designing and implementing effective serious games targeting patients with Alzheimer's disease and different disabilities. The proposed guidelines are based on preliminary research with healthcare professionals from a nursing home, in the context of memory workshops mediated by a robotic interface. They focus on:

(i) Natural interaction helping people to communicate. To deal with that, a "human-friendly" robot, called NAO, is used;
(ii) Capturing attention;
(iii) Encouraging and congratulating users;
(iv) Providing hints through different statement and gestures
(v) Fostering group interactions

In this paper, section 2 address in details these guidelines. Section 3 presents the designed games/exercises, and discusses their benefit for use with Alzheimer's patients before concluding with ongoing work.

2. SERIOUS GAMES USING SOCIAL ROBOTS

2.1 New Design of Serious Games

2.1.1 Humanoid Robot as a Mediation Tool

In coherence with its humanoid aspect, the robot has a participative role defined in the programming of the exercises. By mediation tool, we mean that it has the function of facilitating the therapist-patient-mediation relationship; mediation being the cognitive exercises for the memory workshop.

During workshops, the robot has two roles, animation and self-esteem improvement:

- The animation consists of interacting and communicating directly with the person by giving specific instructions on the exercise to be performed;
- The self-esteem part consists in encouraging the person and highlighting his/her preserved abilities.

To deal with that, the NAO robot was selected for many reasons:

i. **Adequacy to the target population**: the therapeutic workshops for people with Alzheimer's disease or a moderate-stage related illness with moderate psycho-behavioral symptoms aim to offer them psychosocial interventions to promote the use of all the preserved abilities of the person and a well-being in everyday life. Because of its social aspects including the ability to communicate verbally with body expressions and the ability to keep eye contact with people during an interaction, NAO robot is a stimulating mediating object to initiate speaking and to promote exchanges among people whose language is still preserved. These social aspects of the NAO robot are the major elements taken into account in the choice of this robot.
ii. **Adequacy for target use**: As the therapists in our team intended to use the robot for memory exercises and physical exercises, a humanoid robot seemed like the right choice: it would be able to show the physical exercises accurately and it would be more coherent for it to participate in memory exercises than a talking animal or hybrid robot.
iii. **Practicality**: Although its price remains quite high for a nursing home, it is easily available and already used in many different settings in France, with a rather efficient customer service, which is critical to ensure that operations run smoothly during the study that will be conducted.

iv. **Ease of programming**: NAO offers a wide variety of pre-programmed functions and tools that enable researchers to quickly build new applications and iterate on their designs. Thus, although it is not the most technically advanced hardware, it appeared to offer the best mix of performance and ease of use for the work conducted during this study.

This choice was also supported by previous experiments too: some studies show that the NAO robot is judged more positively than other social robots such as the PARO robot, and a dog. In (Valentí S. et al., 2015), a pilot study was conducted in two phases in nursing homes and day care centers, with 201 people with Alzheimer's disease or related diseases to compare the effects on behavior and cognitive abilities of patients of three types interventions: the NAO robot, the PARO robot, and a dog. Groups of patients involved in NAO robot-assisted therapy showed a decrease in apathy as well as a measured decrease in Mini Mental State Examination (MMSE) scores. A recent study was conducted in a geriatric unit with 9 people with cognitive impairment to evaluate the effects of using a NAO-based intervention in psychomotor sessions on commitment, emotional state, well-being of the person as well as the appreciation of this type of session compared to a conventional therapy session. The first results show better commitment, positive answers and a greater satisfaction of patients during sessions attended by the robot (Rouaix, N. et al., 2017).

2.1.2 Natural Interaction

Communication modalities have also been an essential element in designing games in order to facilitate interaction between robot and patients. To deal with that, some natural aspects are considered to design natural human-robot interaction in this work such as:

- *Natural Language (NL)*: Using spoken natural language (NL) has received increasing attention in human-involved robotics research. NL facilitates the communication between users and games/robots and helps to create natural human-environment interaction (Baraglia et al. 2016).
- *Body Language Expression*: Beyond verbal communication, some associated gestures are designed to support the robot's utterances, in order to give more expressive messages, for example, affirmative and negative gestures, or a specific posture for when the robot is waiting for an answer from the patient.

2.1.3 Encouraging and Congratulating Users

Boosting patients' self-esteem is one of the important aspects that should be considered when designing serious games for this population. In this work, two such features are considered: encouragement and congratulation, which consist in highlighting the patient's preserved abilities through statements and gestures of encouragement (e.g. "I know you can find the answer") and congratulation (e.g. "You made it !").

2.1.4 Hinting

In order to create a failure-free experience for patient, game designers should consider a hinting feature.

Indeed, the robot participates as a mediator in strengthening the identity of the elderly, which is often undermined by the experience of old age. To this end, it is important that patients do not experience failures that would highlight their impairments. Thus, in case of a wrong answer, the robot gives clues successively to help the patient in finding the correct answer with the therapist's support.

2.1.5 Fostering Group Interaction

Many studies observed improvements in cognitive functions through social activities. A recent study from the National Social Life, Health, and Aging Project found that participation in "congregant," "diverse," or "friends" social networks was significantly associated with good mental health in later life, especially with avoidance of depressive symptoms and their potential to impair cognition (Nelson LA, et al. 2013). Moreover, emulation helps people to engage more fully in tasks. This justifies designing games for use by a group of patients instead of individual sessions.

2.2 Cognitive Training Games

The memory workshop program is based on three exercises frequently used by psychologists, some of which refer to the psychological assessment from the Mini Mental State Examination.

2.2.1 Temporal Orientation Exercises

The first exercise concerns temporal orientation. It consists in searching the date of the day sequentially, such as, the day, the month and the year with the possibility of referring to clues to find the correct answer. This exercise is related to the alteration of episodic memory, i.e. the storage and retrieval of acquired experiences in a specific spatio-temporal context (Giffard B., et al., 2001). This is the memory of facts personally experienced. In the context of Alzheimer's disease, erasure advances in reverse order: patients forget recent facts first and progressively forget older facts, for example leading them to believe they live in the time of their childhood instead of present times. Patients forget the events that they experience, which generates a disorientation in time and space. Reorienting patients over time, at the beginning of the exercise, helps to limit the feeling of disorientation and anxiety.

2.2.2 Semantic Memory Exercises

The second exercise focuses on semantic memory. Semantic memory is the memory of concepts, ideas and facts that are independent of the spatio-temporal framework. It stores the learned general knowledge and it is queried when humans must give the spelling or the meaning of a word, or to remember a general fact such as the name of the president, capital city of a country, etc. (Giffard B., et al., 2001).

The proposed exercise consists in giving a random letter of the alphabet and asking patients to propose a list of categorized words, such as, first names, verbs or common names, starting with the given letter. Proposing cognitive exercises with the ability to access to semantic memory has many interests such as maintaining this form of memory while promoting a situation of success for the patient.

2.2.3 Short and Long Term Memory Training

The third exercise focuses on working memory. This latter is a short-term memory and is the most requested. It allows (i) to temporarily store data, such as a phone number, a path, etc.; (ii) to carry out mental operations such as calculations; or (iii) to follow reasoning at several stages, for example, applying instructions and following a path (Giffard B., et al., 2001). The alteration of this memory makes patients' attentional resources difficult to mobilize and very labile. The evolution of cognitive neuropsychology has allowed the development of specific and targeted methods of deficit brain function. These new approaches are used especially for verbal working memory. Propose exercises for specific cognitive rehabilitation to this form of memory had been efficiency shown in (Seron X., et al., 2008).

As this memory has a moderate alteration for the monitored population, the goal is to retrain it during a therapeutic workshop. In fact, the proposed exercise consists in listening to four words from the same lexical category, and recalling them after some delay. People gradually learn the words in a repetition task to generate memory encoding. A delayed recall is performed after an interfering task based on mental calculation such as reciting a multiplication table. This exercise also refers to several Mini Mental State Examination items.

3. CONCLUSION

This paper proposes a set of specific guidelines for designing and implementing effective serious games targeting patients with Alzheimer's disease and different disabilities. These guidelines focus on providing natural interaction helping people to communicate with the ability to capture their attention and to improve their self-esteem. The proposed approach is based on introducing a mediator tool such as the humanoid NAO robot. Indeed, the design of this robot has a soothing, attractive and playful effect. Furthermore, introducing a mediator tool is able to improve the interaction and the communication between the therapist and patients with behavioral disorders. It is important that this approach is not intended to suit all kinds of patients: dementia is a progressive disease, and the use of NAO is, in our view, most suited for people with moderate cognitive impairment. For example, animal robots with fewer interactional modalities such as PARO, which are less demanding for participants, seem to be more suited for interventions targeting people with more advanced dementia. A promising use for such robots is pain relief during medical care (Demange et al., 2019), for which NAO would be rather impractical to use. Indeed, the gradual deterioration of language, reasoning and comprehension abilities during the illness need an adaptation regarding to communication modality during the interaction. Focusing, at an advanced stage, on nonverbal communication and sensory stimulation in the relation favors a valorization of the preserved capacities and avoids any form of defeat. In comparison with the NAO robot, the choice of a robot such as PARO at this stage seems more relevant and adapted.

The ongoing works focus on the implementation of the proposed guidelines according to a robot cognitive architecture, introduced in previous work (Ayari et al. 2017), with an emphasis on system aspects according to multi-agents and service-oriented computing paradigms, such as, the natural language understanding model, reasoning mechanisms, and communication capabilities.

REFERENCES

Chibani A. et al. 2015. *Using cognitive ubiquitous robots for assisting dependent people in smart spaces*. In Intelligent Assistive Robots. Springer.

Castilla D. et al., 2013. Process of design and usability evaluation of a telepsychology web and virtual reality system for the elderly: Butler, *International Journal of Human-Computer Studies*, Vol. 71, No. 3, pp 350-362.

Demange, M., et al., 2019. Development of a robot-assisted intervention for the management of acute pain situations in dementia: framework and results from a feasibility study. *Journal of pain research (in press).*

Giffard B., et al., 2001. Le vieillissement de la mémoire: vieillissement normal et pathologique. *Gérontologie et société*, Vol. 24, No. 2, pp 44-45.

Kort H., and Huisman C. 2017; Care Robot ZORA in Dutch Nursing Homes; An Evaluation Study. *Studies in health technology and informatics*, 242, 527-534.

McCallum S. and Boletsis C. 2013. A taxonomy of serious games for dementia. *In Games for Health*, Springer Vieweg, Wiesbaden, pp. 219-232

Nelson LA, et al. 2013. Social engagement and physical and cognitive health among American Indian participants in the health and retirement study. *J Cross Cult Gerontol,* VOL. 28, No. 4, pp 453-456.

Pyae A. et al. 2016, When Japanese elderly people play a Finnish physical exercise game: a usability study. *J. Usability Stud.* Vol. 11, No. 4, pp 131–152.

Rouaix, N. et al. 2017. Affective and engagement issues in the conception and assessment of a Robot-Assisted psychomotor therapy for persons with dementia. *Frontiers in psychology*, 8, pp 950.

Seron, X., Rossetti, Y., Vallat-Azouvi, C., Pradat-Diehl, P., & Azouvi, P. (2008). La rééducation cognitive. *Revue neurologique*, 164, S154-S163.

Valentí S. et al. 2015. Social robots in advanced dementia. *Frontiers in aging neuroscience*, 7, pp. 133.

Ayari, N., et al. 2017. Contextual Knowledge Representation and Reasoning Models for Autonomous Robots. *In AAAI Fall Symposium Series*, Washington USA.

Baraglia J. et al. 2016, Initiative in robot assistance during collaborative task execution, In 11th ACM/IEEE International Conference on Human-Robot Interaction (HRI), 67–74.

Bouchard B. et al, 2012. Developing serious games specifically adapted to people suffering from Alzheimer. *In International Conference on Serious Games Development and Applications,* Springer, Berlin, Heidelberg, pp. 243-254.

Chebaa B. et al. 2016. Multimodal analysis of serious games for cognitive and physiological assessment. In Proceedings of the 9th ACM International Conference on PErvasive Technologies Related to Assistive Environments, p. 30.

Eichhorn C. et al, 2018. Innovative Game Concepts for Alzheimer Patients. *In International Conference on Human Aspects of IT for the Aged Population*, Springer, Cham, pp. 526-545

Tong T. et al. 2017. Serious games for dementia. *In Proceedings of the 26th International Conference on World Wide Web Companion*, Perth, Australia, pp. 1111-111

Posters

WHATSAPP: A TOOL FOR ADHERENCE TO ASYMPTOMATIC CHRONIC-DISEASE DRUG THERAPIES?

Amanda Caroline Sartori[1,a], Julio de Souza Sá[1,a], Marcelo Picinin Bernuci[2,b], Ely Mitie Massuda[2,b], Tiago Franklin Rodrigues Lucena[2,b] and Mirian Ueda Yamaguchi[2,b]

[1]*Unicesumar – Cesumar University Center , Guedner Ave, 1610, Maringá, Paraná, Brazil*
[2]*Unicesumar – Cesumar University Center, Health Promotion Graduate Program and ICETI - Cesumar Institute of Science, Tecnology and Inovation, Guedner Ave, 1610, Maringá, Paraná, Brazil*
[a]*MSc*
[b]*PhD*

ABSTRACT

Studies have shown that mobile health (mHealth) technologies are increasingly being successfully used in control programs for symptomatic diseases, such as cancer and HIV, as methods of behavioral intervention to enhance patient adherence to treatments. The objective of this research was to evaluate the potential of WhatsApp messages as an intervention method to improve adherence to the treatment of two asymptomatic diseases: hypertension and diabetes. A controlled trial was conducted with 499 participants from the basic health units of Maringá, Paraná, Brazil. They received 58 WhatApp messages over three months with information about the importance of adherence to drug therapy. Adherence to the therapy was assessed by the Morisky-Green test (MGT). The analysis of the efficacy of the intervention was performed by comparing the MGT test results before and after the messages were sent. The results showed an expressive difference in adherence to the drug therapy for the group that received the WhatsApp messages (67.5%) compared to those who did not receive them (58.5%). However, in contrast to our expectations, there was no significant association between adherence and the intervention group. WhatsApp can be used as a welfare technology in various health-related areas, however it was concluded that, for asymptomatic diseases, mHealth technology interventions can be useful if associated with other strategic processes.

KEYWORDS

mHealth, Drug-Therapy Compliance, Asymptomatic Diseases, Primary Health Care

1. INTRODUCTION

Chronic non-communicable diseases (CNDs) claim the lives of about 38 million people a year in the world. Among these diseases, diabetes mellitus (DM) and systemic arterial hypertension (SAH) are chronic diseases frequently asymptomatic and associated with low therapeutic adherence (Malta et al, 2017).

Interventions associated with mobile health (mHealth) technologies have the potential to change health behaviors. These technologies are devices easily introduced into individuals' daily lives (Aurora et al, 2014). Among the mHealth devices, smartphones are personal, portable and allow access to WhatsApp, which is a tool with good acceptance, reaching approximately one billion people nowadays (Ganasegeran et al., 2017). Studies have demonstrated that disease self-management interventions can positively influence the behavior in health (Free et al, 2013). In this context, the objective of this study was to carry out an intervention by means of sending messages through WhatsApp in order to improve adherence to antidiabetic and antihypertensive drug therapy.

2. METHODS

A representative sample of 499 diabetic and hypertensive patients registered in the 33 basic-health units of Maringá city - Paraná, Brazil, participated in the present study. Data collection was performed during October 2017 to April 2018. The research was approved by the Research Ethics Committee No. 2,234,211, 8/23/2017. Of the patients, 57 were diabetic, 280 were hypertensive and 162 had both diabetes and hypertension.

This study was conducted in three stages. In the first, the Morisky-Green Test (MGT), which consists of a self-report scale for assessing compliance with drug therapy, was applied to the intervention and control groups. The MGT test considers patients adherent when they reach a score of four in an instrument containing four objective questions, whose answers can be positive (yes - 0 points) or negative (no - 1 point) (Morisky et al, 1986).

In the second stage, 58 messages (text, audio and pictures) were sent by WhatsApp to the intervention group, using the platform BulkService®. The content of the messages reported about protection and risk factors, signs and symptoms of diabetes or hypertension, the importance of their treatment, complications of the diseases, reminders to take medication and motivational content. After the intervention, at the third stage of the study, telephone contact was made with all the 499 patients included in the research (intervention and control groups), and the MGT was applied again, in order to identify the effectiveness of sending the messages via Whatsapp.

3. RESULTS

One-third of the individuals that participated in the survey (33.0%) were between 50 and 59 years old, and the average of all the patients was 56.88 years of age. The mean was lower among those interviewed who had only one disease, 53.05 years for diabetics and 56.34 years for hypertensives, and higher among those who had both diseases: 59.17 years. Regarding sex, the majority (65.0%) were women; however, the distribution between the sexes was approximately equal for those who only had diabetes, but among the hypertensive patients, women (70.0%) predominated. The majority were white (79.0%), married (63.0%) and the most frequent level of schooling was incomplete secondary education, reported by about a third of respondents.

The proportions of adherent patients at the time before the interventions were very similar between the groups, 38.3% and 39.1%, and there was no significant difference in adherence (p-value of 0.927) before the intervention (Table 1). On the other hand, the difference in the proportion of adherence between the groups was greater after the intervention, 67.5% for the intervention group and 58.5% for the control group. However, there was no significant association between adherence and the intervention group. (p value of 0.077).

Table 1. Relative frequencies of adherents to the drug therapy and results of the chi-square test, to compare the responses between the intervention and control groups, before and after sending the messages by WhatsApp. Maringá, PR, Brazil, 2018

Time	Adherence frequency (%)		Result of χ^2 teste	
	Intervention group	**Control group**	χ^2	**p value**
Before intervention	38.3%	39.1%	0.008	0.927
After intervention	67.5%	58.5%	3.117	0.077

4. CONCLUSION

The present study showed an expressive difference in the proportion of adherence to the drug therapy for the group that received the WhatsApp messages compared to those who did not receive them. However, statistically there was not sufficient sample evidence to indicate the association between adherence and the group after the intervention. In view of the complexity of adherence to drug use for generally asymptomatic diseases like hypertension and diabetes, it can be concluded that WhatApp, as a mHealth-technology intervention, may be useful if it is associated with other strategies. Health-education processes to improve literacy in users' health and / or relationships of trust between patient and health professionals, could be associated to combat low adherence to drug therapies. The importance of the control group's result in this study must be highlighted, which could identify a possible phenomenon known as 'social desirability bias', which refers to the fact that interviewees report behavior that they believe represents an acceptable social position (Nolte et al, 2013). This phenomenon may have influenced the interviewees to report greater adherence for both groups (the intervention and control groups). Finally, a future study should be considered, including clinical controls by measuring the blood pressure of hypertensive patients and the blood glycemic concentration for diabetic patients before, during and after sending messages by WhatsApp.

REFERENCES

Arora, S. et al. (2014). Trial to examine text message-based mHealth in emergency departamento patients with diabetes (TExT-MED): A randomized controlled trial. *Annals of Emergency Medicine,* 63(6), 745-754.

Free, C., Phillips, G., Galli, L. (2013). The effectiveness of mobile-health technology-based health behaviour change or disease management interventions for health care consumers: A systematic review. *Plos Med,* 10(1), 1-45.

Ganasegeran, K. et al. (2017). The -Health revolution: Exploring perceived benefits of WhatsApp use in clinical practice. *International Journal of Medical Informatics*, 97(1), 145-151.

Malta, D.C et al. (2017). Doenças crônicas não transmissíveis e a utilização de serviços de saúde: análise da Pesquisa Nacional de Saúde no Brasil. *Revista Saúde Pública,* 51(1), 1-10.

Morisky, D.E., Green, L.W., Levine, D.M. (1986). Concurrent and predictive validity of a self- reported measure of medication adherence. *Medicine Care*., 24, 67-74.

Nolte, S., Elsworth, G.R., Osborne, R.H., (2013). Absence of social desirability bias in the evaluation of chronic disease self-management interventions. *Health and Quality of Life Outcomes,* 11(114), 1-9.

ANALYSIS OF MOBILE APPLICATIONS FOR DIABETES SELF-MANAGEMENT IN THAILAND

Maneerat Rattanamahattana and Palung Rungsa
Faculty of Pharmaceutical Sciences, Khon Kaen University, Khon Kaen, Thailand

ABSTRACT

With advances in mobile technology, use of mobile devices like smartphones, has grown rapidly around the world including Thailand. Many mobile applications (apps) have been launched to help patients with chronic non-communicable diseases to manage their health behavior. However, little is known about the apps available in commercial application stores. Our study aimed to explore and evaluate the content and features of free download mobile apps available for diabetes self-management according to eight categories: (1) information support; (2) subscription requirement; (3) notifications; (4) visual aids; (5) goal setting; (6) export data; (7) medical consultation; and (8) compatibility. We found 22 apps available in the AppStore and 84 apps in the Google Play store. The applications available in the App Store could be categorized as tracking only (23%) as well as tracking and education (77%) with the three most common features; visual aids (95%), export data (82%), and notification (73%). The applications on Google Play store could be categorized into 4 groups; tracking (35%), tracking and education (15%), education alone (29%), and diabetes food recipes (21%). The apps in the tracking and tracking and education groups had a wide range of features, with app scores ranging from one to seven. Apps in the education and diabetes food recipe groups only had one feature - information support. Many apps also contained advertisements, with no references provided. Therefore, it may be practical that healthcare providers explore each application before recommending to their patients. Mobile apps that specifically fit the Thai setting, with quality assessment of the reliability of information as well as assessment of e-health literacy are needed for diabetes self-management among Thai patients.

KEYWORDS

Diabetes Mellitus, Self-Management, Mobile Devices, m-Health, Thai

1. INTRODUCTION

Diabetes mellitus is a non-communicable disease (NCD), of which an alarming increase of diabetes patients has been reported of with over 1.6 million deaths worldwide in 2016 (WHO, 2018) and it is the third leading cause of all deaths among NCDs in Thailand (Ministry of Public Health Thailand, 2016). Previous studies suggest that use of mobile phone apps in diabetes management can modestly improve glycemic control (Wayne and Ritvo, 2014; Kirwan, Vandelanotte, Fenning & Duncan, 2013). In fact, smartphones have gained rapidly popular use across the population. The proportion of people who possess smartphones in Thailand was 82.1% and the trend of use among baby boomers aged 51-69 has been growing significantly. The aim of this study was to explore and evaluate contents and features of diabetes self-management applications available in Thailand.

2. METHODS

This was a descriptive study using online search for diabetes-related applications available in Thailand from two popular platforms i.e. Android Google Play Store and iOS AppStore by using two mobile devices - Sony Xperia XA Ultra and iPhone SE, respectively. The key terms for diabetes self-management applications included both in English ('diabetes', 'diabetes management') and in Thai ('เบาหวาน'). The inclusion criteria included the applications that were specifically designed for patients regarding diabetes self-management with free download; provided in either English or Thai language; used with smartphones; and updated on a regular basis; and with no duplicates. Descriptive information regarding the name of developer, latest update, and operating system of the applications was collected. Based on the exhaustive literature review (Izahar et al, 2017; Demidowich, Lu Tamler, & Bloomgarden, 2012), the applications were then categorized and evaluated the availability of the following features; (1) information support; (2) subscription requirement; (3): notifications; (4) visual aids; (5) goal setting; (6) export data; (7) medical consultation; and (8) compatibility. A point would be given for each available feature for the total score across all applications included in the study.

3. RESULTS

From the above inclusion and exclusion criteria, 22 apps on AppStore and 84 apps on Android Google Play Store were selected. The content found in AppStore apps could be classified in two categories i.e. tracking blood sugar of patients (77%) and the combination of tracking and patient education (23%). The apps from Google Play store contained more varieties as shown in Figure 1. The distribution of features in the apps from both iOS and Android is compared in The most popular features for iOS were visual aids (95%) followed by export data (82%), notification (73%), compatibility (50%), goal setting (41%), subscription (32%), information (23%), and medical consultation was the least popular feature provided in only 9% of the total 22 iOS apps. Information feature was the most popular feature in Android apps (65%) followed by visual aids (46%), export data (38%), notification (33%), compatibility (22%), goal setting (19%), subscription (8%). Similarly to the iOS apps, medical consultation was the least popular feature available (5%) for Android. Considering the distribution of the features available for each Android app, there was a wide range of scores from 1 to 7 features (Figure 2). Most of the apps available on AppStore contained more than 4 features while the apps from Google Play Store contained less than 4 features. However, none of the apps from either platforms obtained a full score of 8. In addition, the apps that provided information in Thai language were still very rare.

Recent evidence suggests the use of mobile technology as a tool for self-management has the potential to fulfill the needs of patients, healthcare providers, or both (Goyal and Cafazzo, 2013). In this study, both the Android and iOS platform provided diabetes self-management apps tended to provide tracking and education, but the Android apps had a wider variety of categories than the iOS apps. For instance, Android apps provided more education features and suggested diabetes recipes. However, across all of the apps, most of the content was provided as information without any evidence-based citations or references. Thus, there is no guidance regarding the reliability of the information. Therefore, lay patients will require high health literacy skills to evaluate the information before they can confidently apply the healthy tips recommended by the apps.

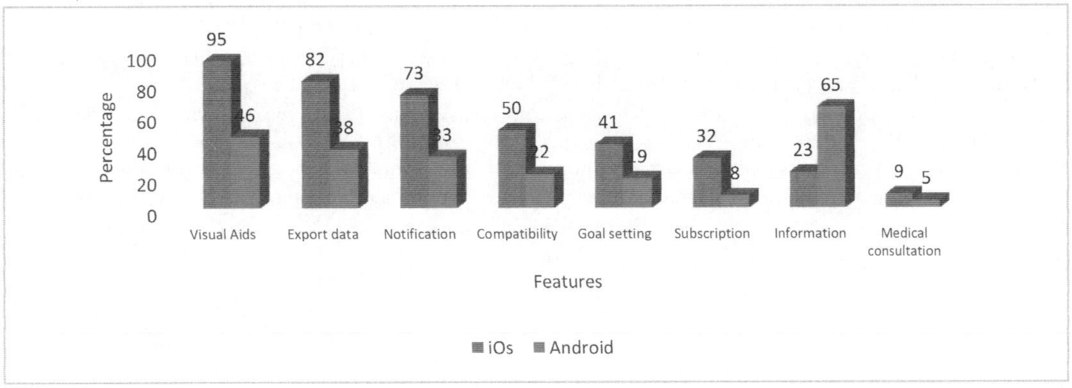

Figure 1. Comparison of available features found on iOS apps versus Android apps

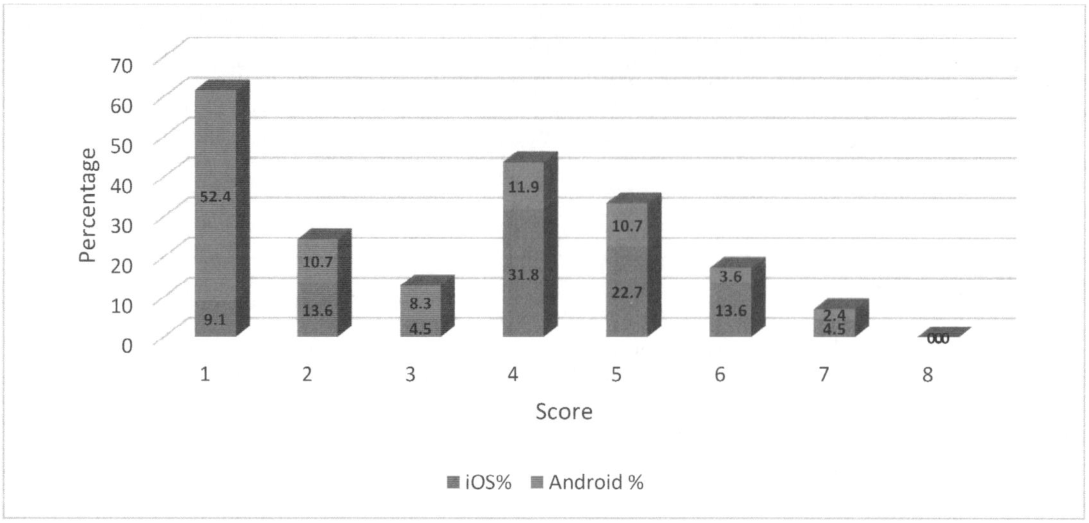

Figure 2. Proportion of score regarding features on iOS apps versus Android apps

4. CONCLUSION

The contents and features of free download mobile apps available for diabetes self-management in both iOS and Android are plethora. However, mobile apps that specifically fit the Thai setting i.e. language and lifestyle should be further developed. Moreover, the apps should be assessed for the quality and reliability of the information they provide and investment in e-health literacy is needed for the effective use of these apps in the self-management of diabetes among Thai patients.

REFERENCES

Demidowich AP, Lu K, Tamler R, Bloomgarden Z., 2012. An evaluation of diabetes self-management applications for Android smartphones. J Telemed Telecare 18(4):235–8.

Izahar, S., Lean, Q. Y., Hameed, M. A., Murugiah, M. K., Patel, R. P., Al-Worafi, Y. M., Wong, T. W., … Ming, L. C., 2017. Content Analysis of Mobile Health Applications on Diabetes Mellitus. Frontiers in endocrinology, 8, 318.

Kirwan M, Vandelanotte C, Fenning A, Duncan MJ., 2013. Diabetes self-management smartphone application for adults with type 1 diabetes: randomized controlled trial. J Med Internet Res.;15(11): e235.

Thailand Digital Economy and Society Development Plan, Ministry of Information and Communication Technology:2016:12.

Wayne N, Ritvo P., 2014. Smartphone-enabled health coach intervention for people with diabetes from a modest socioeconomic strata community: single-arm longitudinal feasibility study. J Med Internet Res.;16(6): e149.

World Health Organization, 2018. *Noncommunicable diseases country profiles 2018*. Geneva: Licence: CC BY-NC-SA 3.0 IGO.

THE ROLE OF THE ICT IN PROCESS MONITORING AND SYSTEM IMPACT ASSESSMENT OF RARE DISEASE COLLABORATIVE NETWORKS

Paolo Locatelli[1], Federica Cirilli[1], Eliana Panno[1], Salvatore Provenzano[2], Roberta Sanfilippo[2] and Paolo G. Casali[2, 3]
[1]*Fondazione Politecnico di Milano, Italy*
[2]*Medical Oncology Unit 2 – Adult mesenchymal tumour and rare cancer, Fondazione IRCCS Istituto Nazionale dei Tumori di Milano, Italy*
[3]*Department of Oncology and Haemato-Oncology, University of Milan, Italy*

ABSTRACT

Rare cancers represent about 20% of the total number of cancers diagnosed in Europe, with an occurrence of less than 6 per 100.000 individuals annually. Despite the impact of each rare disease is limited, rare diseases collectively are a true challenge for public health authorities. In particular, rare cancers face specific challenges that include late/incorrect diagnosis and lack of clinical expertise.
In this context, rare cancers networks play a central role in linking national designed centres of expertise far from each other. These collaborations among clinical centres aim to improve the care quality of rare cancers in terms of patients' survival and of life quality. Moreover, the presence of a national collaborative network should induce a reduction of health migration and social costs.
To foster these networks, it is important to use ICT solutions not only to support network collaborative processes, but also to monitor the overall collaboration process and to assess the related system impact.
This paper describes the project "Italian Rare Cancer Network – Process Monitoring and System Impact Assessment" that aims to strengthen the collaborative Italian network Rete Tumori Rari (RTR) developing its quality system through an information system. The project results can be used as input for the definition of the organizational model and of the information system of the new Italian network of rare tumours (Rete Nazionale Tumori Rari) and to strengthen Italy within the European Reference Networks on Rare Adult Cancers (ERNs-EURACAN).

KEYWORDS

Rare Cancer, Collaborative Networks, Digital Health, Teleconsulting

1. INTRODUCTION

In social sciences, a network is defined as an association of a definite number of people connected through social relations at different levels (Héon-Klin, 2017).

In healthcare, especially for rare diseases, a network includes different clinical centres with two main roles: the providers, who are the experts in that specific rare disease, and the users, who request clinical services (e.g., teleconsultations) to the providers. Users and providers are usually far from each other.

As demonstrated by different EU funding programs (Health Program, the Connecting Europe Facility and Horizon 2020), rare diseases are still one of the priorities in EU. Indeed, there are different initiatives to support virtual networks that involve healthcare providers across Europe - European Reference Networks (ERNs) - aimed at tackling rare diseases and conditions that require highly specialised treatment and concentrated knowledge and resources.

These programs highlight two main lines of action: the exchange of information via existing European information networks on rare diseases, and the development of strategies and mechanisms for information exchange to encourage continuity of work and trans-national co-operation. To foster this information exchange, it is important to use ICT solutions (Aymé et al., 2007), not only to facilitate the collaboration among clinical centres and to limit the patient's travels, but also to monitor the overall process of collaboration and to assess the related system impact (Waks et al, 2013).

This paper describes an Italian project focused on the importance of the ICT in supporting diseases networks, developing a network quality system through an information system. The project results can be also used in the European Reference Networks on Rare Adult Cancers (ERNs-EURACAN), strengthening the Italian role in the network.

2. RARE CANCER NETWORKS: SCENARIOS AND CHALLENGES

According to the Surveillance of Rare Cancers in Europe (RARECARE) definition, rare cancers are best defined as those with an incidence ≤6/100,000 cases per year (Gatta et al, 2011). However, when considering all of them together, rare cancers represent about 20% of the total of new cancer cases in Europe. Three main groups of rare cancers may be singled out: rare adult solid cancers (~15%), hematologic (~5%), and paediatric cancers (~1%). In Italy, professional networks are collaborating with the aim at improving the quality of care for Italian patients suffering from rare cancers. Gruppo Italiano Malattie Ematologiche dell'Adulto (GIMEMA) and Associazione Italiana di Ematologia e Oncologia Pediatrica (AIEOP) deal respectively with hematologic and paediatric cancers. Rete Tumori Rari (RTR) is a professional bottom-up network on rare adult solid cancers, operating since 1997 on a web-based platform. In 2017, Rete Nazionale Tumori Rari (RNTR) was established by the Italian government, as a service of the Italian National Health System, in order to enhance the three aforementioned professional networks and to improve the care quality of rare cancers limiting the health migration and social costs.

Moreover, in 2016, a European Joint Action on Rare Cancers (JARC) was launched to support the growth of the new European reference Networks (ERNs). The Clinical Patient Management System (CPMS) is the web-based application projected within the ERNs to be used for teleconsultations among centres at the European level.

Currently, ongoing efforts from the Italian government and the three partners of the RNTR aim at exploiting RNTR constitution to define the main requirements needed by the network organizational model and by the information system.

3. PROJECT "ITALIAN RARE CANCER NETWORK – PROCESS MONITORING AND SYSTEM IMPACT ASSESSMENT"

The project "Italian Rare Cancer Network – Process Monitoring and System Impact Assessment", funded by the Italian Ministry of Health, focuses on the collaborative Italian network RTR, with the aim of: (1) strengthening it and developing its quality system; (2) monitoring the collaborative process and assessing the system impact.

Indeed, the project aims at improving appropriateness through a systematic matching between individual clinical decisions and updated clinical practice guidelines. Moreover, the project aims at optimizing rare solid cancer patients' access to the most appropriate treatments by distant sharing of clinical cases. This should result in efficacy improvement, in terms of patients' survival and quality of life. The consequent reduction/rationalization of health migration should induce also social costs decreasing. Another by-product of this project is to foster clinical and translational research in rare solid cancers, gathering specific information (e.g. the number of studies related to RTR) and developing a virtual tissue bank.

The project also tests some specific functionalities on current information systems in order to support and monitor the network in its collaborative processes and to assess the related impact (e.g. in terms of healthcare migration).

The project results can be used as input for the definition of the organizational model and of the information system of the RNTR. This is a great opportunity for Italy not only to improve the quality of rare cancer treatment within the National Health System, but also at a European level, strengthening Italy within the European Reference Networks on Rare Adult Cancers (ERNs-EURACAN).

4. CONCLUSIONS

The project "Italian rare care network – Process monitoring and system impact assessment" started in September 2018. The partners are currently focusing on the improvement of a clinical guidelines program, on network rules, on mechanisms by which centres may interact with each other (Hammer, 1990). The solution selected in the project to graphically representing processes is based on the Business Process Modelling Notation (Osterwalder, & Pigneur, 2010) with specific notations to mark process instances and underline the difference between the as-is and to-be model (Weber et al., 2009).

Fondazione IRCCS Istituto Nazionale dei Tumori will evaluate how to implement additional functionalities in the RTR, focusing on its interaction with other users. Finally, the project results will be also used to define a new business model for a rare cancer network to be transferred in other Countries (Pistorio, et al., 2017). The updated results will be published in further papers.

ACKNOWLEDGEMENT

The authors would like to thank the project partners: Fondazione IRCCS Istituto Nazionale dei Tumori di Milano, Istituto Europeo di Oncologia and Fondazione Politecnico di Milano.
This work was supported by the Italian Ministry of Health (Grant No. RF-2016-02363386 to P.C.).

REFERENCES

Aymé S. et al., Networking for rare diseases: a necessity for Europe, Bundesgesundheitsblatt Journal, 50:12 (2007), 1477-1483

Gatta G. et al., Rare cancers are not so rare: the rare cancer burden in Europe, European Journal of Cancer 47:17 (2011): 2493-2511.

Hammer M. (1990). Re-engineering work: don't automate, obliterate. Harvard Business Review, 68(4):104-112.

Héon-Klin V., European Reference networks for rare diseases: what is the conceptual framework?, Orphanet Journal of Rare Diseases, (2017):12-137.

OMG (2009). Business Process Model and Notation, version 1.2.Osterwalder, A., & Pigneur, Y. (2010). Business Model Generation (pp. 1–288). Hoboken, New Jersey: John Wiley & Sons, Inc.

Pistorio, A., Locatelli, P., Cirilli, F., Gastaldi, L., Solvi, S., 2017. A Business Model for Digital Healthcare Environments: An Organic Approach and a Use Case for Handling Cognitive Impairment. Proceedings of the 10th International Joint Conference on Biomedical Engineering Systems and Technologies HEALTHINF, BIOSTEC, pp. 340–347.

Waks Z. et al., Analyzing the 'CareGap': assessing gaps in adherence to clinical guidelines in adult soft tissue sarcoma, Studies in Health Technology and Informatics, 186 (2013): 46-50.

Weber B, et al. (2009). Providing Integrated Lyfe Cycle Support in Process-aware Information Systems. Int'l Journal of Cooperative Information Systems (IJCIS), 18 (1), 115-165.

E-HEALTH: THE WAY TO IMPROVE COORDINATED CARE IN A PRIVATE INSURANCE COMPANY IN COLOMBIA, 2019

Karine Posada, Susana Orrego, Ana M. Ortiz and María M. Yepes
Seguros SURA Colombia
Carrera 64 B # 49 A 30 Medellín Colombia

ABSTRACT

The use of digital technology is a fundamental pillar of health ecosystem. Through virtual healthcare, members can obtain information about their health policies, evaluate their symptoms and obtain recommendations about care according to the risk level identified. The main objective is to coordinate the healthcare of members through Information and Communication Technologies (ICTs). In this paper, the results from a first year implementation of a virtual healthcare program from a private insurance company in Colombia are discussed. There were 3,045 pediatric virtual consultations in 2017 and 9,364 in 2018, equivalent to 10.1% of the total pediatric emergencies attentions. The number of cases that were resolved through the virtual channel was 78.3% and the satisfaction of the users was approximately 94.0%. Virtual healthcare has resulted in about $100,000 US Dollars in cost savings for the insurance company. The implementation of virtual healthcare has had an impact on health of people through immediacy in care, increased member's satisfaction through easy user experience, mobility, savings in out-of-pocket expenses, and optimizes the use of resources in providing health services by reducing investments in physical infrastructure and saving costs when compared to person-to-person healthcare services.

KEYWORDS

Healthcare Quality, Virtual Healthcare, Coordinated Care, Health Ecosystem

1. INTRODUCTION

One of the goals of Information and Communication Technologies (ICTs) in an insurance company in Colombia is to develop applications to facilitate access to healthcare services. Using digital technology is a fundamental pillar of healthcare ecosystem, especially in Colombia where using technology to assist patients is incipient and scarce.

The digital technology team within the insurance company has developed an App, which offer several services one of which is the virtual healthcare. Through virtual healthcare, members can obtain information about their health policies, evaluate their symptoms and obtain recommendations about care according to the risk level identified. The main objective is to coordinate the health of members through ICTs. These include providing comprehensive, opportune care, and facilitating access to virtual or readily available assistance, institutional or domiciliary healthcare. In addition, the insurance company take advantage of each contact with members to solve their health concerns and integrate it with their risk management. In this paper, results from a first year implementation of a virtual healthcare program from a private insurance company in Colombia are discussed.

2. BODY OF PAPER

Virtual healthcare service was developed in house and is available at the insurance company's App, through which both administrative and healthcare services are provided. Administrative services is designed to handle concerns, claims, certifications, reimbursements and other procedures through a ChatBot, based on frequently asked questions. If a question necessitates in-person communication, that is offered. About healthcare services, a pre-classification of symptoms: virtual, domiciliary or institutional medical care – are provided based on the risk level identified.

Virtual healthcare assistance has a model of complete care that incorporates the physicians. Likewise, it allows prescriptions to be sent digitally, recommendations and laboratory exams to be ordered, and is complemented with a visit for blood works, and other domiciliary procedures if needed. In cases where person-to-person assistance is required, home and/or institutional medical care is coordinated through the healthcare network providers.

All services are offered through the digital platform in a coordinated manner, being an end-to-end experience for the insured whose needs are solved by both the virtual channel and face-to-face if required. The insured does not ask for services but once healthcare team prescribes them, this team does all the management with the providers through the platform.

Virtual healthcare assistance implementation began in February 2017 with pediatrics services. There were 3045 pediatric virtual consultations in 2017 and 9,364 in 2018, equivalent to 10.1% of the total pediatric emergencies attentions. The number of cases that were resolved through the virtual consultation service was 78.3% and the satisfaction of the users was approximately 94%. Virtual healthcare has resulted in about $100,000 US Dollars in cost savings for the insurance company.

3. CONCLUSION

The implementation of virtual healthcare has had an impact on health of people through immediacy in care, increased member's satisfaction through easy user experience, mobility, savings in out-of-pocket expenses, and optimizes the use of resources in providing health services by reducing investments in physical infrastructure and saving costs when compared to person-to-person healthcare services. The future goal is to include virtual care services in populations with specific risks such as chronic conditions, and palliative needs.

Virtual healthcare is an innovative complement to the traditional healthcare system. It increases the opportunity for members to receive assistance and get their issues resolved. Additionally, it aligns interests between patients, insurer and providers. The main challenges after implementation include members adherence to the use of ICT, maintain coordination among allies that support the services, have the platforms and technological updates required to maintain availability of care and ensure a good user experience.

Is a challenge to articulate Healthcare providers once a single patient could need more than one medical supplies. Additionally, changing person-to-person assistance to a virtual assistance model is a big goal to achieve in Colombians health culture, were historically the technology use in health is almost nule.

ACKNOWLEDGEMENT

Our acknowledgment to the project team, leaders and developers that makes this project becomes a reality and all the providers that believe in this project to make healthcare affordable and better every day.

REFERENCES

Yara M. Asi. Et al, 2018. The role of digital health in making progress toward Sustainable Development Goal (SDG) 3 in conflict-affected populations. International Journal of Medical Informatics, Vol. 114, pages 114 – 120.

Meskó, B. et al, 2017. Digital health is a cultural transformation of traditional healthcare. mHealth, 3, 38. doi:10.21037/mhealth.2017.08.07.

Abhinav, S. et al, 2018. Using Digital Health Technology to Better Generate Evidence and Deliver Evidence-Based Care. Journal of the American College of Cardiology, Vol. 71, Issue 23, pages 2680 – 2690.

Sanjeev, B. et al, 2016. Mobile technology and the digitalization of healthcare. European Heart Journal, Vol. 37, issue 18, pages 1428 – 1438.

Hamine, S. et al, 2015. Impact of mHealth chronic disease management on treatment adherence and patient outcomes: a systematic review. Journal of medical Internet research, 17(2), e52. doi:10.2196/jmir.3951.

Sanjeev, B. et al, 2017. 2017 Roadmap for Innovation—ACC Health Policy Statement on Healthcare Transformation in the Era of Digital Health, Big Data, and Precision Health: A Report of the American College of Cardiology Task Force on Health Policy Statements and Systems of Care. Journal of the American College of Cardiology, Vol. 70 Issue 21, DOI: 10.1016/j.jacc.2017.10.018.

DESIGNING NO-SQL BASED PERSONAL HEALTHCARE DATA EXCHANGE SCHEMA

Hee Joung Hwang
Dept. of Computer Engineering, Gachon University.
1342 Seongnamdaero, Seongnam-si, Kyeonggi-do, Korea

ABSTRACT

With the advance of micro-service architecture, existing large systems have begun to be deployed as small service units. There is an increasing number of NoSQL-based services that can handle unstructured data in the massive schema structure of large RDBMSs. However, the heterogeneity and diversity of the numerous existing systems impede the well-informed selection of a data store appropriate for a given application context.
In this paper, we design a NoSQL - based data exchange server to integrate Healthcare data and implement Restful API for interworking with various medical data sources. Experimental results show that medical data such as HL7 FHIR, Fitbit and CDM can be exchanged.

KEYWORDS

Data Exchange Server, HIS, NoSQL, Healthcare

1. INTRODUCTION

Due to the rapid development of IT technology in recent years, the amount of data has increased exponentially. As the penetration rate of smartphones increases and the number of wearable devices and IoT technology increases, the amount of data to process has also increased dramatically and the importance of storage and processing power has also increased. In the past, relational database management systems (RDBMS), such as MySQL, Oracle, and Postgre SQL, have begun to show limitations in terms of cost and efficiency when processing large amounts of data, effectively managing and processing large amounts of data [Abadi, 2012]. NoSQL is growing rapidly, although Oracle, MySQL and MSSQL are still at the top of database rankings [Choi et al, 2016].

In this study, we designed and implemented a Health Data Exchange Server architecture that works with various healthcare data sources. Generally, standards such as HL7, HL7 CDA, and HL7 FHIR are required for medical information systems and medical data exchange. However, methods other than HL7 FHIR were expensive to implement with too large a specification, so the standard did not spread.

Our works is theoretically compatible with systems such as HL7 v 2.x, CDA (Clinical Document Architecture) / CCD [Zhang et al, 2015], ASTM CCR [Tseng et al, 2003] and HL7 MLLP, IHE XDS.b. Outbound modules supporting standards-based communication protocols have been developed and can be interchanged with existing medical data systems to ensure interoperability with a variety of medical devices.

In particular, the development of medical and personal healthcare devices and the increase in personal health data require a repository structure to effectively integrate and operate these data.

2. DESIGN AND IMPLEMENTATION

2.1 Data Message Schema

There are two main models of data architecture for health information exchange. One is a federated or distributed model, and the second is a centralized model. There is also a hybrid model with two elements. The centralized HIE has a central (or master) database that holds a complete copy of all records for all patients included in the HIE. The federated HIE does not have a master database.

In the federated model, each health care provider must maintain a record of individual patients. The main function of the HIE in this model is to enable the provider to exchange patient records as needed. For example, when a doctor in the federal HIE requests a record for patient Y, a query is sent to each server in the system to request that records for patient Y be returned. Each federated HIE can do this in a slightly different way. The notable difference is that the affiliate model does not have a central database to store and download previously compiled medical records.

In personal healthcare areas there are various types of data exist like HL7 FHIR, ISO, and OHDSI. Among them, HL7 FHIR is attracting attention because it uses flexible message standards such as JSON and XML. However, server systems that handle these messages require a micro-services architecture based on modular software architecture, unstructured data processing, and open APIs.

Figure 1 shows 3 different personal healthcare data.


```
{
 "resourceType": "Observation",
 "code": {
  "coding": [{
   "system": "http://loinc.org",
   "code": "29463-7",
   "display": "Body weight"
  }]
 },
  "effectiveDateTime": "2017-03-28",
  "valueQuantity": {
   "value": 185,
   "unit": "lbs",
   "system": "http://unitsofmeasure.org",
   "code": "[lb_av]"
  }
}
```


(a) HL7 FHIR


```
{
  "activities-heart": [
    {
      "dateTime": "2017-08-04",
      "value": {
        "customHeartRateZones": [],
        "heartRateZones": [
          {
            "caloriesOut": 740.15264,
            "max": 94,
            "min": 30,
            "minutes": 593,
            "name": "Out of Range"
          },
        ],
        "restingHeartRate": 68
      }
    }
  ]
}
```


(b) Fitbit smart band


```
{
 "DAYS_SUPPLY_DISTRIBUTION": {
  "CATEGORY": "Days supply",
  "MIN_VALUE": 10,
  "P10_VALUE": 10,
  "P25_VALUE": 30,
  "MEDIAN_VALUE": 30,
  "P75_VALUE": 30,
  "MAX_VALUE": 30
 },
 "QUANTITY_DISTRIBUTION": {
  "CATEGORY": "Quantity",
  "MIN_VALUE": 0,
  "P10_VALUE": 10,
  "P25_VALUE": 30,
  "MEDIAN_VALUE": 30,
  "P75_VALUE": 60,
  "MAX_VALUE": 120
 },
```


(c) OHDSI CDM

Figure 1. Personal healthcare data schema

2.2 Data Repository

NoSQL systems are purpose-built solutions, designed to address specific technical requirements. NoSQL systems originated to provide high throughput, fault-tolerant horizontally scalable simple data storage and retrieval with a bare minimum of additional functionality. Specifically, NoSQL systems were created in order to provide:

Horizontally distributed data management of simple structured and unstructured data across a large cluster of commodity storage systems,

- Highly fault-tolerant data management and ability to continue operating even after multiple hardware and system failures,
- Very high throughput for simple read/write operations, with limited or no transaction semantics,
- Schema-less or flexible schema definitions allowing highly variable data and record structures,
- Application and application developer-centric special purpose data models and APIs

The most commonly employed distinction between NoSQL databases is the way they store and allow access to data. Each system covered in this paper can be categorized as either key value store, document store or wide-column store.

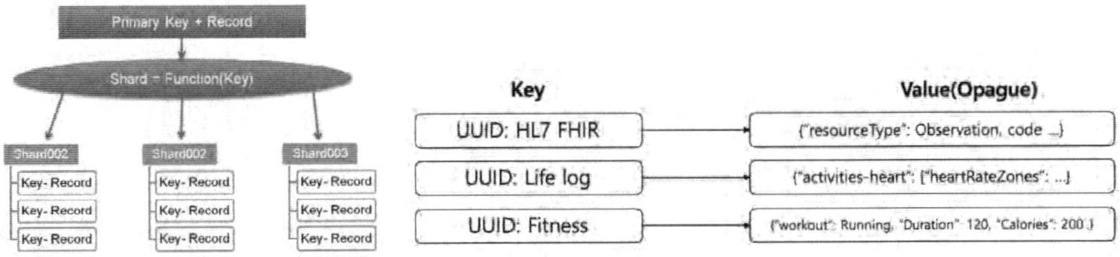

Figure 2. Proposed NoSQL repository data structure

3. CONCLUSION

NoSQL database technology is here to stay. It addresses specific technical requirements that are not as efficiently or cost effectively addressed with other data management technologies, including RDBMS systems. The RDBMS and NoSQL Database are complementary in the sense that they work together to address the overall data management needs of our customers, each providing the technical capabilities required by today's complex and evolving applications.

In summary, we are confident that the proposed schema model is an effective decision to support effective filtering. Centralized large NoSQL database system Requirements. NoSQL Toolbox Mapping between functional and non-functional requirements A common implementation technique for classifying constantly evolving NoSQL spaces.

ACKNOWLEDGEMENT

This research was supported by the MSIT(Ministry of Science and ICT), Korea, under the ITRC (Information Technology Research Center) support program(IITP-2019-2017-0-01630) supervised by the IITP (Institute for Information & communications Technology Promotion)

REFERENCES

Abadi D (2012) Consistency tradeoffs in modern distributed database system design: cap is only part of the story. Computer 45(2):37–422. Attiya H, Bar-N.

Choi, J., Choi, C., Ko, H., and Kim, P., "Intelligent Healthcare Service Using Health Lifelog Analysis," Journal of medical systems, 40(8), 1-10, 2016.

Minkyung Kim, Dong-Wook Lee, Kangseok Kim, Jai-Hoon Kim, "Hierarchical struc- tured data logging system for effective lifelog management in ubiquitous environment," Multimedia Tools and Applications, 2013.

Myung-Kyu Yi and Hee-Joung Hwang," A New Lifelog Management Scheme for In- telligent Healthcare Service", International Journal of Elderly Welfare Promotion and Management, Vol. 1, No. 1, pp. 29-36, 2017.

Reza Rawassizadeh, Martin Tomitsch, Katarzyna Wac, A. Min Tjoa, "UbiqLog: a generic mobile phone-based life-log framework," Personal and Ubiquitous Computing, Vol. 17(4), pp. 621-637, 2013.

Sheldon M. Ross, "Introduction to Probability Models: Eighth Edition," Academic Press, 2002.

Wiese L et al (2015) Advanced Data Management: For SQL. Cloud and Distributed Databases. Walter de Gruyter.

X. Zhang, J. Castellanos, and A. Campbell, "Design and Performance of Mobile IP Paging," ACM Mobile Networks and Applications, pp.127-141, March, 2002.

Yu-Chee Tseng, Lien-Wu Chen, Ming-Hour Yang, Jan-Jan Wu, "A stop-or-move mo- bility model for PCS networks and its location-tracking strategies," Journal of Com- puter Communication, pp. 1288-1301, 2003.

Zhang H, Chen G et al (2015) In-memory big data management and processing: a survey. TKDE.

TRACEABILITY AND MANAGEMENT OF MEDICINES AND MEDICAL DEVICES AT HOME: THE CRUCIAL ROLE OF THE DIGITAL SUPPORT

Paolo Locatelli, Federica Cirilli, Eliana Panno and Antonio Jacopo Scardigno
Fondazione Politecnico di Milano, Italy

ABSTRACT

The continuous growth of the healthcare sector is followed by a corresponding demand to reduce costs and to improve the quality of its management processes. One way to meet these requirements is to leverage on a better management and traceability of medicines and medical devices through ICT, not only at the hospital but also at the patient's home. The project "LIFEMED – Integrated Logistic of Medicines and Medical Devices", aims, indeed, at the realization of an innovative digital model for medicines and medical devices logistic management. In particular, for the management and the traceability of medicines and medical devices at home, the solution will be characterized by a telemedicine platform together with a traceability platform based on RFId and barcode labels and readers. The telemedicine and traceability platforms will allow for a better monitoring of respectively the patient's correct medicine administration and the medicines and medical devices logistic, as demonstrated by the example of the peritoneal dialysis, a domiciliary treatment addressed in the project. Despite all the advantages, attention must be paid during the platform design in considering crucial factors like digital illiteracy of older people and the privacy, legal and ethical issues.

KEYWORDS

Medicines, Medical Devices, Digital Support, Traceability, Peritoneal Dialysis

1. INTRODUCTION: SCENARIO AND CHALLENGES

The healthcare sector is currently going through a significant growth, remaining one of the most popular investment theme in developed markets and Countries, especially those with ever increasing ageing population [Lulin et al, 2016]. Nonetheless this growth is followed by a corresponding demand to reduce costs and a continuous pressure to improve the quality of management processes [Agarwal et al, 2010]. One way to meet these requirements is to leverage on a better management and traceability of medicines and medical devices, not only at the hospital, but also at the patient's home. As a matter of fact, having a trustworthy and monitored provision of health services at a distance can result in the costs reduction thanks to a lower patient hospitalization [Paul, 2013]. Moreover, an efficient traceability at a distance can lead to a better management of unused medicines, avoiding uneconomic waste, and to a better prevention of errors and more controlled deployment of medical devices [Lovis, 2008] [Matta et al, 2013].

In this regard, digital technologies proved to play a crucial role in the processes management and organizational evolution of medicines and medical devices [Geneva: World Health Organization; 2019].

2. LIFEMED DIGITAL SOLUTION

The project "LIFEMED – Integrated Logistic of Medicines and Medical Devices", funded by the Italian Ministry of Education, Universities and Research (MIUR – Ministero dell'Istruzione, dell'Università e della Ricerca), aims at the realization of an innovative organizational-technological model for medicines and medical devices logistic management at the hospital and at home. In particular, the overall solution (Figure 1) is conceived as a digital and modular platform that enables the clinical governance and the monitoring of all the processes related to medicines and medical devices, leveraging on virtual warehouse management, risk management, decision support system and medicines prescription and recovery/administration management in and out the hospital. The solution integrates also external existing systems in order to avoid a double-input of information by the hospital staff.

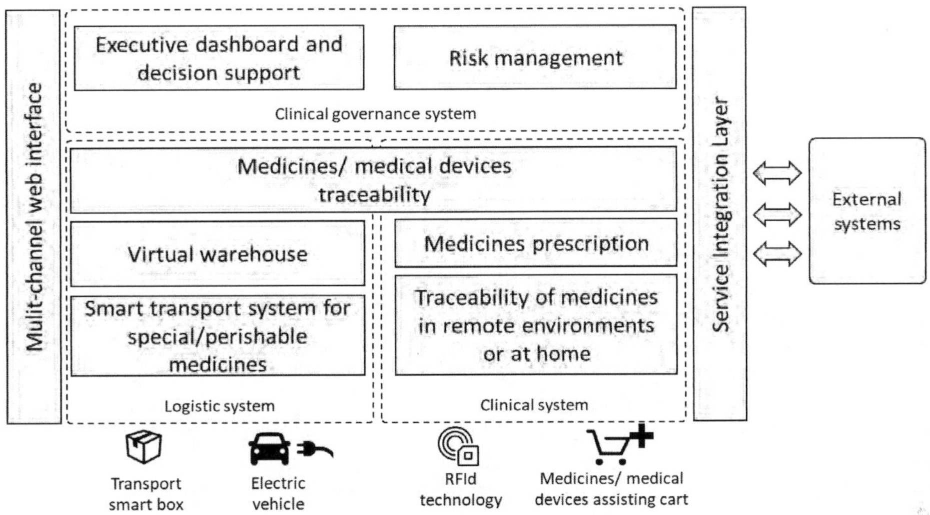

Figure 1. LIFEMED solution

Regarding specifically the management of medicines and medical devices at home, the solution will consist of a client-to-provider telemedicine platform that will enable the patient to correctly perform the medicine administration. The patient, indeed, will be provided with a personal account (accessible through a smartphone, a tablet or a web-based user interface) that will allow him to: receive automatic reminders (e.g. time and dosage) about taking the medicine and notify that the administration has been carried out correctly or enter supporting notes when deciding not to carry it out. Doctors or other members of the medical team will be able to access the monitoring telemedicine platform in order to consult the history of the treatments carried out and view any warnings of non-administration.

Furthermore, the solution will also include a traceability platform so as to monitor all medicines and medical devices movements during transportation to the patient's home. The traceability system will be based on RFId labels technology to track medical devices and special medicines and on barcode technology to track generic medicines, coupled with portable devices provided with RFId and barcode readers.

The LIFEMED platform efficiency in management and traceability of medicines and medical devices at home can be highlighted through the application in a specific treatment: the peritoneal dialysis (PD).

3. CASE STUDY: PERITONEAL DIALYSIS (PD)

PD is a treatment option that enables patients suffering from end-stage renal disease to perform dialysis in their own home thus allowing them to have a higher quality of life [Russo et al, 2006]. Despite its diffusion, PD is still affected by several issues: inadequate and not repeated patients' trainings, trainings require more time than most nephrologist can provide, trainings effectiveness should be periodically tested, high noncompliance rate among patients and waste of domiciliary medicines that are not correctly traced back [Novak et al, 2008] [Kazancioglu et al, 2008]. In many cases, all these problems drive patients not to feel confident in adopting PD or the treatment's inadequate application can cause severe patients' complications leading to their hospitalization as a consequence of the switching from PD to haemodialysis [Fried et al, 1999].

The analysed as-is management process for PD in a clinical centre involved in the LIFEMED project can widely reflect some of the abovementioned issues. In this case, all the medicines are considered as used by the hospital pharmacy as soon as they are delivered to the patient's home. Moreover, patient's education about manual PD is provided mainly at the hospital site. Finally, when a manual PD is provided (without an automated dialysis medical device), there are no reliable information regarding the efficiency and periodicity of the patient's treatment: the patient can only report his/her information in a paper form during periodical visits at the outpatient clinic of the hospital.

LIFEMED solution will provide concrete evidence to tackle these problems allowing for the monitoring of the patient's correct adherence to the treatment. Moreover, the partners are evaluating the chance to introduce some functionalities on the telemedicine platform in order to permit the patient to enter his/her vital signs periodically recorded (e.g. heartbeat, pressure) and consult digital training materials (e.g. digital brochure, video and audio tutorials). Furthermore, RFId and barcode traceability will allow the hospital pharmacy to avoid the waste of unused medicines, that can be transferred from the patient's home to the hospital as warehouse stocks, and will also allow a more efficient tracing of faulty products.

4. CONCLUSIONS

The project is still ongoing and now is focused on analysing the as-is processes and governance in the involved clinical centres. The partners are also currently involved in designing the new process-organizational model and the related support through the integrated digital platform. In particular, they are focusing on the technical design and completion (hardware and software) related to the RFId traceability and to both the telemedicine and traceability platforms. Crucial attention must be paid in considering digital illiteracy of older people, the wide heterogeneity of the processed data, the use of standardized RFId carriers and readers and all the privacy, legal and ethical issues involved in the traceability [Lovis, 2008] [AbuKhousa et al, 2014].

Once the designing is completed, the system applications components will be developed and, finally, the project results will be tested and monitored in the pilot centres in order to have information to define an innovative business model 0 for the medicines and medical devices logistic.

REFERENCES

AbuKhousa E. et al, 2014. Simulation and modeling efforts to support decision making in healthcare supply chain management. *The Scientific World Journal*, Vol. 2014, Article ID 354246, 16pp.

Agarwal R. et al, 2010. Research Commentary—The Digital Transformation of Healthcare: Current Status and the Road Ahead. *Information Systems Research*, 21(4): 796–809.

Fried L. et al, 1999. Hospitalization in peritoneal dialysis patients. *American Journal of Kidney Diseases*, 33(5): 927-933.

Geneva: World Health Organization; 2019. WHO guideline: recommendations on digital interventions for health system strengthening.

Kazancioglu R. et al, 2008. Importance of Home Visits in Peritoneal Dialysis. *Dialysis & Transplantation*, 1-3.

Lovis C., 2008. Traceability in Healthcare: Crossing Boundaries. *IMIA Yearbook of Medical Informatics*, 47 Suppl 1:105-113.

Lulin Z. et al, 2016. The effect of herd formation among healthcare investors on health sector growth in China. *International Journal for Equity in Health*, 15:116.

Matta A. et al, 2013. A Decision-Making Approach Supporting Hospital Drug Logistics. *Proceedings of the International Conference on Health Care Systems Engineering*. Springer Science & Business Media, (19): – 241-252.

Novak M. et al, 2008. Why Is Peritoneal Dialysis Underutilized in the United States? *Dialysis & Transplantation*, 1-2.

Paul D. P., 2013. An Innovation in Healthcare Delivery: Hospital at Home. *Journal of Management Policy and Practice*, 14(6): 73-91.

Pistorio A. et al, 2017. A Business Model for Digital Healthcare Environments: An Organic Approach and a Use Case for Handling Cognitive Impairment. *Proceedings of the 10th International Joint Conference on Biomedical Engineering Systems and Technologies HEALTHINF*, BIOSTEC, pp. 340–347.

Russo R. et al, 2006. Patient re-training in peritoneal dialysis: Why and when it is needed. *Kidney International*, 70: S127–S132.

TELEMONITORING OF USERS WITH PACEMAKERS: INFORMAL CAREGIVERS AT 5 YEARS AFTER IMPLANT

Antonio Lopez-Villegas[1], Remedios Lopez-Liria, Cesar Leal-Costa and Daniel Catalan-Matamoros
Hospital of Poniente, University of Almeria , University of Murcia & University Carlos III of Madrid, Spain

ABSTRACT

Introduction: The use of telemonitoring of pacemakers is limited. Few studies have focused on their informal caregivers at long-term. To evaluate the burden borne to informal caregivers of patients with pacemakers, five years after implantation.
Methods: This is a controlled, non-randomised clinical trial. Data on baseline characteristics, workload and time worked were collected from informal caregivers, five years after implantation of pacemakers. Besides, the Spanish version of the Survey on Disabilities, Personal Autonomy, and Dependency Situations was used to get information on Clinical and social characteristics, levels of professionalism, duration and types of care, difficulties in providing care, health status, economic and job aspects, family and leisure.
Results: Fifty five patients with pacemakers finished the study. Of which, 50 needed to be helped by a caregiver, 18 were included in the telemonitoring and 32 in the hospital monitoring group. Overall, females represented 96.0% of the informal caregivers (88.89% in telemonitoring and 100.0% in hospital monitoring group). The mean ages were 63.17 ± 15.92 and 63.13 ± 14.56 years, respectively (p = 0.83) in the groups of telemonitoring and hospital monitoring group. The majority (88.0%) of the caregivers declared that they had to provide their services between 6 and 7 days per week (83.33% in telemonitoring group.
Conclusions: The results confirm that there were no significant differences between the informal caregivers regarding to baseline characteristics, workload and time worked in both groups of follow-up.

KEYWORDS

Disease Burden, Informal Caregiving, Pacemaker Follow-Up, Remote Monitoring, Telemedicine

1. INTRODUCTION

The use of telemonitoring (TM) of pacemakers is limited. Few studies have focused on their informal caregivers at long-term.

2. OBJECTIVE

To evaluate the burden borne to informal caregivers of patients with pacemakers, five years after implantation.

3. METHODS

This is a controlled, non-randomised clinical trial. Data on baseline characteristics, workload and time worked were collected from informal caregivers, five years after implantation of pacemakers. Besides, the Spanish version of the Survey on Disabilities, Personal Autonomy, and Dependency Situations was used to get information on 1) Clinical and social characteristics; 2) Levels of professionalism, 3) Duration and types

[1]Contact Details Antonio Lopez Villegas: antoniolopezvillegas@andaluciajunta.es

of care; 4) Difficulties in providing care, 5) Health status, 6) Economic and job aspects, 7) Family and leisure.

Patients were included in the study if: 1) they had been implanted with a single [VVI-VVIR] or double chamber [VDD-DDD] Medtronic Carelink® pacemakers, 2) they were over 18 years old, 3) they were capable of understanding and correctly performing the home auto-monitoring or had a caregiver and 4) they agreed to participate in the study by signing the informed consent. Patients were excluded if: 1) they were enrolled in another study, 2) they had any other cardiac device, 3) they refused to participate.

The Almeria Health Research Ethical Committee approved the trial protocol and the study was developed in accordance with the precepts of the Declaration of Helsinki and Spanish laws on data protection and patient rights. All patients signed the corresponding informed consent prior to their enrolment, and appropriate measures were taken to ensure data privacy. The trial protocol was registered in ClinicalTrials.gov (Identifier: NCT02234245).

4. RESULTS

Fifty five patients with pacemakers finished the study. Of which, 50 needed to be helped by a caregiver, 18 were included in the TM and 32 in the hospital monitoring group (HM). Overall, females represented 96.0% of the informal caregivers (88.89% in TM and 100.0% in HM group). The mean ages were 63.17 ± 15.92 and 63.13 ± 14.56 years, respectively ($p = 0.83$) in the groups of TM and HM group. The majority (88.0%) of the caregivers declared that they had to provide their services between 6 and 7 days per week (83.33% in TM versus 90.63% in HM group). No significant differences between the two groups were found in any of questions of the Survey administrated.

5. CONCLUSIONS

The results confirm that there were no significant differences between the informal caregivers regarding to baseline characteristics, workload and time worked in both groups of follow-up.

ACKNOWLEDGEMENT

The study has been financially supported by the General Secretariat for Research, Development and Innovation, Regional Government of Andalusia (Spain), project reference number PI/0256/2017, under the research call "Development and Innovation Projects in the Field of Biomedicine and Health Sciences," 2017.

REFERENCES

López-Villegas A, Catalán-Matamoros D, Robles-Musso E, Peiró S. Workload, time and costs of the informal cares in patients with tele-monitoring of pacemakers: the PONIENTE study. Clin Res Cardiol. 2016;105(4):307-13. doi: 10.1007/s00392-015-0921-5. Epub 2015 Sep 30. PMID:26423396

Instituto Nacional de Estadística. Encuesta sobre Discapacidades, Autonomía personal y situaciones de Dependencia. 2008. http://www.ine.es/daco/daco42/discapa/edad_cui.pdf

AUTHOR INDEX

International Association for Development of the Information Society (IADIS)
Rua Sao Sebastiao Da Pedreira
No. 100, 30, 1050-209
Lisbon, Portugal

ISBN 978-1-5108-9239-2

CURRAN ASSOCIATES INC.
proceedings.com